Secure and Smart Cyber-Physical System.

Cybersecurity is a paramount concern in both Internet of Things (IoT) and Cyber-Physical Systems (CPSs) due to the interconnected and often critical nature of these systems. The integration of AI/ML into the realm of IoT and CPS security has gained significant attention and momentum in recent years. The success of AI/ML in various domains has sparked interest in leveraging these technologies to enhance the security, resilience, and adaptability of IoT and CPS. *Secure and Smart Cyber-Physical Systems* provides an extensive exploration of AI/ML-based security applications in the context of IoT and CPS.

Features

- Presents cutting-edge topics and research in IoT and CPS.
- Includes contributions from leading worldwide researchers.
- Focuses on CPS architectures for secure and smart environments.
- Explores AI/ML and blockchain approaches for providing security and privacy to CPS including smart grids, smart cities, and smart healthcare.
- Provides comprehensive guidance into the intricate world of software development for medical devices.
- Covers a blueprint for the emergence of 6G communications technology in Industry 5.0 and federated-learning-based secure financial services.

This book covers state-of-the-art problems, existing solutions, and potential research directions for CPS researchers, scholars, and professionals in both industry and academia.

Secure and Smart Cyber-Physical Systems

Edited by Uttam Ghosh, Fortune Mhlanga, and Danda B. Rawat

CRC Press
Taylor & Francis Group
Boca Raton London New York

CRC Press is an imprint of the
Taylor & Francis Group, an **informa** business

Designed cover image: © Shutterstock

First edition published 2025
by CRC Press
2385 NW Executive Center Drive, Suite 320, Boca Raton FL 33431

and by CRC Press
4 Park Square, Milton Park, Abingdon, Oxon, OX14 4RN

CRC Press is an imprint of Taylor & Francis Group, LLC

Library of Congress Cataloging-in-Publication Data
Names: Ghosh, Uttam, editor.
Title: Secure and smart cyber-physical systems / edited by Uttam Ghosh,
    Fortune Mhlanga, Danda B. Rawat.
Description: First edition. | Boca Raton, FL : CRC Press, 2025. |
    Includes bibliographical references and index.
Identifiers: LCCN 2024002986 (print) | LCCN 2024002987 (ebook) |
    ISBN 9781032453828 (hardback) | ISBN 9781032453859 (paperback) |
    ISBN 9781003376712 (ebook)
Subjects: LCSH: Cooperating objects (Computer systems)—Security measures. |
    Internet of things—Security measures. | Computer security.
Classification: LCC QA76.9.A25 S3752 2025 (print) | LCC QA76.9.A25 (ebook) |
    DDC 005.8—dc23/eng/20240527
LC record available at https://lccn.loc.gov/2024002986
LC ebook record available at https://lccn.loc.gov/2024002987

ISBN: 9781032453828 (hbk)
ISBN: 9781032453859 (pbk)
ISBN: 9781003376712 (ebk)

DOI: 10.1201/9781003376712

Typeset in Sabon
by Apex CoVantage, LLC

Dedication

To my wife Pushpita and our two children Dhriti and Shriyan—
Uttam Ghosh

To my wife Florah and our two boys (young men) Carl and Craig—
Fortune S. Mhlanga

Dedicated to my family—Danda B. Rawat

Contents

Preface

Introduction

Around the world today, technology continues to become more and more pervasive with dynamic innovation processes increasingly impacting societies and their governments, industries, communities, and individuals. For example, integrating Internet-connected things with other devices creates larger systems of systems that solve the world's most challenging and technological problems. One of the most popular revolutions of technology is the Cyber-Physical System (CPS). CPS is an integration of cyber world (computation and communication systems) and man-made physical world (e.g., utility networks, vehicles, and factories.) formed by using sensors and actuators. Cyber systems make the physical infrastructures smarter, more secure, and reliable, and fully automated systems foster a more efficient, resilient, and sustainable built environment. In the near future (industry 4.0 revolution or 4IR), CPSs will become the new "techno-economic" paradigm.

CPS have become ubiquitous and the core of modern critical infrastructure and industrial applications in recent years. CPSs such as self-driving cars, drones, and intelligent transportation rely heavily on machine learning techniques for ever-increasing levels of autonomy. Further, the deployed sensors generate a massive amount of real-time Big data from the physical infrastructure and send it to the cyber systems using communication infrastructure (such as switches and routers). In turn, the cyber systems also send feedback to the physical devices using the communication infrastructure. On the other hand, these systems provide an appeal to attackers. Cybersecurity is, thus, of prime concern in CPSs. Due to the success of Deep Learning (DL) in a multitude of domains, the development of DL-based CPS security applications has received increased interest in the past few years. However, despite the broad body of work on using DL for ensuring the security of CPSs, to our best knowledge, very little work exists where the focus is on the development of these DL applications. DL based on artificial neural networks is a very popular approach to modeling, classification, and the recognition of complex data including images, voice, and text. The unparalleled precision of DL approaches has made them the cornerstone of new Internet-based AI-based services. Commercial businesses gathering user data on a large scale were the main beneficiaries of this phenomenon since the performance of DL techniques is directly proportional to the amount of data available for training. The vast collection of data required for DL poses clear privacy problems. Highly sensitive personal data such as photos and voice recordings of users, that is collected by some companies, is kept indefinitely. Users cannot uninstall it, nor restrict the purposes for which it is being used. In addition, data stored internally is subject to legal subpoenas and extra judicial monitoring. For example, data owners, such as healthcare institutions and organizations that may want to apply DL methods to clinical records and are prevented from sharing data due to privacy and confidentiality concerns, benefit from DL on a large scale.

Aim and Scope of the Book

The major aim of this book is to explore the advanced technologies used to facilitate Artificial Intelligence and Big data on the Internet of Things and CPSs. This is a much-needed ingredient for researchers/scholars who are about to start their research in this emerging area. The comprehensive coverage of state-of-the-art problems, existing solutions, and open research directions is of paramount importance to research groups, postgraduate programs, and doctoral programs around the world, especially within the dynamic field of computer science and cybersecurity. This book serves as a comprehensive guide that not only addresses the current challenges in the field but also highlights effective solutions and outlines future research paths.

Target Group

This book explores the advanced technologies used to facilitate the transformation of these novel spaces into a pervasive environment with the development of computational and physical infrastructures into a single habitat that facilitates the professional researchers and application developers of solutions. AI and Big data can be offered as an elective course for graduate and postgraduate computer science engineering students to impart knowledge of current trends, scope, and technology in mobile applications and the physical structure of pervasive management. Furthermore, the system designers, industry personnel, and policy-makers working for the transformation of the ubiquities of computing with the employment of security professionals will benefit.

Contents of the Book

After careful selections, we have been able to put together a total of ten chapters in this book. Here, we discuss the chapters in brief.

Chapter 1: *Machine Learning and Deep Learning Approaches in Cyber-Physical Systems* presents a detailed discussion of machine learning (ML) and DL approaches to address the various challenges posed in CPS, such as Big data analytics and ways to enable ML/DL capability on the resource constraint devices of CPS. It discusses how ML and DL algorithms are used as the solutions for addressing various security issues in CPS including anomaly detection, cybersecurity, fault prediction, predictive maintenance, process optimization, QoS analysis, and resource allocation.

Chapter 2: *Securing Cyber-Physical Systems Using Artificial Intelligence* focuses on how AI-enabled CPS addresses a variety of cybersecurity threats including identifying and forecasting malware incursions, network-based assaults such as Distributed Denial of Service attacks, and other pattern obfuscation-based incursions. It discusses the specifics and endpoints of cyberattacks in a CPS device in addition to various AI-based approaches for increased CPS performance and security and provides case studies and proof of concepts in simulated environments in addition to the current research direction and trend in this domain.

Chapter 3: *Toward Fast Reliable Intelligent Industry 5.0—A Comprehensive Study* provides a comprehensive exploration of the key principles, methodologies, and emerging technologies essential to realizing the vision of Industry 5.0. In the context of Industry 5.0, it delves into the convergence of various cutting-edge technologies, including IoT, AI, Big data analytics, and advanced robotics to create an ecosystem of interconnected, intelligent, and adaptive industrial systems. It highlights the pivotal role of real-time data acquisition, processing, and decision-making in optimizing industrial operations. The chapter further investigates the challenges and opportunities associated with Industry 5.0, addressing concerns related to data security, privacy, and scalability. It emphasizes the need for robust, fast, and reliable communication networks, such as 5G, as a critical enabler for the seamless integration of CPSs. This study provides insights into the adoption of digital twins, predictive maintenance, and autonomous manufacturing in Industry 5.0, illustrating how these concepts enhance reliability and performance while reducing downtime and operational costs. It also explores the concept of human–machine collaboration and the evolving role of the workforce in an intelligent manufacturing environment.

Chapter 4: *Software Development for Medical Devices: A Comprehensive Guide* provides a comprehensive guidance into the intricate world of software development for medical devices, offering invaluable insights for developers, engineers, and stakeholders in the healthcare industry. It explores the evolving landscape of the Internet of Medical Things (IoMT), where interconnected devices gather and exchange real-time health data. The study underscores the transformative potential of AI/ML in predictive analytics, disease diagnosis, and personalized treatment. Furthermore, the guide examines Software as a Medical Device (SaMD) and its multifaceted applications across various platforms. It provides a comprehensive overview of the challenges, regulations, and significance of software development within the medical device sector, emphasizing the importance of quality, compliance, and patient well-being.

Chapter 5: *6G Communication Technology for Industry 5.0: Prospect, Opportunities, Security Issues, and Future Directions* provides a blueprint for the emergence of 6G communications technology in Industry 5.0 verticals mitigating the implementation challenges and security issues along with its opportunities and advancements. 6G could be an ideal communication technology choice in Industry 5.0 owing to its ample advancements such as huge bandwidth, high data rate, and low latency. Since 6G communication technology assures high-quality experiences (QoE) and high-quality services (QoS), the amalgamation of Industry 5.0 and 6G communication technology could be proven as a game-changer in the modern smart world that can influence many contemporary applications, mainly focusing on holographic communication, UAV, virtual/augmented reality, and so on. However, the data exchanged in various heterogeneous networks and spanning diverse authoritative domains of the Industry 5.0 ecosystem are vulnerable to security and trust issues. This chapter discusses the implementation challenges and security issues of 6G communication technology in Industry 5.0.

Chapter 6: *Cyber-Physical System in AI-Enabled Smart Healthcare System* elaboratively discusses various CPS technologies required for its physical and cyber components and continuous health-monitoring techniques. Here, it compares several existing works through theoretical and tabular methods to cover all aspects in this era for improving the overall performance of CPS in AI-enabled Smart Healthcare System. It presents cyber-physical systems in smart healthcare, their characteristics, different technologies incorporated in the growth of CPS, and major challenges encountered during the successful implementation of cyber physical medication systems.

Chapter 7: *Service-Oriented Distributed Architecture for Sustainable Secure Smart City* underpins several challenges in the process of adopting an urban space as a smart city. Given the importance of urban data accumulation, storage, retrieval, and the simultaneous deployment of networked infrastructure, it proposes a high-level framework to better represent the various subsystems of this computing model. The framework is essentially a hierarchical model that represents the flow and storage of urban data, as well as the participation of different actors in the provision of services to citizens. It ensures an incremental implementation of a smart city and provides a seamless integration of new services as they are needed.

Chapter 8: *A Comprehensive Security Risk Analysis of Cliff Edge on Cyber-Physical Systems* discusses the protection threats including concerns in a CPS and recognizes the reasonable vulnerabilities, attack arguments, opponent's features, and an assemblage of provocations that necessitate being inscribed. Further, the chapter discusses some recommendations and failures in CPS.

Chapter 9: *Securing Financial Services with Federated Learning and Blockchain* gives an in-depth look at how Blockchain and federated learning (FL) are used in financial services. It starts with an overview of recent developments in both use cases. It explores and discusses existing financial service vulnerabilities, potential threats, and consequent risks. The chapter focuses on addressing the issues present in financial services and how the integration of Blockchain technology and FL can contribute to solving these challenges. These issues include data protection, storage optimization, and making more money in financial services. The chapter presents various Blockchain-enabled FL methods and provides possible solutions to solve several challenges including cost-effectiveness, automation, and security control in financial services. It also provides future research directions at the end of the study.

Chapter 10: *A Comprehensive Survey on Blockchain-Integrated Smart Grids* provides an overview of Blockchain technology and a number of recent research works presented in different literature on Blockchain integration into smart grid systems for energy management, energy trading, security and privacy, microgrid management, and electric vehicle management. It also presents the limitations and future directions of applying Blockchain in smart grids.

Acknowledgments

First and foremost, praises and thanks to God, the Almighty, for His abundant blessings and guidance, which have illuminated our path and enabled us to successfully complete this book. We extend our heartfelt appreciation to all the contributors, whose valuable insights and dedication have enriched the content and depth of this project. We are indebted to the dedicated publishing team at CRC Press, whose expertise and commitment have played an important role in shaping and bringing this book to fruition. Last but not the least, we would like to acknowledge our families and friends for their patience, understanding, and encouragement during the countless hours spent on this project.

It should be specifically noted that for this project, Uttam Ghosh has been partly supported by the US NSF under grants award numbers 2219741 and 2334391, also Visiting Faculty Research Program (VFRP) with the Information Assurance Branch of the AFRL, Rome, NY, United States; and the Information Institute (II).

Uttam Ghosh, PhD
Meharry Medical College, United States

Fortune S. Mhlanga, PhD
Meharry Medical College, United States

Danda B. Rawat, PhD
Howard University, United States

About the Editors

Uttam Ghosh joined Meharry Medical College, Nashville, Tennessee, United States, as Associate Professor of Cybersecurity in the School of Applied Computational Sciences in January 2022. Earlier, he worked as Assistant Professor in the Department of Computer Science at Vanderbilt University, where he was awarded the 2018–2019 Junior Faculty Teaching Fellow (JFTF). Dr. Ghosh earned his MS and PhD degrees in Electronics and Electrical Communication Engineering from the Indian Institute of Technology (IIT) Kharagpur, India, in 2009 and 2013, respectively. He had post-doctoral experiences at the University of Illinois in Urbana-Champaign, Fordham University, and Tennessee State University. He received research funding from the US National Science Foundation (NSF), US National Security Agency (NSA), National Aeronautics and Space Administration (NASA), and Thurgood Marshall College Fund (TMCF). Dr. Ghosh has published over 120 papers at reputed international journals including IEEE Transactions, Elsevier, Springer, IET, and Wiley, and in top international conferences sponsored by IEEE, ACM, and Springer. He has coedited and published seven books: *Internet of Things and Secure Smart Environments, Machine Intelligence and Data Analytics for Sustainable Future Smart Cities, Intelligent Internet of Things for Healthcare and Industry, Efficient Data Handling for Massive Internet of Medical Things, Deep Learning for Internet of Things Infrastructure, How COVID-19 is Accelerating the Digital Revolution: Challenges and Opportunities,* and *Security and Risk Analysis for Intelligent Edge Computing.* Dr. Ghosh is listed in the Stanford-Elsevier list of the world's most-cited scholars, placing him among the top 2% of cited scientists. He is a senior member of the IEEE and a member of ACM.

Fortune Mhlanga is a proven leader with a strong track record of developing and managing organizations in the academic, private, and public sectors with a specific focus on computing and technology. He joined Meharry Medical College in August 2020 as executive director of the Data Science Institute (DSI). In February 2021, he became the founding dean of the School of Applied Computational Sciences (SACS), the fourth school in Meharry's proud history, after leading the transition of the DSI to the school level. He also serves as founding senior vice president of the Enterprise Data and Analytics (EDA) division whose establishment, in March 2022, posited an advanced technology-driven enterprise data ecosystem that plays a critical role in achieving success in programs and initiatives across the entire Meharry enterprise as it caters for all data storage, harmonization, management, stewardship, and business intelligence in Meharry's clinical, research, academic, and business enterprises. Previously, Mhlanga served as Founding Dean of the College of Computing and Technology and Professor of Computer Science, Data Science, and Software Engineering at Lipscomb University. His tenure at Lipscomb began in August 2011 as Director of the then newly established School of Computing and Informatics. He led the school through a period of growth that resulted in its transformation into the College of Computing and Technology in 2014, offering 11 bachelor's degrees and 3 master's degrees. Before his appointment at Lipscomb, Mhlanga was Professor of Computer Science in the School of IT & Computing (SITC) from August 2007 to July 2011. He also served as Director of the SITC during his last year at ACU. From 2002 to 2007, he served as Associate Professor and subsequently Professor and Founding Chair of the computer science department at Faulkner University. Mhlanga was the Founding Director of the Informatics and Electronics Institute at the Scientific and Industrial Research and Development Centre in Harare, Zimbabwe, from 1998 to 2002. He was Senior Lecturer in the computer science department at the University of Zimbabwe from 1993 to 1994 and subsequently served as Chair of the department until December 1997. Mhlanga has published widely in computing, technology, and computational sciences including modeling and simulation. His experience also includes several domestic and internationally based academic and research fellowships. Mhlanga earned his bachelor's degree in computer science from Harding University in 1984, and his master's and PhD degrees in computer science from the New Jersey Institute of Technology in 1989 and 1993, respectively. His PhD work focused on database systems and resulted in the conception of a Data Model and Query Algebra for Office Documents.

Danda B. Rawat is Executive Director, Research Institute for Tactical Autonomy (RITA)—a University Affiliated Research Center (UARC) of the US Department of Defense; Associate Dean for Research & Graduate Studies; a Full Professor in the Department of Electrical Engineering & Computer Science (EECS); Founding Director of the Howard University Data Science & Cybersecurity Center; Founding Director of DoD Center of Excellence in Artificial Intelligence & Machine Learning (CoE-AIML); Director of Cyber security and Wireless Networking Innovations (CWiNs) Research Lab, Graduate Program Director of Howard CS Graduate Programs; and Director of Graduate Cybersecurity Certificate Program at Howard University, Washington, DC, United States. Dr. Rawat is engaged in research and teaching in the areas of cybersecurity, machine learning, Big data analytics and wireless networking for emerging networked systems including CPSs (e-Health, energy, transportation), Internet-of-Things, multi domain operations, smart cities, software-defined systems, and vehicular networks. Dr. Rawat is the recipient of NSF CAREER Award in 2016, Department of Homeland Security (DHS) Scientific Leadership Award in 2017, Provost's Distinguished Service Award 2021, Researcher Exemplar Award 2019 and Graduate Faculty Exemplar Award 2019 from Howard University, the US Air Force Research Laboratory (AFRL) Summer Faculty Visiting Fellowship 2017, Outstanding Research Faculty Award (Award for Excellence in Scholarly Activity) at

GSU in 2015, the Best Paper Awards (IEEE CCNC, IEEE ICII, IEEE DroneCom and BWCA), and Outstanding PhD Researcher Award in 2009. He has delivered over 50 keynotes and invited speeches at international conferences and workshops. Dr. Rawat has published over 300 scientific/technical articles and 11 books. He has been serving as Editor/Guest Editor for over 100 international journals including being Associate Editor of IEEE Transactions on Cognitive Communications and Networking, Associate Editor of IEEE Transactions of Service Computing, Editor of *IEEE Internet of Things Journal*, Associate Editor of *IEEE Transactions of Network Science and Engineering,* and Technical Editor of *IEEE Network*. Dr. Rawat earned a PhD from Old Dominion University, Norfolk, Virginia. Dr. Rawat is Senior Member of IEEE and ACM, a member of ASEE and AAAS, and Fellow of the Institution of Engineering and Technology (IET). He is ACM Distinguished Speaker (2021–2023).

Contributors

Sourav Banerjee
Department of Computer Science and Engineering, Kalyani Government Engineering College, Kalyani, West Bengal, India

Sudip Barik
Department of Computer Science and Engineering, Kalyani Government Engineering College, Kalyani, West Bengal, India

Asmita Biswas
Department of Computer Science and Engineering, Institute of Engineering and Management, Kolkata, West Bengal, India

Pushpita Chatterjee
Deaprtment of Electrical Engineering and Computer Science, Howard University, Washington DC, United States

Debashis Das
Department of Computer Science and Data Science, Meharry Medical College, Nashville, TN, United States

Priyanka Das
Deaprtment of Computer Science and Engineering, National Institute of Technology, Ravangla, Sikkim, India

Agni Datta
Department of Computer Science and Engineering, VIT Bhopal, Madhya Pradesh, India

Uttam Ghosh
Department of Computer Science and Data Science, Meharry Medical College, Nashville, TN, United States

Mahesh Chandra Govil
Deaprtment of Computer Science and Engineering, National Institute of Technology, Ravangla, Sikkim, India

Pon Harshavardhanan
Department of Computer Science and Engineering, VIT Bhopal, Madhya Pradesh, India

Charles A. Kamhoua
Network Security Branch, US Army Research Lab, Adelphi, MD, United States

Yash Kartik
Department of Computer Science and Engineering, VIT Bhopal, Madhya Pradesh, India

Sumathi Lakshmiranganatha
Los Alamos National Laboratory, Los Alamos, NM, United States

Jerry Chun-Wei Lin
Department of Computer Science, Western Norway University of Applied Sciences, Bergen, Vestland, Norway

Koustav Kumar Mondal
Department of Computer Science and Engineering, Indian Institute of Technology (IIT) Jodhpur, Rajasthan, India

Koushik A. Manjunatha
Department of Instrumentation Controls and Data Science,University of Alabama, California, United States

Manash Kumar Mondal
Deaprtment of Computer Science and Engineering, University of Kalyani, Kalyani, West Bengal, India

Pravin Mundra
Medtronic, Minneapolis, MN, United States, the United States

Laurent L. Njilla
Cyber Information Branch, Air Force Research Lab, Rome, NY, United States

Akhilesh Pokale
Department of Computer Science and Engineering, VIT Bhopal, Madhya Pradesh, India

Danda B. Rawat
Deaprtment of Electrical Engineering and Computer Science, Howard University, Washington DC, United States

Sangram Ray
Deaprtment of Computer Science and Engineering, National Institute of Technology, Ravangla, Sikkim, India

Sidheswar Routray
Department of Computer Science and Engineering, Indrashil University, Gujarat, India

Deepsubhra Guha Roy
Department of Computer Science and Engineering, Institute of Engineering and Management, Kolkata, West Bengal, India

Dipanwita Sadhukhan
Department of Computer Science and Engineering, National Institute of Technology, Ravangla, Sikkim, India

Arijit Sil
Department of Information Technology, Meghnad Saha Institute of Technology, Kolkata, West Bengal, India

Yuvraj Singh
Department of Computer Science and Engineering, VIT Bhopal, Madhya Pradesh, India

Jyoti Srivastava
Department of Computer Science and Engineering, Indrashil University, Gujarat, India

Chandan Thota
Department of Computer Science and Engineering, VIT Bhopal, Madhya Pradesh, India

Yudong Zhang
Department of Medical Social Sciences, University of Leicester, Leicester, United Kingdom

Abbreviations

DNS: Domain Name System
LDAP: Lightweight Directory Access Protocol
mssql: Microsoft Structured Query Language
NetBIOS: Network Basic Input/Output System
NTP: Network Time Protocol
SNMP: Simple Network Management Protocol
SSDP: Simple Service Discovery Protocol
TFTP: Test and Performance Tools Platform
UDP: User Datagram Protocol

Chapter 1

Machine Learning and Deep Learning Approaches in Cyber-Physical Systems

Sumathi Lakshmiranganatha and Koushik A. Manjunatha

Chapter Contents

1.1   Introduction

Cyber-Physical System (CPSs) consist of physical systems interconnected with multiple software algorithms to form an intelligent system. With the evolution of the Internet, there has been a huge transformation of how the objects in a system are connected. Nowadays, the interconnected traditional objects are transformed to smart objects, allowing people to interact with the engineered systems CPS is an amalgamation of sensors, control, computation, and networking integrating into the physical systems that connect each other and also connect them to Internet. Technological advancements are not confined to a particular area when it comes to CPS. The advancements can be observed in the embedded sensors, software frameworks for data analytics, communication protocols, and computing resources to name a few. CPS contains devices or

DOI: 10.1201/ 9781003376712-1

things connected to the Internet to form a network to communicate and exchange data among each other. This network of devices in a CPS is called the Internet of Things (IoT). The things in IoT that are the physical objects are an integral part of our daily life which can be as simple as a smartwatch tied to the wrist, smart bulb controlled by voice command, a smart refrigerator that remembers weekly groceries to more sophisticated devices like wearable heart monitor, Radio Frequency Identification (RFID) devices, smart parking lots, self-driving cars, smart building, and many more. The heart of all these devices are various sensors that are embedded which collect data based on the time interval set for the device. Therefore, IoT is driving the world toward sensor-rich data that are continuously streamed and exchanged with each other through better computing resources, sensing devices, and communication capabilities.

CPS applications can be applied to a wide range of applications namely: Consumer applications like a smart home; critical infrastructure applications like smart grid and smart city; industrial applications like agriculture and manufacturing; and organizational applications like healthcare, transportation, and military applications.One of the main objectives of CPS is to provide intelligent services.

According to a forecast by International Data Corporation, there will be 41.6 billion IoT devices that will be connected by 2025 which will inherently become a part of CPS. As the network becomes large scale, new challenges arise with privacy and security, data storage, communications, computations, and management of the devices in the network. As IoT generates a huge amount of data, traditional data processing and storage will not work for such networks. The data obtained from IoT devices can be used for improving the user experience or enhancing the IoT framework performance. The data can be studied for pattern behavior, predictions, and assessments of the IoT network. Also, the data representation in an IoT network is heterogeneous, and the devices generate a continuous stream of raw data. Therefore, a single data-processing framework cannot be applied to the IoT-generated data which paves the way for a need for new computational methods instead of traditional methods.

Also, the decision-making process and control actions of CPS in certain computing layers demand real-time solutions. Traditional analytical methods are not suitable for achieving real-time solutions. Therefore, data-driven methods are explored to minimize the computations and achieve real-time solutions. The most commonly used data-driven method is ML. ML methods can extract useful information from the data generated. This enables in better automation process, and the devices can change their behavior or take necessary actions based on the knowledge inferred from the data which is a vital characteristic of a CPS. The convergence of ML also improves communication and computations. While DL is a subset of ML, they both have different capabilities. Therefore, in this chapter, we distinguish ML and DL methods separately. ML and DL approaches have seen tremendous advancements in the last decade. ML and DL methods have been implemented to a wide spectrum of applications from image processing, stock market analysis, behavioral analysis, and natural language processing to mention a few. They have shown promising results with close to human performance. ML and DL have proven to be strong tools for huge data volumes.

1.2   Cyber-Physical Systems

CPS intertwines the physical components and the software components to perform operations at different time scales. It provides the combination and coordination of the physical, computation, and software elements of the systems. The components in a CPS interact and exchange information at different levels to maintain the synchronism of the system. The main goal of CPS is to improve the quality of human life by developing a smarter environment and providing a simplified lifestyle with a reduction in cost, energy, and time. CPS also has potential applications which can benefit multiple industries. In CPS, the physical and networking devices are connected and exchange data with or without any human intervention or inputs. The layer in which the devices of the CPS are connected through the Internet or network is known as IoT. Therefore, in many applications, IoT can be treated as a subset of CPS as shown in Figure 1.1.

The models that consist of CPS and IoT have a certain degree of overlap based on the requirements of the application. Based on the percentage of overlap between CPS and IoT, the models are divided in four categories as described in Figure 1.2. In this chapter, the main focus is the model in which IoT is the subset of CPS.

CPS applications involve Internet, and inherently, the devices connected in the application become IoT devices. The aim of IoT is similar to the CPS to achieve a smarter environment except without any human interaction, making it a completely automated process. Therefore, majority of ML and DL models are implemented and tested at the IoT layer as it is well known for Big data and fast streaming data generation. The ML/DL models which are implemented for IoT are inherently part of CPS. Depending on the application and needs, appropriate data processing and preparation methods need to be performed. The flow of data can be summarized as follows: the raw data of the environment is collected from the IoT devices and sensors; the important features or knowledge from the data is extracted; the data is sent to other objects, servers, or devices via the Internet. In recent

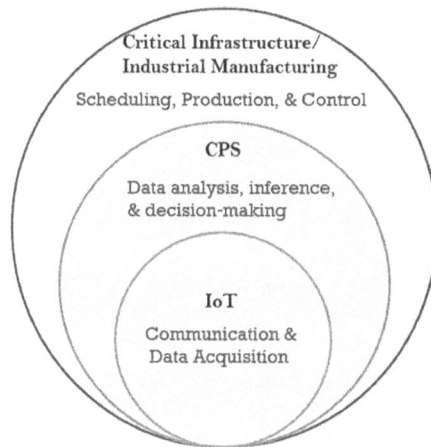

Figure 1.1 CPS–IoT connection.

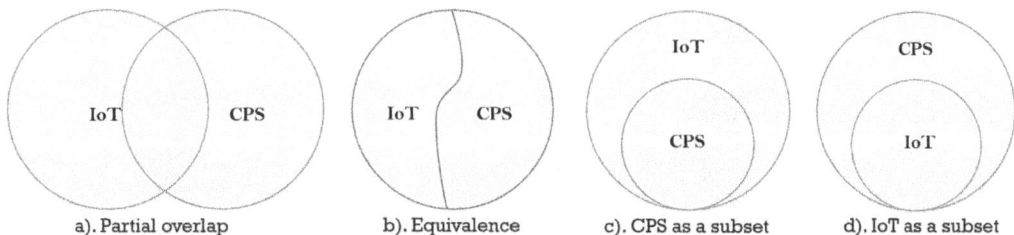

a). Partial overlap b). Equivalence c). CPS as a subset d). IoT as a subset

Figure 1.2 CPS/IoT overlap models.

years, Big data and data analytics for IoT have gained immense research interests due to the following reasons:

1. **Huge volume of data generated:** A huge number of devices are deployed for various IoT applications in a CPS, and these devices/sensors will generate continuous streams of data. For example, in a smart grid, phasor measurement units (PMU) are placed to capture the phasor values of the grid in real time. The phasor data generated by PMU is a huge volume of continuous data.
2. **Heterogeneous data:** IoT network consists of a large variety of devices due to the convergence of multiple applications and their requirements. Therefore, the data acquired by different applications have different parameters sampled at varying times that result in heterogeneous data. For example, IoT devices/sensors used for smart homes are different from the devices/sensors used in military surveillance.
3. **Inconsistency and noisy data:** Inconsistency in IoT data is an inevitable problem due to uncertainties associated with the devices/sensors. The data captured can be noisy due to faulty devices or cyber intrusion. This leads to noisy data or gross errors in the data during transmission and acquisition. For example, the data from PMU could be modified by cyberattacks during transmission from a device to the cloud.

1.2.1 Data Characteristics

The Big data characteristics can be described through the 6Vs presented in (1): volume, variety, velocity, veracity, variability, and value. Based on the reasons described, the data generated by a CPS application can be characterized in terms of Big data characteristics.

1. **Volume:** The devices in CPS generate huge amounts of data, clearly indicating Big data.
2. **Velocity:** The data generated by the devices in CPS are sampled at very high velocity in general. However, the velocity of the data generated changes over time and is dependent on the application.
3. **Variety:** The data is heterogeneous in nature. The data representation acquired from the devices consists of various data types like time-series data, images, audio signals, video, sensor measurements in different system of units, and many more.
4. **Veracity:** It refers to the trustworthiness, quality, and reliability of the data generated from the devices. It plays an important factor in an intelligent decision-making approach.
5. **Variability:** This refers to the data flow at different rates. CPS consists of various devices that generate data at different time intervals. Also, the data flow in IoT networks can be at specific times like the exchange of data in wearable devices during a particular mode.
6. **Value:** It is the crucial characteristic of Big data. It refers to information that can be obtained from the data collected in business terms. In IoT, the value of the data can be derived on the basis of the application.

1.2.2 Computing Framework

It is also important to understand the different computing frameworks in an IoT and the methods of data processing for deploying intelligent models at each level. Three major computing frameworks constitute an IoT network as shown in Figure 1.3.

1. **Cloud computing:** In this computing architecture, the computing resources are located in remote locations, and the data is sent over the Internet. The resources are used to

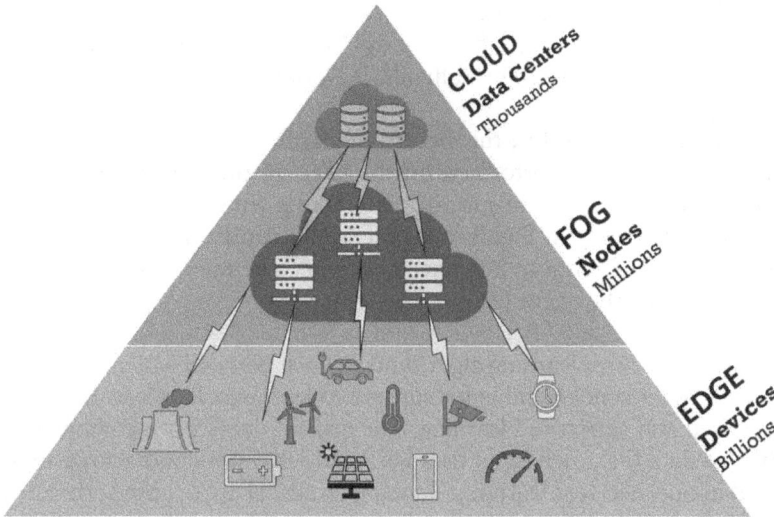

Figure 1.3 IoT computing framework.

store, gather, analyze, and process the data. Cloud computing supports different types of services based on the requirements like i) infrastructure as a service (IaaS), ii) platform as a service (PaaS), iii) software as a service, iv) mobile backend as a service (MBaaS). The cloud architectures vary depending on the user scenario as public, private, hybrid, or community-based architecture (2).

2. **Edge computing:** In this architecture, the computing and data processing happen at the source of the data or close to it in terms of physical location. In this computing framework, the devices are not continuously connected to the network; therefore, the only inference can be performed at the edge. The major advantages of edge computing are i) low latency (there is no data transfer between the central server and the device) and ii) increased analytical efficiency. Edge computing is more suitable for time-sensitive problems that demand real-time operations (2). Edge computing can be adhoc in nature and does not require extensive planning like cloud computing. In the near future, it has been predicted that around 45% of the IoT data will use edge architecture (3).

3. **Fog computing:** Here, the computing architecture can be categorized between edge computing and cloud computing. This was proposed by Cisco to address the latency issues caused by time-sensitive applications (4). Fog computing provides limited computing, storage, and networking facilities through virtual platforms.

1.3 Machine Learning

The ML approach is a data-driven approach that learns the features from the data sample provided to make predictions or decisions. There are different types of learning methods that ML uses to learn and extract the features from the data which are discussed as follows:

1. **Supervised learning:** In this type of learning, each input has its corresponding output labels. The objective of the learning algorithm is set. The learning algorithms try to learn the relationship between the input dataset and the output labels that are used for predicting the output label of the new set in similar input data.

2. **Unsupervised learning:** In this type of learning, the ML learning algorithm receives only the input data, and the output labels are not specified. Therefore, defining a common objective for this type of learning is difficult. It is commonly used to classify the data into different groups.
3. **Reinforcement learning:** In this, the algorithm learns an appropriate action or sequence of actions that need to be performed on the basis of the reward function. The agents learn through feedback mechanisms after interacting with the environment. These algorithms are applied widely in highly dynamic applications like robot control where the tasks are accomplished without a defined outcome. The reward function determines the success and failure of the agent (5).

Along with the above three learning algorithms, semi-supervised ML algorithm is used when the majority of the data is not suitable for supervised learning. This learning method falls between unsupervised and supervised learning. The general dataset is a combination of a small amount of labeled data (i.e., each input data has a labeled output and a large amount of unlabeled data used in unsupervised learning). In many practical applications, the labeled data acquisition cost for training is high due to the requirement of human experts; whereas unlabeled data acquisition is relatively inexpensive. Therefore, this method can be used for modeling when there is a large set of unlabeled dataset combining with a small set of labeled dataset.

Most of the applications use ML models which are based on the above learning mechanisms. In supervised learning, the type of tasks is dependent on the labels. The labels can be discrete category values or continuous values. If the labels belong to a single discrete category, the task is called a classification task; if the labels consist of one or more continuous values, then the tasks are classified as regression tasks. First, different types of ML models are explained followed by DL models.

1.3.1 ML Models

1. **Support vector machines (SVM):** SVM is one of the ML models that is widely used in both classification and regression tasks using supervised learning. It is a non-probabilistic binary classifier in which the major objective is to find a hyperplane in an N-dimensional space that separates both classes with a maximum margin. The data points falling on either side of the hyperplane can be denoted as different classes. The number of hyperplanes depends on the number of input features. For example, if the number of input features is three, then the hyperplane is 2D. As the number of input feature increases, the hyperplane dimension also increases. SVMs are among the best supervised learning models that can deal with high-dimensional datasets and have efficient memory usage. In (6), the SVMs can be trained in an online fashion. SVM can be extended to solve regression problems using the support vector regression (SVR) process. SVR model depends only on a subset of the training data since the model only cares about training points within the margin, and all the data points beyond the margin are rejected. The mathematical equations to compute y and loss function are shown below. **Model:**

$$y = \text{sign}\left(\sum_{i=1}^{n} \alpha_i y_i \langle x_i, x \rangle + b\right)$$

Loss Function:

Hinge Loss $= \max(0, 1 - y \cdot f(x))$

2. **Naive Bayes (NB):** NB is a probabilistic classification model based on Baye's theorem with an assumption of independence between the predictors. NB first creates a frequency table of all the classes and then creates a likelihood table from it. Based on the likelihood, it calculates the posterior probability. NB is highly scalable and requires a small number of data points for training. They can also deal with high-dimensional data points and trains fast.

3. **K-nearest neighbors:** KNN is a non-parametric classification model that classifies an unseen data point by looking at the K given data points in the training set that are closest to the feature space. The commonly used distance metrics are Euclidean distance, Hamming distance, or L_∞ norm. Since the algorithm relies on distance for classification, if there are multiple data represented in different physical quantities, it is suggested to normalize the data before training. One of the major drawbacks of KNN is scalability. KNN requires storing the entire training dataset which is cumbersome for large datasets.

4. **Decision trees (DT):** DT is a non-parametric supervised learning model that can be used for classification and regression tasks. It is a decision-support tool with the main objective to predict the value of the target variable based on simple decision tools based on features. DT splits into multiple branches until a decision is made. The DT is drawn in an upside-down fashion with its root at the top. Based on the decision it splits, a DT will have multiple branches. The terminal node is called a leaf node which cannot be split any further since it is the end of the branch. The performance of the trees can be increased by pruning which involves the removal of branches that make use of features with low importance. DT is simple to implement and understand. DT is independent of data features; therefore, data preparation is minimal. However, DT is unstable because if there is a small change in the data, it can lead to a completely different tree generation. DT can create complex trees that cause overfitting. The mathematical method to compute decision trees is given as

Model:

$$\mathcal{Y} = \text{Tree}(x)$$

Loss Function:

$$\text{Gini Index} = 1 - \sum_{i=1}^{C} (p_i)^2$$

5. **Random forests (RF):** RF is constructed by combining multiple DT that can be used for classification and regression tasks. Each tree is trained on a subset of data that is chosen randomly with replacement. If there are M input variables, then each node gets m variables selected randomly out of M where m «M, and the best split is used on these m values. The value of m remains unchanged during forest growing. RF has high accuracy, but it causes overfitting. A large number of trees can make the algorithm slow and infeasible for real-time applications. The below equation is the mathematical representation for computing random forests.
Model:

$$\mathcal{Y} = \text{MajorityVote}(\text{Tree}_1(x), \text{Tree}_2(x), \dots, \text{Tree}_n(x))$$

6. **K-means clustering:** It is an unsupervised learning model which involves dividing the data points into K clusters; each described by the mean of the sample in the cluster. The mean is generally defined as the cluster center or centroids. It assigns the data point to a cluster based on the computed distance between the data point and the cluster centroid

which is minimum. K-means clustering is a very fast and scalable algorithm. However, K-means assign one data point to one cluster which may result in inappropriate clusters. Also, it is not robust for the outliers.

7. **XGboost:** XGBoost is an ML algorithm that has gained popularity in recent years. This algorithm works well for structured and tabular data. It is a gradient-boosted DT that is been developed for faster execution speed and higher performance. XGBoost algorithm implements optimizations at hardware, software, and algorithmic levels. In XGBoost, the process of sequential tree building is parallelized due to the loop interchangeable property. The hardware resource is efficiently utilized by allocating internal buffers in each thread to store gradient statistics. The overfitting of the model is prevented by penalizing through L1 and L2 regularization.

In Table 1.1, a summary of different ML models discussed above is presented with potential applications that can be implemented.

1.3.2   Feature Extraction Techniques

Feature extraction techniques are widely used in ML algorithms due to their inability to handle a large number of variables in high-dimensional spaces. Some of the variables in the dataset may be redundant or do not have dominating features that can be extracted by the ML algorithm. Therefore, the variables which are redundant or have features that do not distinguish significantly from others are removed, and only the dominating features are retained using feature extraction techniques. It reduces the dimensions of the input dataset. Hence, the number of resources required to describe the dataset is reduced and aids in ML model efficiency. Feature extraction techniques were widely implemented in the early 1990s and 2000s due to hardware resource limitations. The common feature extraction technique is principal component analysis (PCA) and canonical correlation analysis (CCA). In PCA, the dimension of the data is reduced by projecting each data point onto only the first few principal components while preserving as much variation in the data as possible. There are other similar techniques that are used for feature extraction like linear discriminant analysis and independent component analysis. CCA is a linear dimensionality reduction technique that infers the information from the cross-covariance matrices. CCA is closely related to PCA. CCA deals with two or more vector of features, unlike PCA which deals with one variable or a feature. Table 1.2 shows a brief comparison between CCA and PCA.

1.3.3   DL Models

DL model consists of multiple layers of artificial neural networks (ANNs) that are trained using supervised or unsupervised learning. DL which is a subset of ML can be classified into three categories: i) Generative models that are used for unsupervised learning, ii) discriminative models that are used for supervised learning, and iii) the hybrid models that can be classified as semi-supervised learning. The different DL models are discussed briefly in this chapter.

1. Fully connected deep neural network (FC-DNN) or multi layer perceptron (MLP): A fully connected deep neural network consists of one input layer, multiple hidden layers, and an output layer. Each layer has multiple neurons that are the processing units of the network. The neurons receive multiple input signals coming from the external environment denoted by a

Table 1.1 Summary of Different ML Models

Model	Learning Category	Input Data	Characteristics	Sample Applications
SVM	Supervised	Various	• It is a non-probabilistic model with low-computational complexity • It aims to classify the input data into n dimensions by drawing a n–1 hyperplanes • It is suitable for high-dimensional datasets with low memory usage	• Attack detection (7) • Intrusion detection (8) • Stability classification in smart grids (9)
NB	Supervised	Various	• It calculates the posterior probability • It uses Bayes' theorem to calculate the probability of a particular feature set • Can deal with high-dimensional data	• Securing DDoS attack (10) • Smart agriculture (11)
KNN	Supervised	Various	• The objective is to classify the new data point by looking at the K closest points in the training set • For good performance, a new data point needs to be closely associated with the training set • It requires storing the entire training set which makes it non-scalable for large datasets	• Attack detection for small networks (7) • Anomaly detection (12)
DT	Supervised	Various	• The training data is represented as trees and branches for the learning process • The model is pretrained and then used to predict for new samples • The trees are used to split the dataset into branches based on a set of rules • Due to its constructive nature, it requires large storage	• Smart healthcare scheduling (13) • Intrusion detection (14)
RF	Supervised	Various	• It is constructed combining several DTs • It is robust to over-fitting • It is not suitable for some real-time applications with large datasets since it is constructed with multiple DTs	• Detection of unauthorized IoT devices (15) • Crop yield prediction (16) • Traffic accident detection (17) for smart cities
K-means clustering	Unsupervised	Various	• It is an unsupervised learning method that classifies the input data into K clusters based on the feature similarities • It is very fast and highly scalable • The cluster centers do not perform well for outliers	• Privacy protections in wireless sensor networks (18) • Smart cities—traffic network assignment (19)
XGBoost	Supervised	Various	• It works for tabular and structured data • It has faster performance compared to other ML algorithms • Can handle missing data	• Anomaly detection (20) • Fall detection in the elderly using wearable devices (21)

Table 1.2 Comparison of CCA and PCA

Method	CCA	PCA
Objective	Maximizes correlation between two datasets (multivariate analysis)	Maximizes variance in a single dataset (univariate analysis)
Data type	Used for two sets of continuous variables (e.g., feature vectors, time series)	Used for a single set of continuous variables (e.g., feature vectors, images)
Output	Canonical variates (linear combinations) representing correlated patterns	Principal components (orthogonal axes) representing most significant variation
Applications	Multimodal data integration (e.g., neuroscience, genetics)	Dimensionality reduction (e.g., data visualization, feature extraction)

set of input values or from other neurons connected to them. The relationship of each input is calculated by multiplying the synaptic weight to validate all the information received by that neuron. The weighted sum is then passed through an activation function to limit the output of the neuron. Depending on the task defined, the output of the neuron can be a continuous or discrete value. The neural network is supposed to learn the relationship between input and output provided to the network. Once the network learns the relationship, it is capable of predicting the output for any given input. To generalize the output, the weights associated with the neurons need to be tuned by a training process using a suitable learning algorithm.

The training process begins with the initialization of random synaptic weights for the network. The weights are adjusted or changed at the end of each iteration in a particular direction to reduce the error between the desired and the actual outputs. An iteration to complete the adjustment of the synaptic weights and the threshold for all the input patterns presented to the network is called an epoch. The network is trained after going through the defined number of epochs with the given training data; the network can be used for testing other values. The error between the actual output and the desired output is termed as the cost or loss function. The training algorithm aims at achieving the minimum cost function to train the network by adjusting the weights for each layer to classify the data. It is crucial to have correct weights at each layer to train the network. Therefore, the general method for obtaining the weights updates of the networks is done by optimization methods. There are several optimization methods like gradient descent, momentum, and RMSProp to update the weights. The commonly used optimizer to train the network is the gradient descent optimizer. This optimizer implements a full-batch fundamental backpropagation algorithm which updates the weights for the entire dataset using its gradient. The weights must have the negative direction of the gradient to find the global minimum. The cost function is minimized to find its global minimum. One of the widely used training algorithms for FC-DNN is the backpropagation algorithm. The training process using the backpropagation algorithm is done by the successive application of two specific stages: a) forward stage and b) backward stage as shown in Figure 1.4. The forward stage is the first stage, where the inputs from the training data are inserted into the network inputs and are propagated layer-by-layer until the output is calculated. In the second stage, differences between the actual and desired outputs are calculated and used to adjust the weights and thresholds of the neurons in the network. Therefore, the successive application of forward and backward stages allows the synaptic weights and thresholds of the neurons to be adjusted automatically in each iteration.

Neural Network

Single Neuron

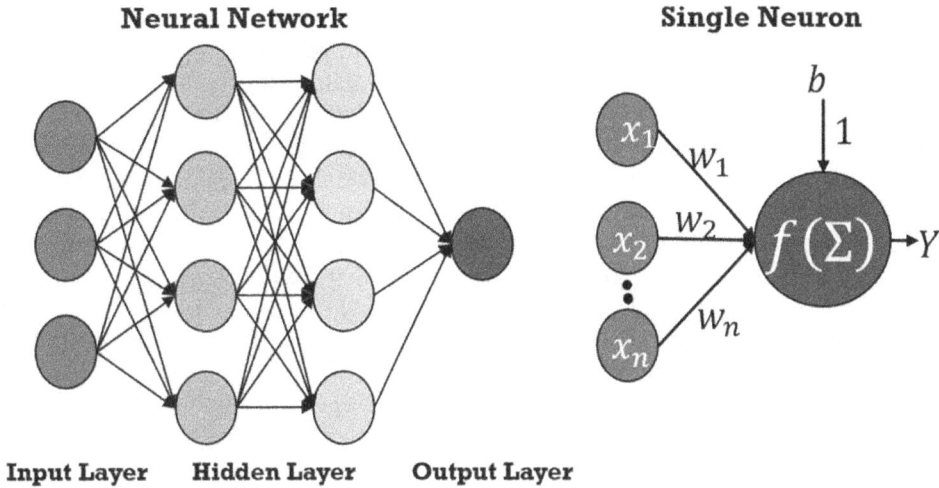

Input Layer   Hidden Layer   Output Layer

Figure 1.4 Backpropagation algorithm for FC-DNN.

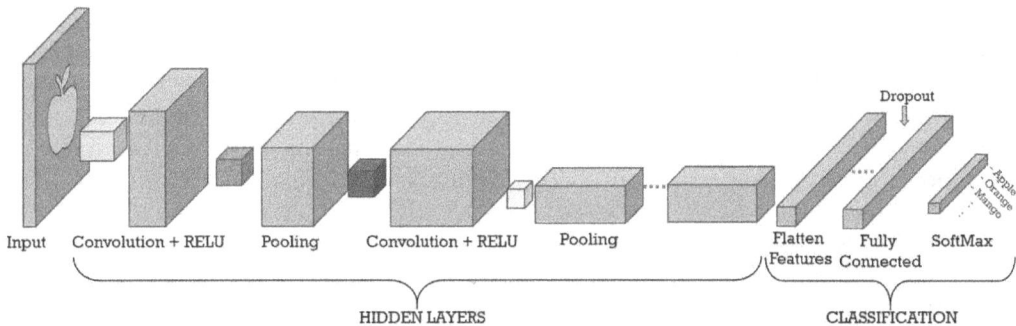

Figure 1.5 CNN architecture.

2. **Convolution neural networks (CNN):** FC-DNN is a dense network due to connections between each neuron in every layer limiting the network to scale. For images or vision-related tasks, where the image contains the translation invariance characteristics that are not learned in the FC-DNN networks. However, CNN solves this problem using convolutional kernels that capture the translation invariance charateristics in the image. A typical CNN architecture is shown in Figure 1.5. The input to a CNN typically is a grid topology. The data is represented either in a 1D grid or 2D grid which can be an image, audio signal, or video frame. CNN extracts the features of the image using convolutional kernels in each hidden layer. The output of the convolutional layer is the feature map. The feature map is obtained by calculating the inner products of the input and the filter. Each convolutional kernel can be different sizes starting from 1 × 1. The general rule of thumb is to have few feature maps with a large kernel size at the beginning of the CNN network so the large features are extracted. As the network gets deep, more kernels with smaller sizes are used to extract nonlinear and smaller features. In many CNN implementations, a small filter size is used to have deeper networks. The convolution layer is followed by a detection layer which has an activation function.

Usually, the activation function is rectified linear unit (ReLU) for the CNN. Following the detection layer, the pooling layer is applied. It downsamples the input data to reduce the number of learnable parameters and number of computations in the network (usually by half at every layer). The pooling layer operates over each activation map independently. Max pooling and average pooling are used in the CNN architecture. The last layer of the CNN architecture is the fully connected layer which converts the matrix to a 1D vector. The output of the fully connected layer is the number of desired outputs. One of the major advantages of using CNN over DNN is that each neuron in the DNN connects to all input dimensions; whereas in CNN, only a small portion of the input is connected to a neuron. This helps in the great reduction of the parameters that need to be trained. Given an input image or feature map X with dimensions $N \times H \times W \times C$, where N is the number of samples, H is the height, W is the width, and C is the number of channels. The CNN applies convolutional filters or kernels K, nonlinear activation functions σ, and pooling operations Pool.

The output feature map of a CNN layer Y can be obtained using the following mathematical formulation:

$$Y = \sigma(\mathrm{Conv}(X, K) + b)$$

3. **Recurrent neural networks (RNN):** While CNN is suitable for a spatially oriented application, RNN is suitable for time series applications. RNNs carry the information time step in their internal memory which is known as hidden state h_t making it suitable for time series or sequential applications. A standard RNN unrolled in time is shown in Figure 1.6. The output of h_t is calculated from the input x_t and the history h_{t-1} which is the output at time $t - 1$. Block A in the figure represents a layer of n neurons. Each neuron is equipped with a feedback loop that returns the current output as an input for the next step. However, for the standard RNN, the number of hidden states to store temporal information is one. Therefore, standard RNN is not suitable for long-term dependency of the data. Also, standard RNN implementation results in vanishing gradients and gradient explosion. This problem affects the weight updates during the training process leading to erroneous learning.

To overcome the problems posed by the standard RNN network, the variant called long short-term memory network (LSTM) is developed. LSTM network is made up of LSTM cell

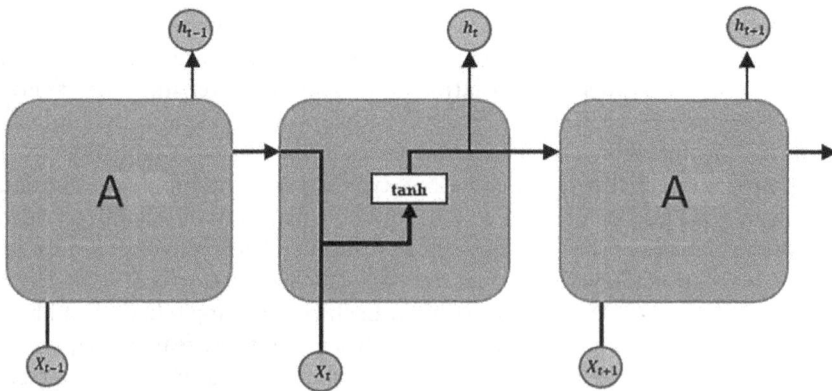

Figure 1.6 RNN architecture.

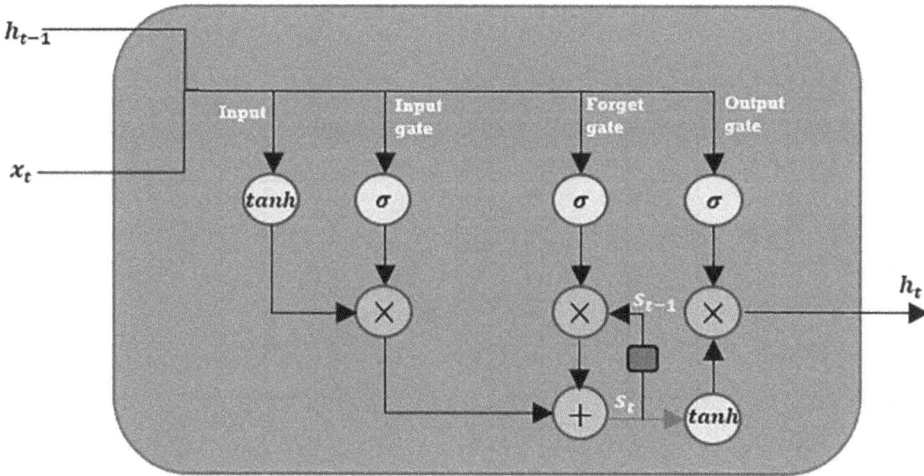

Figure 1.7 LSTM cell.

block shown in Figure 1.7 instead of standard RNN layer. The backbone component of the LSTM network is the cell state that has a memory that helps to remember the past values by accumulating the state information *st* at every time step t. The LSTM cell block consists of three gates namely: input gate, forget gate, and output gate. The sigmoid activation function is applied to all three gates to accumulate the information and control the information flow to trap the gradient in the cell. Based on which gate is activated, the input information is accumulated, forgotten, or passed on to the next state. LSTM networks have proven to perform better than the standard RNN networks.

Given an input sequence $X = \{x_1, x_2, \ldots, xT\}$, the LSTM computes hidden states H and cell states C using input gate i, forget gate f, output gate o, and a candidate cell state $\tilde{C}$.

The LSTM equations are as follows:

$$i_t = \sigma\left(W_{xi}x_t + W_{hi}h_{t-1} + b_i\right)$$
$$f_t = \sigma\left(W_{xf}x_t + W_{hf}h_{t-1} + b_f\right)$$
$$o_t = \sigma\left(W_{xo}x_t + W_{ho}h_{t-1} + b_o\right)$$
$$\tilde{C}_t = \tanh\left(W_{xc}x_t + W_{hc}h_{t-1} + b_c\right)$$
$$C_t = f_t \cdot C_{t-1} + i_t \cdot \tilde{C}_t$$
$$h_t = o_t \cdot \tanh\left(C_t\right)$$

4. Autoencoders (AE): AE is a type of DL model that falls under the unsupervised category. AE network has two parts. First, the encoder learns a lower-dimensional feature representation from an unlabeled dataset. Second, in the decoder, the code or the representation features of the data try to reconstruct the original input data. The encoder and decoder can have multiple hidden layers. AE has the same number of input and output units. The basic architecture is shown in Figure 1.8. Due to their ability to reconstruct the input at the output layer, autoencoders have a wide range of applications in industries for

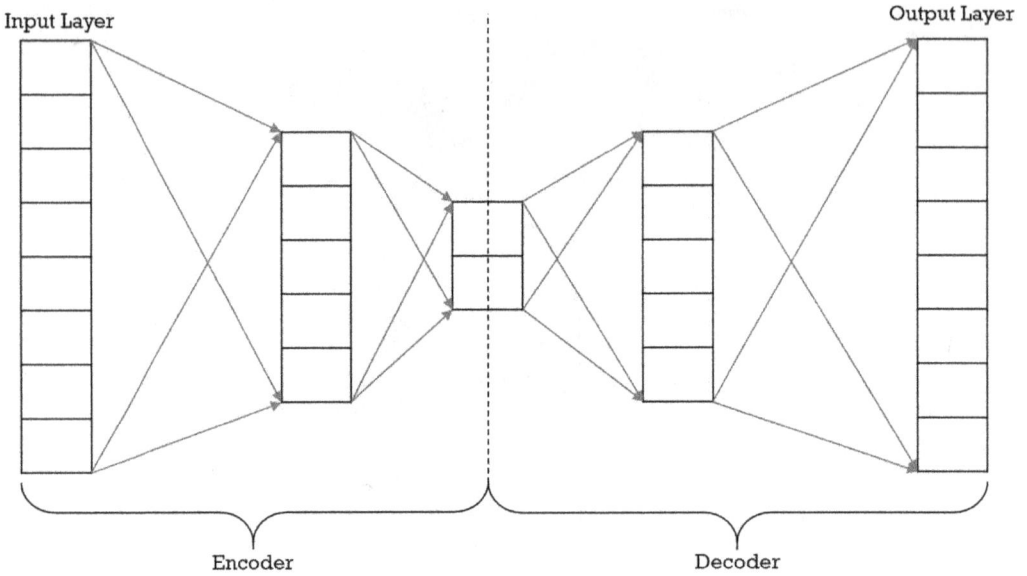

Figure 1.8 Autoencoder architecture.

fault detections and anomaly detection. There are variants of AE namely, contractive autoencoder (CAE), variation autoencoder (VAE), sparse autoencoder (SAE), and de-noising autoencoder (DAE).

The autoencoder objective function minimizes the reconstruction error between the input X and the output $\hat{X}$:

$$\min_{E,D} \mathcal{L}(X, \hat{X}) = \|X - D(E(X))\|^2$$

And VAE objective function combines a reconstruction loss and a regularization term using the Kullback–Leibler (KL) divergence:

$$\min_{E,D} \mathcal{L}(X, \hat{X}, z) = \mathbb{E}_{q(z|X)}[-\log p(X \mid z)] + \mathrm{KL}(q(z \mid X) \| p(z))$$

5. Generative Adversarial Networks (GANs): GANs are used for generative modeling using DNN like CNN and RNN. For a given training set, GANs learn to generate new data with the same statistics as the training set. GANs consist of two neural networks: i) generative network and ii) discriminative network. The two networks work together to produce high-quality and synthetic outputs. The generative networks learn from the input dataset to generate new data. The discriminative model tries to classify the date if it is real (actual input data from the domain) or fake (generated from the generative model). The network is trained until the discriminative network cannot distinguish between the actual and fake data. In GANs, two networks compete with each other based on zero-sum games in which one network tries to maximize the value function, and the other network tries to minimize it. The generative network generates a batch of samples in an unsupervised manner. The discriminative network receives two inputs: i) output of the generative network and ii) the real example from the domain to classify them as real or fake. The discriminative network gets updated to get better at classifying the real and fake samples in the next round. The generative networks get updated on the basis of how well the generator samples deceived the discriminative network. In Figure 1.9, the GANs architecture is shown.

GANs have gained immense popularity due to the capability of generative high-quality realistic examples in a wide range of applications. It has shown promising results in cases of image-to-image translation. The GAN objective function can be formulated as a minimax game:

$$\min_{G} \max_{D} V(D,G) = \mathbb{E}_{x \sim p_{\text{data}}(x)}[\log(D(x))] + \mathbb{E}_{z \sim p_z(z)}[\log(1 - D(G(z)))]$$

6. **Restricted Boltzmann Machine (RBM):** RBM is a stochastic ANN that learns the probability distribution from a given training dataset. RBMs consist of two layers: a) visible layer and b) hidden layer. The architecture of RBMs is shown in Figure 1.10. In RMBs, the neurons are called nodes, and each node is connected to each other across the layers, but

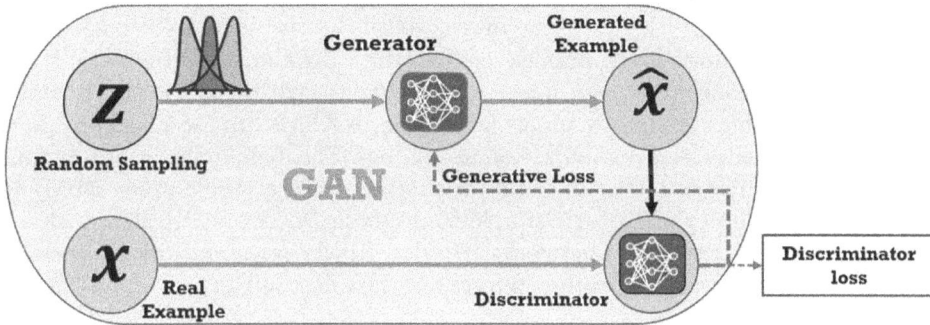

Figure 1.9 GAN structure.

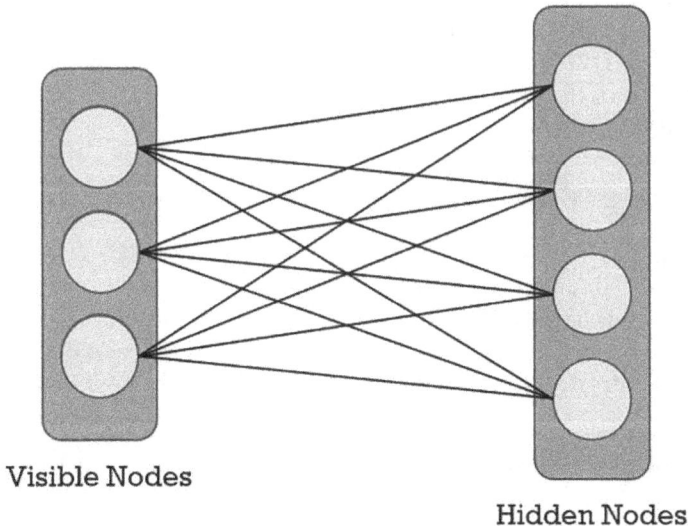

Figure 1.10 RBM structure.

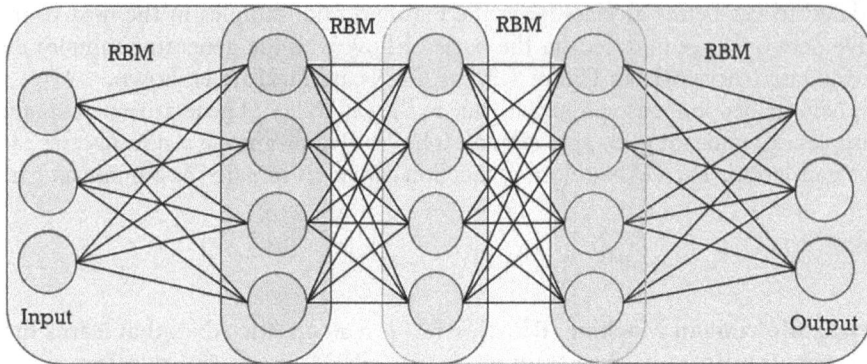

Figure 1.11 DBN structure.

no two nodes are connected in the same layer. The training data is assigned to the visible node that needs to be learned. Then, the output of all the visible nodes is passed to each hidden node in the hidden layer. Because of this, the RBMs form a symmetric bipartite graph. The results from the hidden node are passed through the activation function to produce one output for each hidden unit. The training of RBMs can use a backpropagation algorithm and different weight optimization techniques. The main objective is to minimize the error and maximize the product of all probabilities of the visible units. RBMs function in a similar fashion as AEs for a simple feed-forward network. RBMs are also used as building blocks in deep belief networks (DBNs). RBMs are useful for dimensionality reduction problems, regression, feature extraction, classification, and topic modeling.

7. **Deep Belief Networks (DBNs):** DBNs are generative models which consist of multiple hidden layers which have interlayer connections but not intra-layer connections. DBNs are composed of unsupervised networks like RBMs and AEs. DBNs learn to reconstruct the input as well as extract the features in a hierarchical manner. Once the DBN is trained in an unsupervised manner, DBNs can be further trained to perform classification tasks by adding a softmax layer. The DBNs are trained in a greedy fashion with one layer at a time. This makes the DBNs one of the most efficient and fast algorithms in DL. The architecture of the DBNs is shown in Figure 1.11.

In Table 1.3, a summary of different DL models discussed above is presented with potential applications that can be implemented.

1.4   CPS Applications

ML and DL algorithms have been researched for multiple CPS applications like smart grids, smart healthcare, and smart homes to name a few. A brief discussion on ML and DL methods applied to IoT infrastructure and security is provided in the next section.

1.4.1   CPS Infrastructure Security

One of the major concerns in the CPS is the privacy and security of the data generated by numerous devices in the system. A major threat for the security of CPS arises from the IoT

Table 1.3 Summary of Different DL Models

Model	Learning Category	Input Data	Characteristics	Sample Applications
FC-DNN/ MLP	Supervised	Various	• It is a fully connected network • The error function is minimized using a backpropagation algorithm • It is a dense network with all neurons connected to neurons in the next layer	• Detection of distributed denial of service (DDoS) (22) • Brain tumor detection in healthcare (23)
CNN	Supervised	1D (sound), 2D (images, video frames, etc.)	• It is the robust supervised DL method • It has fewer connections between layers compared to FC-DNN and good for spatial dataset • With new features, the CNNs are scalable and their training time complexity is improved • It has a high computational cost • It requires a large training set to achieve good performance	• Traffic sign detection (24) • Plant detection (25) • Food analysis (26)
RNN	Supervised	Various, time-series data	• It considers the temporal dimension of the sequential data and learns them using hidden units of recurrent cell • The main issue of RNN is vanishing or exploding gradients • To address the vanishing or exploding gradient issue, a variant of RNN called LSTM is used • LSTM is suitable for long-time lag data • The memory cell is accessed by gates depending on the input data	• Activity prediction (27) • Energy demand prediction in smart grid (28) • Authentication and access control (29)
AE	Unsupervised	Various	• It is an unsupervised learning model that has two parts: i) an encoder that converts the input into an abstraction code and ii) a decoder that takes the abstraction code to reconstruct the input data • It is suitable for feature extraction and dimension reduction with no prior data knowledge • It is computationally expensive	• Botnet attack detection (30) • Real-time traffic speed estimation (31)

(Continued)

Table 1.3 (Continued)

Model	Learning Category	Input Data	Characteristics	Sample Applications
GANS	Semi-supervised	Various	• It simultaneously trains two models: i) a generative model that learns the data distribution and generates data samples and ii) a discriminative model that predicts the possibility that a sample originates from the training dataset via an adversarial process • It is suitable for noisy data • It is unstable and has difficulty in learning to generate discrete data	• Clinical decision support in healthcare (32) • Objection detection (33) • Cybersecurity of IoT systems (34)
RBM	Unsupervised, supervised	Various	• It is an unsupervised learning algorithm that uses deep generative models • It uses a feedback mechanism that allows vital feature extractions • Training is computationally expensive	• Intrusion detection (35) • Anomaly detection in IoT networks (36)
DBNs	Unsupervised, supervised	Various	• It consists of stacked RBMs that are suitable for hierarchical feature extractions • It has greedy layer-wise training to improve performance for an unsupervised learning • Training is computationally expensive	• Traffic prediction (19) • Stress monitoring in healthcare (37)

layer since the data is communicated over the Internet, and the data is visible to the entire world and easily accessible to hackers. This opens up a new research interest in implementing ML and DL models in areas like preventing malicious attacks by detecting an intruder or an unauthorized device/user and providing access control and authentication and detection of malware attacks to name a few. Research in (38) gives a survey of learning techniques for network intrusion detection systems (NIDS) for IoT. The survey in (39) gives a comprehensive survey of ML approaches for intrusion detection systems. The performance and implementation details of various ML models such as SVM, DT, NB, and K-means clustering were discussed. The research in (22) demonstrated the detection of DDoS attacks in IoT networks using MLP networks. (40) uses ensemble-based ML algorithms to mitigate botnet attacks against DNS, HTTP, and MQTT protocols that are utilized in IoT networks. The AdaBoost algorithm was used to distribute the network data. NB, DT, and ANN are used for the detection of botnet attacks. The proposed method had a high detection rate and low positive rate compared to the state-of-the-art techniques. The research in (41) demonstrates a multivariate correlation analysis to detect the DoS attack. The study was conducted for various DoS attacks for original data and normalized data. The detection rate of 99.95% was achieved for normalized data for various attacks, whereas 95.25% detection rate was achieved for the original data. In (42), the researchers propose a DNN approach for the user authentication technique for IoT-based human physiological and behavioral characteristics inherited from their daily activities. The proposed authentication method achieves 94% and 91% accuracy for 11 subjects for walking and stationary behaviors, respectively. The research in (7) demonstrates cyberattack detection in smart grids using ML models. Based on the results presented, SVM performs better in large-scale systems compared to other algorithms. However, KNN performs better for small-scale systems. The traditional ML algorithms like SVM, NB, and DT have a limitation when dealing with big network traffic data due to the shallow network architecture (43) for malware detection and classification. DL methods like CNN and RBMs are suitable for big network traffic data like IoT network data. Edge intelligence in CPS (44, 45) is another vital research area that is being explored to avoid data breaches and maintain the quality of the data in CPS. Processing and inferring the data at the edge would reduce the data exchange between various layers and devices and minimize the exposure to cyber threat.

1.4.2   Critical Infrastructure

ML and DL methods applied to a wide spectrum of applications like smart grids, healthcare, industries, agriculture, and transportation can be categorized as the critical infrastructure of CPS applications for a smart city. A brief literature review of ML and DL methods applied to these applications are presented.

The power grid is no longer an unidirectional communication with the advanced digital technologies for real-time analysis along with the integration of renewable energies to the grid. Various ML and DL models are implemented to forecast the load requirements in real-time. The research in (46) compares various ML models for load forecasting in a smart grid. In this research, they used data-imaging conversion-based CNN for the time-series data with temperature, weather, and date for load forecasting. The implementation of SVM for predicting the stability of the grid using synchrophasor data collected during the post-fault period of the system subjected to a large disturbance is shown in (9). Overall, 97% prediction accuracy was obtained for simulated cases containing PMU data. They achieve the highest performance for the proposed method with SVM being the second-best method. The researchers in (28) use a combination of CNN- and LSTM-based neural networks for residential energy consumption prediction. The CNN layer can extract the features between several variables impacting the energy consumption, and the LSTM layer is used to obtain the temporal information of the irregular trends in the time-series components. The proposed CNN–LSTM network method achieves the prediction performance very close to the actual data that was previously difficult to predict using the DNN method. The research in (47) uses the RBM-based DL method to predict a building's energy flexibility in real time. The research in (48) implemented the CNN for transient stability assessment and instability mode prediction. CNN is implemented to predict the stability of the system and classify the data into stable, aperiodic unstable, and oscillatory unstable. The best performance was achieved using the CNN when compared to other ML models with 97% accuracy. Along with CNN, a new class of DNN called ConvLSTM network is used to predict the parameters of the critical values like PMU voltage phasors to perform stability analysis is being researched in (49). This research explores the application of DNN for system-wide parameter prediction. A survey on various frameworks, performance, and challenges for the smart grid implementation can be found in (50).

Various methods are explored to produce healthy crops and efficient farming techniques to provide a healthy and sustainable environment. There are several uses of ML and DL models for smart agriculture for different stages of farming from soil testing to harvesting. In (16), the research demonstrates the use of the RF method for yield predictions of sugarcanes. The prediction helps in making farm decisions like how much nitrogen fertilizer to apply, maintenance of the miller, and labor schedules for milling. The research in (51, 52) demonstrates the use of CNN-based DL method for plant disease detection. The research in (25) used CNN for the identification of citrus trees using unmanned aerial vehicle imagery. The authors in (53) propose a multilevel DL architecture land cover and crop-type classification from multi-temporal, multi source satellite imagery. CNN implementation outperforms other ML models like RF and MLP with 85% accuracy in detecting major crops. The research in (54) used SVM for pest detection. DL is widely used in fruit detection and determining the stage of fruit for automatic harvesting. The research in (55, 56) uses region-based CNN (R-CNN) for the analysis of the fruit. A comprehensive review of ML and DL methods applied to agriculture is provided in (57–59).

CPS, IoT, and ML methods are infused to bring better healthcare solutions and well-being practices among individuals and communities. ML-based models are widely used in healthcare for various ailments' detection. In (23), the authors propose the use of MLP in detecting brain tumors using MRI images. The proposed model outperforms the NB method with 98.6% accuracy. CNN is implemented in assessing cardiovascular disease based on mammograms (60). The results demonstrate that the DL approach achieves a level of detection similar to the human experts. An overview of different ML algorithms applied in cardiovascular disease is provided in (61). Models have been implemented for fitness devices to analyze the data and make predictions (62). The data collected from wearable smartwatches such as participants, terrain, calories, and steps. DT and RF have the best performance for participant prediction, whereas NB has the best performance for steps. CNN-based framework is developed to provide the nutritional value of the food in real time (63). The top-1 accuracy achieved is 85% in classifying the food. The research in (26) provides visual support of calorie intake of the food using R-CNN-based DL model. Various DL models are implemented for the early diagnosis and study of Alzheimer's disease (64, 65). A brief survey of ML models for cancer detection is studied in (66). Along with traditional DL and ML methods with IoT, the recent research advancements are focused on Federated Learning (FL) which is a distributed AI paradigm. FL is more beneficial in smart healthcare as they don't share any raw data for training. This addresses the main concern of using any AI-related frameworks which is data privacy. The use of FL and its challenges are provided in (67), and a survey of smart healthcare is summarized in (68). The research in (69) discusses about the effective use of cloud-based healthcare system using DL methods for IoT healthcare in smart cities.

Transportation system is one of major CPS applications where most of the data are generated. As the automotive industry is making technologically advanced vehicles, it is important to integrate the intelligent system. The research in (24) used CNN for real-time traffic sign detection. This approach has 99.96% of accuracy in classification. The model is run on an embedded GPU platform. With autonomous driving, driver assistant systems, mobile mapping services, and reliable services are demanded. Self-driving cars use DNN for real-time analysis of various tasks like detecting the speed limit and identifying the pedestrians, traffic signs, etc (70). (71, 72) researches driver behavior identification using DL models and challenges associated with it. ML and DL algorithms have been explored for transportation systems in various aspects. (73) discusses the use of CNN and LSTM networks for traffic flow prediction of smart city planning using intelligent transportation systems. This method incorporates the spacial and temporal features for better prediction of traffic flow. In (74, 75), various ML algorithms applied for smart transportation are provided. Industries are incorporating advanced technologies with IoT and CPS called Industry 4.0. The industries must have high-accurate intelligence to work efficiently and productively. ML and DL models have been used in many industrial applications making them intelligent with the reduction in operation and maintenance costs. There are numerous research surveys conducted for ML and DL algorithms implemented in industrial applications like smart machine process (76), fault diagnosis (77), and autonomous order dispatching (78). A survey of current state-of-the-art CPS and IoT technologies implemented for Industry 4.0 is discussed in (79). The progress of incorporating the technological advancements in Industry 4.0 to the need for more explanability in AI is summarized in (80).

1.5 Adapting ML models

ML and DL models have shown promising results like image, speech, and video applications. Majority of these applications are executed on IoT devices of the CPS. Transforming these IoT devices into smart devices requires some computationally inexpensive intelligence frameworks for deployment. ML models, particularly DL models, are not feasible for resource constraint devices for the training process since DL models need a large amount of computing resources, memory, and power. Studies in (81) show that resources available on these devices are insufficient to perform inference from a pretrained model. Many researchers have explored and studied fundamental methods of DL models to adapt to low-power resource constraint devices with modification to the DL models while maintaining high accuracy (82). Here, various approaches for the DL model deployment for IoT devices are briefly discussed.

1.5.1   Model Compression Techniques

A typical DL model consists of millions of trainable parameters that require large computational and memory resources. To adapt to a resource constraint problem, research interests in model compression have grown over the past few years.

1. **Pruning:** This is one of the model compression methods, in which the redundant and useless weights of the network are removed to reduce the network complexity and over-fitting (83, 84) called the pruning technique. Recently, (85) demonstrated that pruning does not impact the accuracy when applied to the state-of-the-art CNN models. In this method, the dense network is converted into a sparse network, and the sparse values are stored using a compressed sparse row (CSR). This method does not reduce the number of layers in the original network. In Figure 1.12 , the concept of DNN pruning is demonstrated.
2. **Quantization:** A typical representation of weights and activations in deep networks are 32-bit float point numbers. Quantization reduces the number of bits that represent these weights and activations which in turn reduce the memory required for storing the values. Quantization applied for the pruned network reduces the memory storage and computations significantly. The authors in (85) demonstrate the quantization technique for pruned AleXNet with eight-bit representation without any compromise in accuracy.
3. **Knowledge distillation (KD):** In this method, two types of networks are trained: 1) teacher network and 2) student network as shown in Figure 1.13. KD trains the student network (i.e., a smaller network that mimics the large teacher network). The student network utilizes the information contained in the soft targets from the teacher's network to aid the training of the student network. KD along with quantization enables model compression which reduces the memory footprint and complexity of the networks. The number of layers can reduce using KD in the student network (86). In research (87), the KD can be applied for networks with unlabeled datasets.

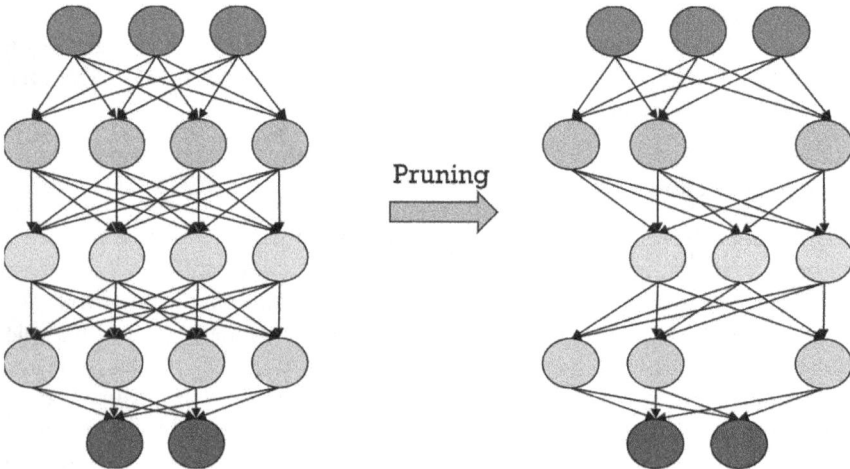

Figure 1.12 Pruning technique.

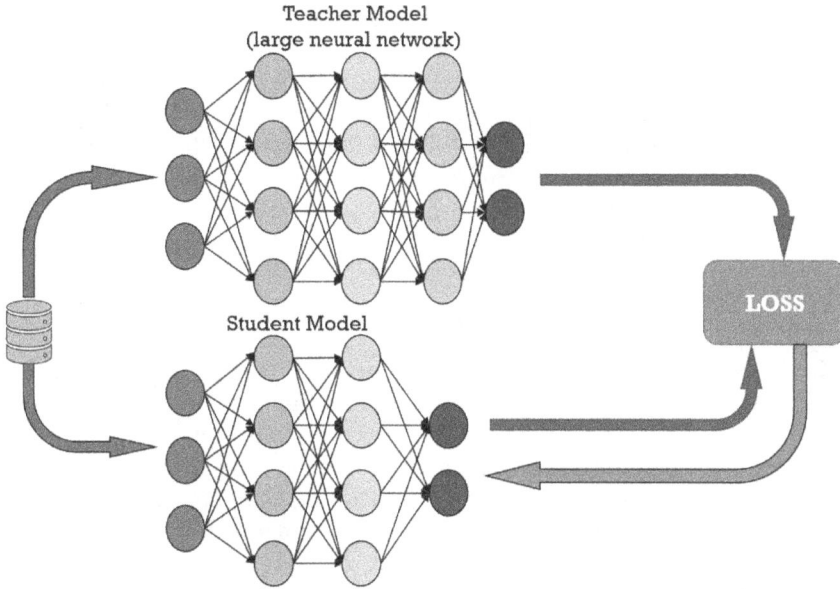

Figure 1.13 Knowledge distillation.

1.5.2   *Accelerators*

With the growth of DL models, suitable hardware designs are developed to optimize the memory footprint and power consumption in IoT devices. The main focus of developing the accelerators for the device is on inference rather than training the DL models. Various researches (88, 89) have been conducted to develop accelerators to adapt to IoT devices. Along with hardware accelerators, various researches have been conducted using software frameworks.

1.6   Open-Ended Research Challenges and Future Directions

Even though ML and DL approaches have shown promising results, and efforts are made to incorporate the ML and DL approaches into CPS, there are several challenges associated with it. In this section, a brief discussion of various challenges is presented followed by future directions and research opportunities.

1.6.1   *Data Scarcity and Scalability for DL Models*

One-size-fits-all solution may not be achieved with DL models, as the model trained for one domain might not be able to perform well for another problem in a similar domain. The model needs to be retrained to fit into another domain. This might not be an issue for static domains but challenging for real-time CPS applications as collecting and processing data, as well as retraining, can be difficult due to limited computational resources. Besides, the representative data for each behavior are rare events (e.g., machine failure) causing an imbalanced dataset for model retraining and thus negatively impacting on the inference and decision-making. Heterogeneity in considered environments, as well as sensors, also impacts ML–DL model scalability across the same domains. On the other hand, due to the heterogeneity of the data collected from devices with different modality and granularity, there could be ambiguity and spuriousness. Future work need to consider addressing these challenges.

1.6.2 Sensitivity and Security of ML and DL Models

The data in CPS are typically collected from sensors, and sensors have different sensitivity and operation levels in different environments. Thus, trained ML and DL models will also be susceptible to change in data patterns leading to the butterfly effect. Such a swing in model results can further pose additional challenges. First, the trained model cannot track or detect issues in the sensors and change their operating characteristics. Second, the ML–DL models become prone to security attacks such as Trojan attacks. An adversary can inject false data and easily mislead the model inferences. More investigation is needed to tackle the privacy and integrity of the models.

1.6.3 Adversarial ML and Cyber Threats

ML and DL models nourish the data, but adversaries can use ML and DL models for malicious purposes. Such a branch of ML and DL is called adversarial machine learning (AML). With AML, adversaries can perform spoofing attacks, generate realistic fake data, and understand inferencing patterns of other ML and DL models by playing with the training parameters, thus misleading the learning system. GANs and other AML models with perturbation can be used by adversaries to attack IoT networks and operations. Hence, ML and DL models can backfire if they are not safeguarded from their vulnerabilities. On the other hand, in critical infrastructure deployment, compromised data can be a threat to human lives and device safety. Furthermore, the compromised data can also be another approach of adversarial attacks. Hence, it is important to ensure the security of IoT devices and the health of the generated data for ML–DL training, and it is extremely important to investigate the role and impact of AML approaches on IoT networks.

1.6.4 Legislative Challenges for ML in Security

The influx of CPS services and applications is surging and spurring legislative discussion. Most CPS applications are struggling with efficient and acceptable legislative policies. Considering ML and DL models are black box solutions, it is very challenging to convince or sell ML and DL solutions and develop technology-oriented legislative policies. As ML and DL models require data for training, the privacy of the users and data leakage become top concerns while making policies. Furthermore, each country will have its own policies for different applications; hence, one security framework might not work for other regions and applications. It is apparent that the legislative policies decide the course for the success, deployment, and adaptation of new technologies among consumers.

1.6.5   Explainability of ML and DL Methods in CPS

Every device in CPS generates data, and that data is processed as needed to train the different ML/DL models. The results inferred from these models are based off from a black box calculations. In order to completely rely on the ML/DL results, the user must be completely confident of the data that has been used for training and testing the models; if not, it would simply be treated as a garbage-in garbage-out scenario. Given the scarcity and quality of the data for certain areas of critical infrastructure, one cannot rely just on the ML/DL results to take the necessary actions. There is a pressing need to more than ever to understand how the ML/DL model concludes the results based on the data that is been used to train and interpret the calculations at each layer. This is one of the major challenges that is hindering from completely deploying AI-driven models in real-world scenarios. There are several software like LIME, Shapley values to name a few that translate the ML/DL models to human understanding which could help us to implement ML/DL models in confidence. In the future, more tools need to be developed in understanding how the ML/DL models work to benefit their implementation.

1.7   Summary

ML and DL approaches have gained great attention in the last decade for various applications. With the growth of CPS for various applications and the number of devices connected in the CPS, there is a need for efficient methods to process, analyze, and predict from the data generated by these devices. One of the main objectives of CPS is to provide intelligent services. Various researches have demonstrated that it is possible to achieve intelligent services by combining the ML and DL approaches with CPS mainly with IoT. In this chapter, we reviewed the characteristics of data generated in a CPS and a brief overview of various ML and DL algorithms that are used for CPS applications. We also provided a brief literature review of ML and DL methods implemented for CPS security and critical infrastructure. From the literature, it is observed that ML and DL approaches have shown promising results for various CPS applications. A brief discussion of adopting ML and DL methods for IoT devices that are resource constraint and the recent advancements in the field was also carried out this chapter. Finally, we concluded the chapter by discussing the challenges posed and future directions for using ML and DL approaches to IoT networks in CPS.

Acknowledgment

The LA-UR for this document is LA-UR-21–29411.

Bibliography

[1] M. H. ur Rehman, C. S. Liew, A. Abbas, P. P. Jayaraman, T. Y. Wah, and S. U. Khan, "Big Data Reduction Methods: A Survey," *Data Science and Engineering*, vol. 1, no. 4, pp. 265–284, 2016.
[2] S. K. Sharma and X. Wang, "Live Data Analytics with Collaborative Edge and Cloud Processing in Wireless IoT Networks," *IEEE Access*, vol. 5, pp. 4621–4635, 2017.
[3] W. Shi, J. Cao, Q. Zhang, Y. Li, and L. Xu, "Edge Computing: Vision and Challenges," *IEEE Internet of Things Journal*, vol. 3, no. 5, pp. 637–646, 2016.
[4] F. Bonomi, R. Milito, J. Zhu, and S. Addepalli, "Fog computing and its role in the internet of things," In *Proceedings of the first edition of the MCC workshop on Mobile cloud computing (MCC '12)*, Association for Computing Machinery, New York, USA, pp. 13–16, 2012.

[5] C. Wirth, R. Akrour, G. Neumann, and J. Fürnkranz, "A Survey of Preference-Based Reinforcement Learning Methods," *Journal of Machine Learning Research*, vol. 18, no. 136, pp. 1–46, 2017.

[6] G. Cauwenberghs and T. Poggio, "Incremental and Decremental Support Vector Machine Learning," *Advances in Neural Information Processing Systems*, pp. 409–415, 2001.

[7] M. Ozay, I. Esnaola, F. T. Y. Vural, S. R. Kulkarni, and H. V. Poor, "Machine Learning Methods for Attack Detection in the Smart Grid," *IEEE Transactions on Neural Networks and Learning Systems*, vol. 27, no. 8, pp. 1773–1786, 2015.

[8] E. Hodo, X. Bellekens, A. Hamilton, P.-L. Dubouilh, E. Iorkyase, C. Tachtatzis, and R. Atkinson, "Threat Analysis of IoT Networks Using Artificial Neural Network Intrusion Detection System," in *2016 International Symposium on Networks, Computers and Communications (IS-NCC)*. IEEE, 2016, pp. 1–6.

[9] F. R. Gomez, A. D. Rajapakse, U. D. Annakkage, and I. T. Fernando, "Support Vector Machine-Based Algorithm for Post-Fault Transient Stability Status Prediction Using Synchronized Measurements," *IEEE Transactions on Power Systems*, vol. 26, no. 3, pp. 1474–1483, 2010.

[10] A. Mehmood, M. Mukherjee, S. H. Ahmed, H. Song, and K. M. Malik, "NBC-MAIDS: Naïve Bayesian Classification Technique in Multi-Agent System-Enriched IDS for Securing IoT Against DDoS Attacks," *The Journal of Supercomputing*, vol. 74, no. 10, pp. 5156–5170, 2018.

[11] P. Padalalu, S. Mahajan, K. Dabir, S. Mitkar, and D. Javale, "Smart Water Dripping System for Agriculture/Farming," in *2017 2nd International Conference for Convergence in Technology (I2CT)*. IEEE, 2017, pp. 659–662.

[12] W. Li, P. Yi, Y. Wu, L. Pan, and J. Li, "A New Intrusion Detection System Based on KNN Classification Algorithm in Wireless Sensor Network," *Journal of Electrical and Computer Engineering*, vol. 2014, 2014.

[13] R. Manikandan, R. Patan, A. H. Gandomi, P. Sivanesan, and H. Kalyanaraman, "Hash Polynomial Two Factor Decision Tree Using IoT for Smart Health Care Scheduling," *Expert Systems with Applications*, vol. 141, p. 112924, 2020.

[14] K. Goeschel, "Reducing False Positives in Intrusion Detection Systems using Data-Mining Techniques Utilizing Support Vector Machines, Decision Trees, and Naive Bayes for Off-line Analysis," in *SoutheastCon 2016*. IEEE, 2016, pp. 1–6.

[15] Y. Meidan, M. Bohadana, A. Shabtai, M. Ochoa, N. O. Tippenhauer, J. D. Guarnizo, and Y. Elovici, "Detection of Unauthorized IoT Devices Using Machine Learning Techniques," *arXiv preprint arXiv:1709.04647*, 2017.

[16] Y. Everingham, J. Sexton, D. Skocaj, and G. Inman-Bamber, "Accurate Prediction of Sugarcane Yield Using a Random Forest Algorithm," *Agronomy for Sustainable Development*, vol. 36, no. 2, p. 27, 2016.

[17] N. Dogru and A. Subasi, "Traffic Accident Detection Using Random Forest Classifier," in *2018 15th Learning and Technology Conference (L&T)*. IEEE, 2018, pp. 40–45.

[18] G. Han, H. Wang, M. Guizani, S. Chan, and W. Zhang, "KCLP: A k-Means Cluster-Based Location Privacy Protection Scheme in WSNs for IoT," *IEEE Wireless Communications*, vol. 25, no. 6, pp. 84–90, 2018.

[19] J. Yang, Y. Han, Y. Wang, B. Jiang, Z. Lv, and H. Song, "Optimization of Real-Time Traffic Network Assignment Based on IoT Data Using DBN and Clustering Model in Smart City," *Future Generation Computer Systems*, vol. 108, pp. 976–986, 2020.

[20] X. Wang and X. Lu, "A Host-Based Anomaly Detection Framework Using XGBoost and LSTM for IoT Devices," *Wireless Communications and Mobile Computing*, vol. 2020, 2020.

[21] D. Cahoolessur and B. Rajkumarsingh, "Fall Detection System Using XGBoost and IoT," *R&D Journal*, vol. 36, pp. 8–18, 2020.

[22] M. Wang, Y. Lu, and J. Qin, "A Dynamic MLP-based DDoS Attack Detection Method Using Feature Selection and Feedback," *Computers & Security*, vol. 88, 2020.

[23] K. Sharma, A. Kaur, and S. Gujral, "Brain Tumor Detection Based on Machine Learning Algorithms," *International Journal of Computer Applications*, vol. 103, no. 1, pp. 7–11, 2014.

[24] A. Shustanov and P. Yakimov, "CNN Design for Real-Time Traffic Sign Recognition," *Procedia Engineering*, vol. 201, pp. 718–725, 2017.

[25] O. Csillik, J. Cherbini, R. Johnson, A. Lyons, and M. Kelly, "Identification of Citrus Trees from Unmanned Aerial Vehicle Imagery Using Convolutional Neural Networks," *Drones*, vol. 2, no. 4, p. 39, 2018.

[26] J. Mejía, A. Ochoa-Zezzatti, R. Contreras-Masse, and G. Rivera, "Intelligent System for the Visual Support of Caloric Intake of Food in Inhabitants of a Smart City Using a Deep Learning," *Applications of Hybrid Metaheuristic Algorithms for Image Processing*, vol. 890, p. 441, 2020.

[27] M. Z. Uddin, "A Wearable Sensor-Based Activity Prediction System to Facilitate Edge Computing in Smart Healthcare System," *Journal of Parallel and Distributed Computing*, vol. 123, pp. 46–53, 2019.

[28] T.-Y. Kim and S.-B. Cho, "Predicting Residential Energy Consumption Using CNN-LSTM Neural Networks," *Energy*, vol. 182, pp. 72–81, 2019.

[29] R. Das, A. Gadre, S. Zhang, S. Kumar, and J. M. Moura, "A Deep Learning Approach to IoT Authentication," in *2018 IEEE International Conference on Communications (ICC)*. IEEE, 2018, pp. 1–6.

[30] R. Alhajri, R. Zagrouba, and F. Al-Haidari, "Survey for Anomaly Detection of IoT Botnets Using Machine Learning Auto-Encoders," *International Journal of Applied Engineering Research*, vol. 14, no. 10, pp. 2417–2421, 2019.

[31] J. J. Q. Yu and J. Gu, "Real-time Traffic Speed Estimation with Graph Convolutional Generative Autoencoder," *IEEE Transactions on Intelligent Transportation Systems*, vol. 20, no. 10, pp. 3940–3951, 2019.

[32] Y. Yang, F. Nan, P. Yang, Q. Meng, Y. Xie, D. Zhang, and K. Muhammad, "GAN-Based Semi-Supervised Learning Approach for Clinical Decision Support in Health-IoT Platform," *IEEE Access*, vol. 7, pp. 8048–8057, 2019.

[33] C. Wang, S. Dong, X. Zhao, G. Papanastasiou, H. Zhang, and G. Yang, "Saliency-GAN: Deep Learning Semisupervised Salient Object Detection in the Fog of IoT," *IEEE Transactions on Industrial Informatics*, vol. 16, no. 4, pp. 2667–2676, 2019.

[34] A. Arora and Shantanu, "A Review on Application of GANs in Cybersecurity Domain," *IETE Technical Review*, pp. 1–9, 2020.

[35] A. Elsaeidy, K. S. Munasinghe, D. Sharma, and A. Jamalipour, "Intrusion Detection in Smart Cities Using Restricted Boltzmann Machines," *Journal of Network and Computer Applications*, vol. 135, pp. 76–83, 2019.

[36] U. Fiore, F. Palmieri, A. Castiglione, and A. De Santis, "Network Anomaly Detection with the Restricted Boltzmann Machine," *Neurocomputing*, vol. 122, pp. 13–23, 2013.

[37] S.-H. Song and D. K. Kim, "Development of a Stress Classification Model Using Deep Belief Networks for Stress Monitoring," *Healthcare Informatics Research*, vol. 23, no. 4, pp. 285–292, 2017.

[38] N. Chaabouni, M. Mosbah, A. Zemmari, C. Sauvignac, and P. Faruki, "Network Intrusion Detection for IoT Security Based on Learning Techniques," *IEEE Communications Surveys & Tutorials*, vol. 21, no. 3, pp. 2671–2701, 2019.

[39] A. L. Buczak and E. Guven, "A Survey of Data Mining and Machine Learning Methods for Cyber Security Intrusion Detection," *EEE Communications Surveys & Tutorials*, vol. 18, no. 2, pp. 1153–1176, 2015.

[40] N. Moustafa, B. Turnbull, and K.-K. R. Choo, "An Ensemble Intrusion Detection Technique Based on Proposed Statistical Flow Features for Protecting Network Traffic of Internet of Things," *IEEE Internet of Things Journal*, vol. 6, no. 3, pp. 4815–4830, 2018.

[41] Z. Tan, A. Jamdagni, X. He, P. Nanda, and R. P. Liu, "A System for Denial-of-Service Attack Detection Based on Multivariate Correlation Analysis," *EEE Transactions on Parallel and Distributed Systems*, vol. 25, no. 2, pp. 447–456, 2013.

[42] C. Shi, J. Liu, H. Liu, and Y. Chen, "Smart User Authentication through Actuation of Daily Activities Leveraging WiFi-enabled IoT," In *Proceedings of the 18th ACM International Symposium on Mobile Ad Hoc Networking and Computing (Mobihoc '17)*. Association for Computing Machinery, New York, NY, USA, Article 5, 1–10, 2017.

[43] S. Mahdavifar and A. A. Ghorbani, "Application of Deep Learning to Cybersecurity: A Survey," *Neurocomputing*, vol. 347, pp. 149–176, 2019.

[44] M. M. H. Shuvo, "Edge AI: Leveraging the Full Potential of Deep Learning," in *Recent Innovations in Artificial Intelligence and Smart Applications*. Springer, 2022, pp. 27–46.

[45] R. Zhu, A. Anjum, H. Li, and M. Ma, "Edge Intelligence-Enabled Cyber-Physical Systems," *Concurrency Computat Pract Exper*, 35(13):e7500, 2023.

[46] X. Liu, Z. Xiao, R. Zhu, J. Wang, L. Liu, and M. Ma, "Edge Sensing Data-Imaging Conversion Scheme of Load Forecasting in Smart Grid," *Sustainable Cities and Society*, vol. 62, p. 102363, 2020.

[47] D. C. Mocanu, E. Mocanu, P. H. Nguyen, M. Gibescu, and A. Liotta, "Big IoT Data Mining for Real-Time Energy Disaggregation in Buildings," in *2016 IEEE International Conference on Systems, Man and Cybernetics*. IEEE, 2016, pp. 003765–003769.

[48] Z. Shi, W. Yao, L. Zeng, J. Wen, J. Fang, X. Ai, and J. Wen, "Convolutional Neural Network-Based Power System Transient Stability Assessment and Instability Mode Prediction," *Applied Energy*, vol. 263, p. 114586, 2020.

[49] S. Lakshmiranganatha, "HPC and Machine Learning Techniques for Reducing the Computation Burden of Determining Time-Evolution of Complex Dynamic Systems," Order No. 28323789 ed. University of Wyoming; 2021.

[50] M. A. Judge, A. Khan, A. Manzoor, and H. A. Khattak, "Overview of Smart Grid Implementation: Frameworks, Impact, Performance and Challenges," *Journal of Energy Storage*, p. 104056, 2022.

[51] J. G. A. Barbedo, "Plant Disease Identification from Individual Lesions and Spots Using Deep Learning," *Biosystems Engineering*, vol. 180, pp. 96–107, 2019.

[52] S. Sladojevic, M. Arsenovic, A. Anderla, D. Culibrk, and D. Stefanovic, "Deep Neural Netwroks Based Recognition of Plant Diseases by Leaf Image Classification," *Computational Intelligence and Neuroscience*, vol. 2016, 2016.

[53] N. Kussul, M. Lavreniuk, S. Skakun, and A. Shelestov, "Deep Learning Classifications of Land Cover and Crop Types Using Sensing Data," *IEEE Geoscience and Remote Sensing Letters*, vol. 14, no. 5, pp. 778–782, 2017.

[54] M.-A. Ebrahimi, M.-H. Khoshtaghaza, S. Minaei, and B. Jamshidi, "Vision-Based Pest Detection Based on SVM Classification Method," *Computers and Electronics in Agriculture*, vol. 137, pp. 52–58, 2017.

[55] I. Sa, Z. Ge, F. Dayoub, B. Upcroft, T. Perez, and C. McCool, "DeepFruits: A Fruit Detection System Using Deep Neural Networks," *Sensors*, vol. 16, no. 8, p. 1222, 2016.

[56] P. Chu, Z. Li, K. Lammers, R. Lu, and X. Liu, "DeepApple: Deep Learning-based Apple Detection Using a Suppression Mask R-CNN," *arXiv preprint arXiv:2010.09870*, 2020.

[57] M. Pathan, N. Patel, H. Yagnik, and M. Shah, "Artificial Cognition for Applications in Smart Agriculture: A Comprehensive Review," *Artificial Intelligence in Agriculture*, vol. 4, pp. 81–95, 2020.

[58] M. Altalak, M. Ammad Uddin, A. Alajmi, and A. Rizg, "Smart Agriculture Applications Using Deep Learning Technologies: A Survey," *Applied Sciences*, vol. 12, no. 12, p. 5919, 2022.

[59] V. K. Quy, N. V. Hau, D. V. Anh, N. M. Quy, N. T. Ban, S. Lanza, G. Randazzo, and A. Muzirafuti, "IoT-Enabled Smart Agriculture: Architecture, Applications, and Challenges," *Applied Sciences*, vol. 12, no. 7, p. 3396, 2022.

[60] J. Wang, H. Ding, F. A. Bidgoli, B. Zhou, C. Iribarren, S. Molloi, and P. Baldi, "Detecting Cardiovascular Disease from Mammograms with Deep Learning," *IEEE Transactions on Medical Imaging*, vol. 36, no. 5, pp. 1172–1181, 2017.

[61] S. J. Al'Aref, K. Anchouche, G. Singh, P. J. Slomka, K. K. Kolli, A. Kumar, M. Pandey, G. Maliakal, A. R. van Rosendael, A. N. Beecy et al., "Clinical Applications of Machine Learning in Cardiovascular Disease and Its Relevance to Cardiac Imaging," *European Heart Journal*, vol. 40, no. 24, pp. 1975–1986, 2019.

[62] T. Reichherzer, M. Timm, N. Earley, N. Reyes, and V. Kumar, "Using Machine Learning Techniques to Track Individuals & Their Fitness Activities," In CATA 2017, pp. 119–124, 2017.

[63] R. Yunus, O. Arif, H. Afzal, M. F. Amjad, H. Abbas, H. N. Bokhari, S. T. Haider, N. Zafar, and R. Nawaz, "A Framework to Estimate the Nutritional Value of Food in Real Time Using Deep Learning Techniques," *IEEE Access*, vol. 7, pp. 2643–2652, 2018.

[64] A. Ortiz, J. Munilla, J. M. Gorriz, and J. Ramirez, "Ensembles of Deep Learning Architecture for the Early Diagnosis of the Alzheimer's Disease," *International Journal of Neural Systems*, vol. 26, no. 07, p. 1650025, 2016.

[65] F. Falahati, E. Westman, and A. Simmons, "Multivariate Data Analysis and Machine Learning in Alzheimer's Disease with a Focus on Structural Magnetic Resonance Imaging," *Journal of Alzheimer's Disease*, vol. 41, no. 3, pp. 685–708, 2014.

[66] T. Saba, "Recent Advancement in Cancer Detection using Machine Learning: Systematic Survey of Decades, Comparisons and Challenges," *Journal of Infection and Public Health*, vol. 13, no. 9, pp. 1274–1289, 2020.

[67] D. C. Nguyen, Q.-V. Pham, P. N. Pathirana, M. Ding, A. Seneviratne, Z. Lin, O. Dobre, and W.-J. Hwang, "Federated Learning for Smart Healthcare: A Survey," *ACM Computing Surveys (CSUR)*, vol. 55, no. 3, pp. 1–37, 2022.

[68] R. Verma, "Smart City Healthcare Cyber Physical System: Characteristics, Technologies and Challenges," *Wireless Personal Communications*, vol. 122, no. 2, pp. 1413–1433, 2022.

[69] S. M. Nagarajan, G. G. Deverajan, P. Chatterjee, W. Alnumay, and U. Ghosh, "Effective Task Scheduling Algorithm with Deep Learning for Internet of Health Things (IOHT) in Sustainable Smart Cities," *Sustainable Cities and Society*, vol. 71, p. 102945, 2021.

[70] C. Badue, R. Guidolini, R. V. Carneiro, P. Azevedo, V. B. Cardoso, A. Forechi, L. Jesus, R. Berriel, T. M. Paixao, F. Mutz et al., "Self-Driving Cars: A Survey," *Expert Systems with Applications*, vol. 165, p. 113816, 2021.

[71] C. Ravi, A. Tigga, G. T. Reddy, S. Hakak, and M. Alazab, "Driver Identification Using Optimized Deep Learning Model in Smart Transportation," *ACM Transactions on Internet Technology*, vol. 22, no. 4, pp. 1–17, 2022.

[72] R. A. Zaidan, A. H. Alamoodi, B. Zaidan, A. Zaidan, O. S. Albahri, M. Talal, S. Garfan, S. Sulaiman, A. Mohammed, Z. H. Kareem et al., "Comprehensive Driver Behaviour Review: Taxonomy, Issues and Challenges, Motivations and Research Direction Towards Achieving a Smart Transportation Environment," *Engineering Applications of Artificial Intelligence*, vol. 111, p. 104745, 2022.

[73] B. Vijayalakshmi, K. Ramar, N. Jhanjhi, S. Verma, M. Kaliappan, K. Vijayalakshmi, S. Vimal, Kavita, and U. Ghosh, "An Attention-Based Deep Learning Model for Traffic Flow Prediction Using Spatiotemporal Features Towards Sustainable Smart City," *International Journal of Communication Systems*, vol. 34, no. 3, p. e4609, 2021.

[74] F. Zantalis, G. Koulouras, S. Karabetsos, and D. Kandris, "A Review of Machine Learning and IoT in Smart Transportation," *Future Internet*, vol. 11, no. 4, p. 94, 2019.

[75] Z. Karami and R. Kashef, "Smart Transportation Planning: Data, Models, and Algorithms," *Transportation Engineering*, vol. 2, p. 100013, 2020.

[76] D.-H. Kim, T. J. Kim, X. Wang, M. Kim, Y.-J. Quan, J. W. Oh, S.-H. Min, H. Kim, B. Bhandari, I. Yang et al., "Smart Machining Process Using Machine Learning: A Review and Perspective on Machining Industry," *International Journal of Precision Engineering and Manufacturing-Green Technology*, vol. 5, no. 4, pp. 555–568, 2018.

[77] N. Amruthnath and T. Gupta, "A Research Study on Unsupervised Machine Learning Algorithms for Early Fault Detection in Predictive Maintenance," in *2018 5th International Conference on Industrial Engineering and Applications (ICIEA)*. IEEE, 2018, pp. 355–361.

[78] A. Kuhnle, N. Röhrig, and G. Lanza, "Autonomous Order Dispatching in the Semiconductor Industry Using Reinforcement Learning," *Procedia CIRP*, vol. 79, pp. 391–396, 2019.

[79] P. K. Malik, R. Sharma, R. Singh, A. Gehlot, S. C. Satapathy, W. S. Alnumay, D. Pelusi, U. Ghosh, and J. Nayak, "Industrial Internet of Things and Its Applications in Industry 4.0: State of the Art," *Computer Communications*, vol. 166, pp. 125–139, 2021.

[80] I. Ahmed, G. Jeon, and F. Piccialli, "From Artificial Intelligence to Explainable Artificial Intelligence in Industry 4.0: A Survey on What, How, and Where," *IEEE Transactions on Industrial Informatics*, vol. 18, no. 8, pp. 5031–5042, 2022.

[81] N. D. Lane, S. Bhattacharya, P. Georgiev, C. Forlivesi, and F. Kawsar, "An Early Resource Characterization of Deep Learning on Wearable, Smartphones and Internet-of-Thiings Devices," in *Proceedings of the 2015 International Workshop on Internet of Things towards Applications*, Association for Computing Machinery, pp. 7–12, 2015.

[82] L. J. Ba and R. Caruana, "Do Deep Nets Really Need to Be Deep?" *arXiv preprint arXiv:1312.6184*, 2013.

[83] N. Ström, "Phoneme Probability Estimation with Dynamic Sparsely Connected Artificial Neural Networks," *The Free Speech Journal*, vol. 5, no. 1–41, p. 2, 1997.

[84] Y. LeCun, J. S. Denker, and S. A. Solla, "Optimal Brain Damage," in *Advances in Neural Information Processing Systems*, 1990, pp. 598–605. https://citeseerx.ist.psu.edu/document?repid=rep1&type=pdf&doi=17c0a7de3c17d31f79589d245852b57d083d386e

[85] S. Han, H. Mao, and W. J. Dally, "Deep Compression: Compressing Deep Neural Networks with Pruning, Trained Quantization and Huffman Coding," *arXiv preprint arXiv:1510.00149*, 2015.

[86] G. Hinton, O. Vinyals, and J. Dean, "Distilling the Knowledge in a Neural Network," *arXiv preprint arXiv:1503.02531*, 2015.

[87] H. Li, "Exploring Knowledge Distillation of Deep Neural," 2018 [Online]. Available: http://cs230.stanford.edu/files_winter_2018/projects/6940224.pdf

[88] S. Shivapakash, H. Jain, O. Hellwich, and F. Gerfers, "A Power Efficiency Enhancements of a Multi-Bit Accelerator for Memory Prohibitive Deep Neural Networks," *IEEE Open Journal of Circuits and Systems*, vol. 2, pp. 156–169, 2021.

[89] Y.-H. Chen, T. Krishna, J. S. Emer, and V. Sze, "Eyeriss: An Energy-Efficient Reconfigurable Accelerator for Deep Convolutional Neural Networks," *IEEE Journal of Solid-State Circuits*, vol. 52, no. 1, pp. 127–138, 2016.

Chapter 2

Securing Cyber-Physical Systems Using Artificial Intelligence

Pon Harshavardhanan, Agni Datta, Yuvraj Singh, Yash Kartik, Chandan Thota, and Akhilesh Pokale

Chapter Contents

2.1 Introduction

CPS is the next generation of intelligent systems that combine the inherent and comprehensive embedding of communication and control technologies to integrate computational resources with hardware facilities [5]. CPS combines computational and physical processes at the micro-level by embedding computer and telecommunications kernels in hardware. CPS is a reactive

DOI: 10.1201/9781003376712-2

hybrid structure comprising distributed asynchronous heterogeneous networks operating in diverse spatial and temporal ranges, as well as numerous resources and configurable modules such as sensing prediction and evaluation. For the implementation of lightweight symmetric cryptography in CPS, difficulties generally include finding a balance between industrial requirements and publicly available research data. While CPS is a huge facilitator of better living standards across the globe, the coupling of physical systems and the virtual realm poses vulnerabilities that if not addressed can have disastrous repercussions. The CPS modules are more vulnerable to an attacker, raising the possibility of physical attacks that might compromise security. The tremendous lifetime of CPS also needs the capacity of CPS components to withstand new threats that are continually emerging. What makes CPS security so much more difficult is that all these risks and security difficulties are frequently to be managed in severely limited contexts due to multiple variables such as low-cost and low-energy requirements [23].

Definition 1.10 Cyber-physical systems (CPS) are a new class of computing systems that combine complex computational and physical functionalities, and they may interact and communicate with individuals in novel forms.

CPSs are computing networking and physical process integration [31]. Physical processes in these systems are usually monitored and controlled by sophisticated embedded computing and networking, which include feedback loops where physical activities affect calculations and vice versa. The commercial and sociological capability of such platforms is far higher than has been realized; therefore, heavy resources are primarily being invested globally to advance this modern technology. The technology draws on the earlier yet quite new field of embedded systems which are processors and software installed in devices besides just computers including automobiles, toys, healthcare products, and laboratory equipment. CPS mixes physical process dynamics with computer software and networking dynamics, giving frameworks as well as modelling design and analysis approaches for the connected total system [31].

The capacity to communicate with and augment the possibilities of the surrounding environment through computing communication and automation is a crucial component for prospective technological advancements in this field. The merging of engineering and physical application domains with computer science engineering devices and cyberspace worlds is the core idea behind CPS development. Physics simulation and realistic intangible attributes like uncertainties in nature and volatility of the business are illustrations of fundamental natural concepts. At the same time, computer science and engineering fundamentals concentrate on embedded applications, communications computing, and programming methods. CPS has five tiers of technological application, and the structure of such a system is as described in the following:

- **Smart Connection Level**: Gathering data from machines or components in an expedient, consistent, and precise manner and delivering it to the information transformation layer.
- **Data-to-Information Conversion Level**: Gathering and transforming data transferred from the smart connection level into relevant data for prediction and business applications.
- **Cyber Level**: A centralized data centre for the system infrastructure and information from each of the interconnected devices constructs a machinery network after accumulating a large amount of data specialized analysis is necessary to retrieve more data to fully comprehend the status of each device; these evaluations could also provide devices with the potential to self analyse with other similar devices or try comparing and analyzing devices at different points in time which facilitates in a thorough understanding of the status of each device in the cluster.

- **Cognition Level:** The device is analysed after gathering data from other devices via the network stack and contrasting it to other occurrences. Certain particular prediction algorithms are used to forecast or determine the time of component failure based on previously gathered information.
- **Configuration Level:** Technicians and industrial managers may base decisions on input from cyberspace to physical space, as well as system monitoring and administration. Simultaneously, the device by itself can decrease the damage by system failures, permitting the rectification of mistakes as well as the implementation of precautionary measures.

CPSs are becoming increasingly prevalent in our daily lives, with the integration of physical processes with computing and communication technologies. CPS has revolutionized several industries, including healthcare, transportation, and energy management. However, this integration has also made these systems vulnerable to cyberattacks, which can have severe physical consequences, such as equipment damage, power outages, and even loss of life. Securing CPS has, therefore, become a critical concern for industries, governments, and society as a whole. CPS security involves the protection of these systems from cyber threats, ensuring that they operate as intended and maintaining their confidentiality, integrity, and availability. CPS security is a multidisciplinary field that requires the collaboration of experts in computer science, engineering, mathematics, and other related fields. However, the integration of physical processes with computing and communication technologies makes these systems vulnerable to cyberattacks [13], which can have severe physical consequences. For instance a cyberattack on an industrial control system can result in equipment damage, power outages, and even injuries or loss of life. Thus, securing CPS has become a critical concern for industries, governments, and society as a whole. This chapter aims to explore the importance of securing CPS, the potential risks and consequences of cyberattacks on these systems, and the various techniques and approaches that can be used to ensure their security. Through this research, we hope to raise awareness about the need for effective CPS security measures and encourage further efforts to safeguard these systems [44]. The remainder of this project is organized as follows: Section 2 talks about the description about CPS followed by Section 3 which discusses the architecture of CPS. Section 4 is focused on the security IN CPS, and Section 5 explores the security methodologies in CPS, and next Section 6 about the relevance of AI for the security challenges of CPS. Section 7 talks about the different cyberattack methods in CPSs followed by conclusion in Section 8 and references at last.

2.2 CPS Description

A system that integrates the cyber and physical worlds is referred to as a "cyber-physical system." CPSs might be used in a variety of industries such as space transportation, logistics, healthcare, home automation, malls, and communication systems. Let's conceive a scenario where someone forgets to utilize milk and other food that they store in a smart refrigerator. The smart refrigerator will start to warn when the milk or curd is near to expire. As the CPS has expanded, the number of people using the Internet has surpassed billions, and it continues to rise daily.

CPS has an impact on how the Internet and wireless sensor networks (WSNs) operate as intelligent systems that automate tasks that were previously heavily dependent on human labour unnaturally connected to physical and technical systems which have been described in a number of ways and in which the monitoring cooperation controlling and integration of the activities are carried out by a computer and communication core. By including a control action, CPS increases the IoT's processing and networking capabilities. Due to the capacity to employ feedback control to tell a selection to act based on physical measurements collected from the detectors, CPSs have a higher level of automation than IoT systems. They have the great capacity to examine practically every aspect of mortal endeavour and conquer difficulties. This has led to a great deal of interest in CPS from the academic community, the government, and individuals with astuteness.

One of the defences against IoT and CPS when employed in crucial structures like agriculture, health service, transportation, home automation, and power systems is the vast volume of data that is created. This is due to the fact that the majority of gadgets are continually connected and switched on. As a result, the creation of Big data analytics (BDA), IoT and CPSs, becomes important because they make data-based information available for decisions on fault prediction diagnosis and preventative maintenance. Data analytics is becoming less and less common as a technique to extract value from the generated data by revealing regression patterns correlations and other relationships as they add new functionality to the systems under stress as well as providing perception from vast volumes of data. Despite all the hoopla surrounding CPS, its actual implementation to tackle practical challenges is hampered by highly strict safety and security requirements such as the need to function in real time and sensitivity to network difficulties like latency.

Additionally, the impact a failure brings to human life and structure is more serious than it is for typical information technology systems. Cybersecurity is now a hot topic in computer science and information technology. Malware, adware, spyware, and ransomware were used at first to provide security, followed by firewall's intrusion detection systems and antivirus software (IDS). The growing interconnectedness is problematic since it expands the attack surface and exposes these systems to adversarial conditioning. An important factor in the increase of cyberattacks is the failure of CPS's detectors selectors and regulators. Machine learning (ML) and artificial intelligence (AI) algorithms have lately been employed to improve the effectiveness of various systems.

Therefore, a prevalent type of cyberattack in CPS and other systems is the development of techniques to tamper with the data or the input. As a result, the model is forced to generate the erroneous labour. Particularly vulnerable to this are deep neural networks (DNNs) which have grown to be widely employed to protect CPSs. The possibility of utilizing system protection techniques against it has also grown into a topic of concern. Adversaries may attack systems using AI and ML algorithms that are designed to secure those systems to initiate aggressive offences. It has recently been proven that comparable assaults have a higher implicit value. They utilize the perspiration of the defence systems to develop stronger harder-to-detect and harder-to-check weapons, making them more advanced rapid and relatively affordable.

2.3 CPS Architecture

The web and mobile applications will increasingly dominate the Internet which will eventually permeate every part of the real world. A dynamic and complex physical world is essentially different from a cyber world which opens a lot of advanced challenges in the development of CPS. CPS differs from conventional embedded systems. Integration of networking computations and physical systems is called "cyber-physical systems" while embedded systems track the physical processes that affect the computation and vice versa.

Better Network Performance: CPSs have applications such as intelligent transportation and CPS in medicine needs meticulous real-time and secure services that lead a way for time-spatial and security affirmation. The network in CPS requires more strategic amplifications in handling time and security.

Heterogeneous: In CPS, many end devices have different processing potential security measurements and communication procedures. Embedded end devices in CPS have the potential to loop to process the information because they have finite memory and computing power. The processing potential of the device is distinct and brings a cluster of challenges to system configuration.

Adaptability: The CPS end device's application scheme is capable of change. The commuting of surrounding domain coordinates leads to differences in sensing data. An appropriate and convenient design of CPS can acclimate to the environment.

Edward A. Lee argues that the networking requirements of CPSs which combine networking and computation with physical dynamics make the case that CPS requires networks where time is a semantic attribute rather than just being a quality criterion [30]. The design of the CPS treats time as a semantic coordinate and not just a quality factor. If we talk about the design of CPS, time is a semantic property not a factor for quality assurance. Don Kang et al. presented an approach to subsidizing these requirements. In his proposed approach, he discussed network-enabled real-time embedded databases (nRTEDBs) that can interact and can be controlled with wireless sensors. This varies significantly from a traditional database as it deals with both prediction and raw data. Three significant issues with system security that must be taken into account are confidentiality, integrity, and availability.

We split CPS into two groups: Systems that are essential to security and systems that are not. Then for the first, there is a strong emphasis on physical security as well as other needs like confidentiality integrity and availability. However, the latter emphasizes integrity and availability while somewhat weakening security. In various settings there are various needs. For instance although the demand for real-time performance is prioritized in smart home systems, the secrecy aspect is crucial for military applications. The three components of CPS security are: Perception security (ensuring the security and accuracy of data collected from physical environments), transport security (preventing data from being lost during transmission processes), and processing centre security (including physical security and safety procedures on servers or workstations).

A typical CPS process was divided into four parts by Eric Ke Wang et al. [58].

(1) Monitoring: A key duty of CPS is to keep an eye on the environment and physical processes. It is also used to provide feedback on any previous CPS activities in order to maintain proper operations in the future.

(2) Deal with data aggregation and distribution through networking.Numerous sensors may provide a large amount of data that must be combined or distributed for further processing by analysers.

(3) Computing, this stage is used to reason about and examine the data gathered during monitoring to see if the physical process complies with predetermined standards; execute the steps decided upon during the computing stage.

(4) A context-aware security architecture for CPS was suggested in light of this.

Numerous security-related tactics including authentication or encryption are included in the security architecture mentioned above. Context-aware frameworks can adapt to changing physical environments, which is why they are so named as shown in Figure 2.1. Additionally, they scale well and adapt well to changing environmental conditions. The CPS

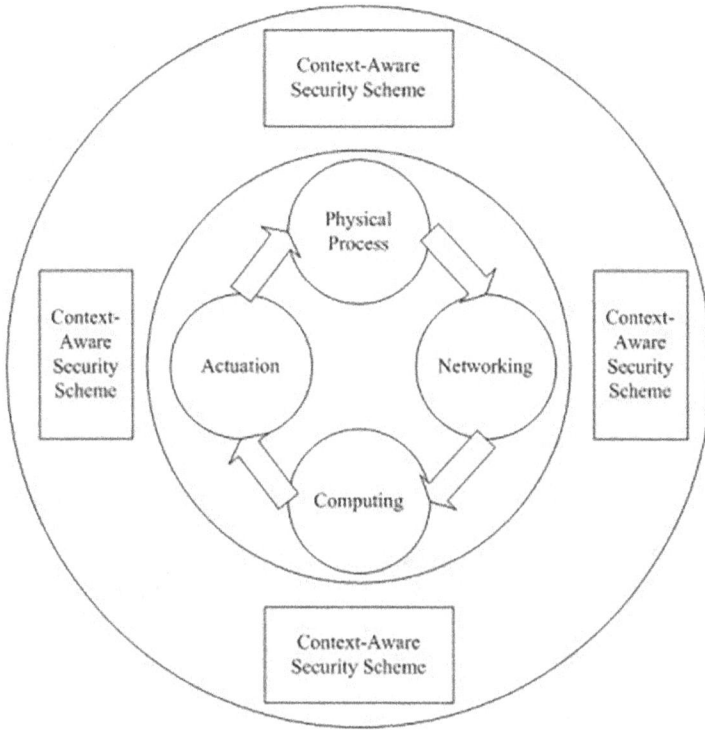

Figure 2.1 Context-Aware security scheme in CPS.

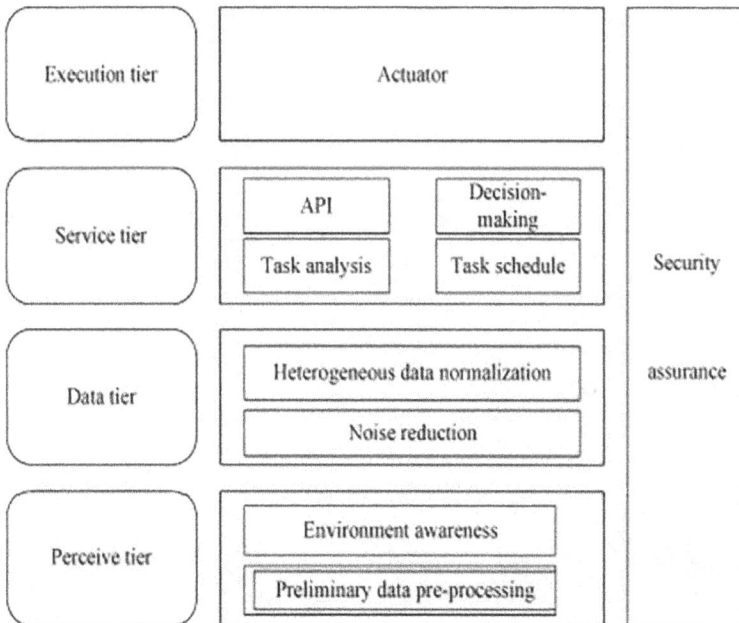

Figure 2.2 CPS architecture based on SOA.

security architecture covers mitigating current attacks as well as predicting potential attacks and making necessary modifications. As part of the system's demand for self-adaptation, the security framework in a changing environment should maintain the system's safety.

Using a standards-based software component technology, SOA (service-oriented architecture) is a dynamically integrating paradigm used to combine loosely connected services into a single workflow. We recommended the generic CPS architecture, which is based on the SOA design as shown in Figure 2.2.

This architecture consists of five tiers [19]:

The data source for the tiers mentioned above also known as the sensor tier is the perceive tier. Environment awareness which is mostly accomplished by sensors and preliminary data pre-processing which transfers the data to the data-processing layer are the duties of this tier. One of the fundamental methods used in this sensor tier is the WSN (Wireless Sensor Network). The computational and storage components that make up the data tier perform heterogeneous data processing, including normalization noise reduction data storage and other related tasks. Between the Producer and the Service lies this layer. The service layer offers the standard system operations such as decision-making task scheduling task analysis and consumer-facing APIs. Several services are installed in this tier that communicate with one another.

The two layers that interact with the environment are the execution and perceived tiers; the actuator might be a real object like an automobile or a bulb. It takes directives from the system and puts them into action. Against access security to data security to device security, the security assurance component is present across the whole system. Require a number of measures to ensure system security from unauthorized access or malicious assaults. The real-time distribution reliability scalability and other properties of CPS create a number of design and implementation issues. The integration flexibility of services or components is one of this suggested CPS architecture's primary benefits. Because diverse needs call for different realizations of these layers, the architecture simplifies the intricacies of each tier [1].

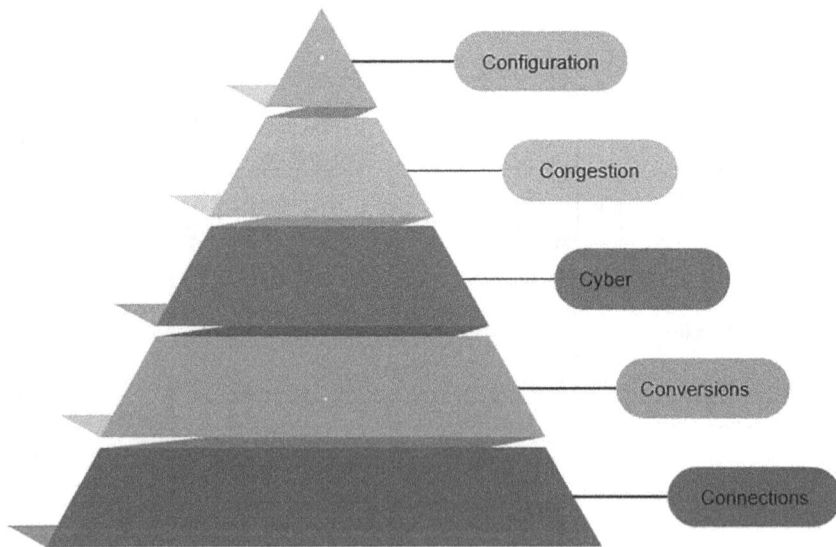

Figure 2.3 5C Architecture for CPS.

2.3.1 *Layers of Attack Vector*

Predictive analytics and CPSs are at the heart of industrial Big data [33, 34]. The physical entities in cyberspace are connected and synthesized by CPSs [35, 38]. A digital twin model is created in the CPS environment using information gathered from the IoT infrastructure to represent a physical entity such as a machine or component in cyberspace [36]. Figure 2.3 illustrates a "5C" architecture that serves as a blueprint for the creation of CPS in industrial applications [37]. We will extrapolate the "5C" levels in detail as follows.

(1) **Connections:** Finding a trustworthy, effective, and secure method of data acquisition is the crucial next step. Typically, sending data from the edge to the central server requires the use of a local agent and communication protocol. The current industrial communication protocols available change based on the manufacturers of the machines the suppliers of the controllers and the data-gathering systems. In order to synchronize and offer an efficient data-collection solution for various equipment [21] including MT-Connect and SmartBox from Mazak and Cisco, a common tether-free communication protocol is encouraged [38].

(2) **Conversion Level:** Data is collected from a variety of sources in the industrial setting, including controllers, sensors, production systems (ERP MES etc.), and maintenance records. Each data point or signal might represent a distinct feature of the under surveillance machine systems. Users can only make the best choice at the correct moment to increase productivity when the data is transformed into "useful" information such as health indices and fault diagnosis findings. For instance all data to information approaches for rotary equipment, employing data-driven methods were included in the review study [34].

(3) **Cyber Level:** The machine health information may be created through the data-to-information conversion level to describe the state of the system. Different info-graphs at the cyber level make the health status of assets evident. Each physical asset may be benchmarked against peers in various time periods once it has a digital twin at the cyber level, a concept known as fleet dimension. The data from each asset is used to generate cyber avatars that represent each machine or system in more detail, giving consumers a better understanding of system variance and life forecast [29, 24, 60]. Thus, fleet modelling in the cyberspace can benefit from a vast scale of comparable assets and improve the robustness and dependability of the algorithms for asset health monitoring. Cyber level algorithms on the other hand will be more adaptable to support dynamic operating regime changes and increase robustness, for example the algorithm can learn new failure modes adaptively and enrich itself to be more robust and dependable. [61].

(4) **Cognition Level:** The machine can assess its probable failure and be aware of its potential degeneration, can assess its probable failure and be aware of its potential degeneration in advance by utilizing the online monitoring system. The system may use the prediction algorithms to forecast possible failure and determine the remaining usable life after learning from the past health assessment. Users may improve manufacturing, operations, maintenance, scheduling, and logistics planning at this level with the use of the predictive information to assist improved decision-making [6,40,47,17].

(5) **Configuration:** The machine's health status may be monitored and visualized online at the first four levels, and the operation level can then get the health information. As a result of the machine's gradual deterioration, timely maintenance will significantly minimize downtime and boost production. In order to limit the loss from machine malfunction and eventually create a resilient system, operators and factory managers

may make the best decision based on maintenance information while also allowing the machine itself to lower its workload and modify the manufacturing schedule.

The "5C" architecture provides a methodical way to turn data into knowledge, explain-production process risks, and make better "informed" decisions. Managers will have the necessary data to calculate the facility-wide OEE with manufacturing transparency; with predictive capabilities, all assets may be handled efficiently with just-in-time maintenance. In order to create a closed loop life cycle redesign lifespan deterioration information may also be sent back to the equipment designer. As a result, an Industry 4.0 plant will move away from traditional TPS management and towards self-aware, self-predictive, and self-configuring processes.

2.4   Security in CPS

Cyber-physical security is just an enhancement of traditional cyber security in which the operation of all physical system is also factored in. Password cracking which entails the procedure of recovering a service's password is one of the most critical cybersecurity threats in the traditional cyber security area, owing to the possibility of personal information breaches. In cyber-physical security, basic information exposure by cracking passwords could indeed harm the CPS; rather, a physical process modification via illegal access via a password might affect the characteristics of the physical systems.

As a result, a wide range of cyber-physical security studies have been carried out by modelling physical dynamics with control theory. Unfortunately, because CPSs are influenced by a diverse range of parameters including quick changes in the environment and unpredictable occurrences, physically model-based cyber-physical security devices suffer from false alarms, reducing detection capability against cyber-physical threats as shown in Table 2.1. Furthermore, when the CPS grows in size and the interactions between its components become more complicated, the degree of accuracy demonstrated by a traditional CPS model and a genuine CPS drops resulting in the generation of new attack vectors.

(1) **Secure Constrained Device:** Constrained systems are incapable of completing complicated encryption and decryption swiftly in real time. These devices are in responsibility of attacks such as power analysis. The data is usually transmitted without encryption or decryption. This can be mitigated by implementing lightweight cryptography and numerous layers of systems such as isolating devices on a septate network.
(2) **Secure Communication:** The security problem with low-power devices is to guarantee network connection. Low-power systems on the other hand do not encrypt data before

Table 2.1 Techniques for the Detection of Malware and AttackPatterns Using Learning Methods

Machine Learning	Deep Learning
SVM	RNN
Logistic Regression	GAN
GTB	CNN
J48 Decision Trees	MLP
Random Forests	LSTM

transferring them over the network. To keep data confidential, employ a separate network and segregate the device which aids in encrypted personal communications.

(3) **Data Integration and Privacy:** It is critical that information is primarily safely kept and handled once it has transmitted across the network. It is critical to establish data protection in low-power systems, which involves forming sensitive material before it has been recorded. Data management within a lawful and consistent framework is likewise a significant difficulty.

Data-driven anomaly-based methodologies (where anomalous data is obtained through various simulations and controlled experiments) are employed in cyber-physical security to circumvent the constraints of legacy model-based cyber-physical security. Machine learning (ML), in particular, which illustrates correlation coefficients between being an input and output using huge quantities of data without modelling based on physical principles, is being used in cyber-physical security to address high-level safety and dependability problems. Besides, ML technologies allow the generation of a prototype for the huge and intricate connections of each constituent of the CPS which include various physical systems operating in the real entire globe of heterogeneous networking technologies and complex software applications in the cyber world in which the resulting model can improve the CPS's security level.

2.5   Methodology

2.5.1   General Approach for Security in CPS

The ISO/IEC 270012013 standard states that troubles might be purposeful, unintentional, or environmental. The following are some exemplifications of common pitfalls: Physical detriment, natural disasters, the interruption of vital services, radiation, malfunctions, concession of information (similar to wiretapping software tampering, etc.), specialized lapses, unauthorized conduct (similar to data corruption), and concession of functions for illustration, forging. and abuse of rights). Using the findings of the study of recent security exploration (Figure 2.4), a chart of pitfalls and attacks grounded on the functional model of CPS is suggested. The chart branches come in the following: Kinds of attacks on detectors (seeing); assaults on selectors (actuating); and attacks against selectors (actuation), computer factors (computing), dispatches (communication), and feedback (E) are all exemplifications of cyberattacks [18].

(1) Threats and failings that vitiate CPS detectors have been uncovered by experimenters (similar to edging in fake radar signals bedazzling cameras with light GPS spoofing, etc.). The trustability and delicacy of the data collecting process must be guaranteed because CPSs are explosively tied to the physical process in which they're integrated. To ensure that any data obtained from a physical process can be trusted, detector security requires methods to enhance physical authentication [18, 2].

(2) The Finite Energy Attack which includes for illustration the loss and modification of particular packets the Finite Time Attack and Impulse attacks as well as the Bounded Attack which results in the suppression of the control signal were two classes that covered a wide range of implicit attacks in Djouadi et al. [12] analysis of the goods of cyberattacks on selectors. The actuation control security refers to the fact that no action may be taken during a unresistant-active or active mode of operation without the necessary authorization. Since the CPS specifications vary over time, the warrants' specification must be dynamic.

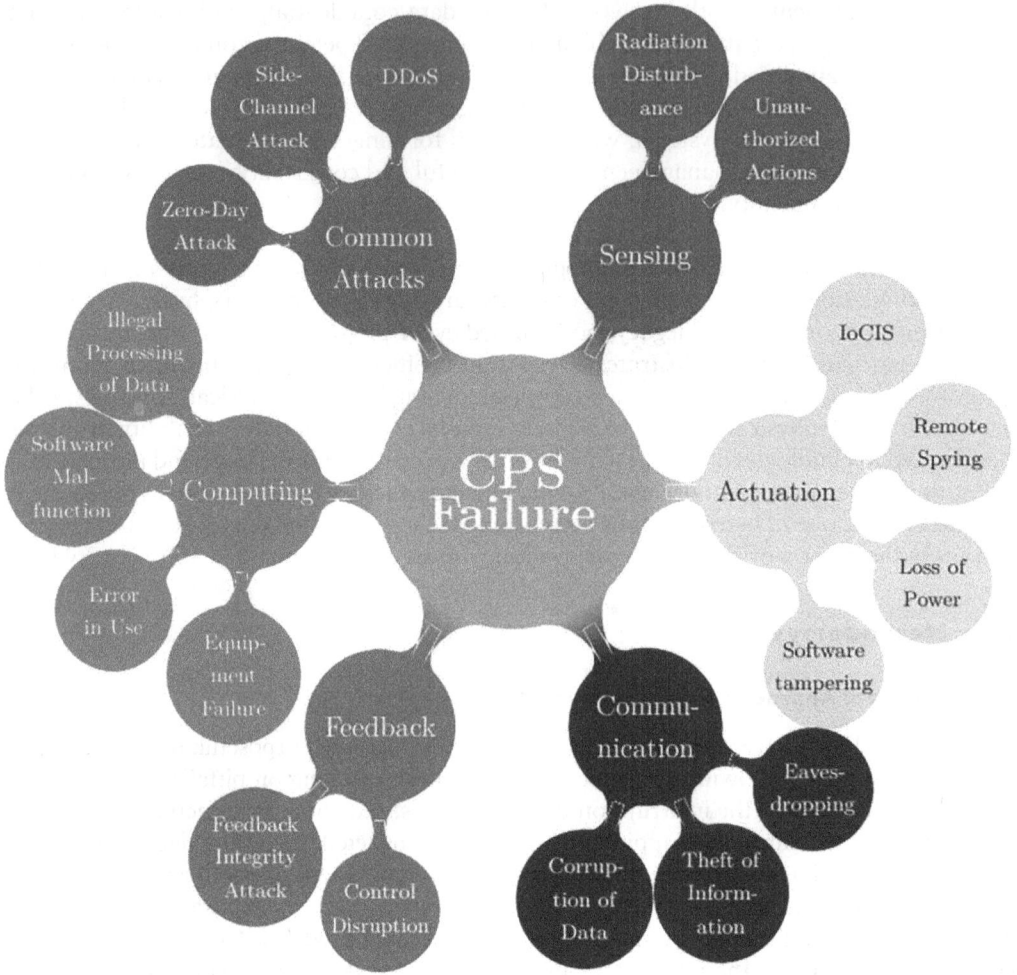

Figure 2.4 Map of attacks and threats on cyber-physical systems.

(3) The specifics of attacks on computer resources are covered in the document and comprise DoS attacks, worms, viruses, and Trojan horses. In details on techniques for data mining (DM) that can be applied to boost cybersecurity [54, 46]. The CPS might sustain harm from a sneaky strike, underscoring the importance of robust security measures and vigilance in safeguarding critical systems. The detecting methods must make sure that the violations and measurement mistakes that are common in control systems won't result in a false alert. The attacker now has a place to hide. An integrity attack occurs when a hacker alters actual data by obtaining a communication key for secure communication or by capturing some network devices. The creation of solutions to protect stored data in CPS platforms from physical or digital intrusions falls under the category of storage security. From the attacker's perspective, developing a false attack strategy typically takes into account a variety of variable resources and security restrictions.

(4) Selective forwarding, packet spoofing, packet replaying, sybil attacks, and other communication attacks that subvert system package routing can be employed to impede resource distribution between nodes in favor of malware. Any revision to the data might

affect miscalculations when recycling it in the future. An effective attack strategy is to record some "normal" data and play it back to avoid discovery if the bushwhacker can only capture and further real data packets [2]. The creation of protocols is necessary for communication security because they link active hindrance sources and resistant listeners between and within CPS.

(5) A three-concentrated logical model of CPS and a cyberattack metamodel were presented, wherein the system was subjected to a feedback integrity assault, whereby only a portion of the control signals retained their integrity. Feedback Security describes the safekeeping of the CPS control systems, which deliver the needed feedback for actuation. Ultramodern security measures are only concerned with defending data, but in order to adequately defend CPSs, it is important to understand how they affect evaluation and operation algorithms [18, 4].

2.5.1.1   Security Implementation Challenges

CPS includes embedded controllers in digital systems (cyberspace) and physical settings. This integrated system is nevertheless vulnerable to several threats. Choo et al. research a variety of security issues including both internal and external assaults. Potential issues can arise with various systems. These include CPS integrated with IoT for healthcare-embedded systems and green cryptography in smart building security. Privacy concerns in CPS trust and security, emerging security schemes for embedded security, and handling sensitive data in CPS are also among these potential issues. In addition, unresolved issues are brought on by the diversity of technologies used in IoT-integrated CPS. The study by Wang et al. looked at time synchronization attacks in CPS [9, 59]. Different attacks are found in CPS environments according to research by Ding et al. They are referred to as deception replay and DoS attacks. Different attacks connected to CPS were identified by Mahmoud et al. Jamming attacks, deception attacks, DoS attacks, and replay attacks are among them [11, 45].

According to Sarker et al., data wisdom is playing a critical part in accelerating the recent technological transition in cybersecurity. Data wisdom was developed to identify attacks through data analysis as the quantum of data relating to cybersecurity events increased. The quantum of training data is growing along with the number of circumstances, which helps to automate the monitoring of cyberspace irruptions with the necessary intelligence. Machine learning is essential in achieving this thing, since it excerpts useful information from numerous cybersecurity data sources. It is known that the number of IoT use cases has sometimes increased,; according to exploration by Sarker et al. [51], IoT use cases that include thousands of linked resource-constrained devices present security enterprises in dispersed surroundings. In such a case, an automated and intelligent system of covering cyberspace is needed to be suitable to take prompt action when circumstances call for it. They also delved several data sets employed for cybersecurity study. Among the data sets are KDD'99 Cup [22], NSL-KDD [56], CAIDA [22], ISOT'10 [22], ISCX'12 [22], CTU-13 [14], UNSW-NB15, CIC-IDS2018 and CIC-IDS2017 [22], MAWI [25] CERT [42], DGA [15], Malware [63], and Bot-IoT [28].

While each of these distinctions are significant, we think that the way in which control systems interact with the physical environment sets them apart from other IT systems. Generally speaking, information security has produced established technologies and design tenets (authentication, access, control, message, integrity, separation of privilege, etc.) that can aid in the prevention and response to assaults on control systems. However, information protection has always been the main emphasis of computer security research. The impact of assaults on estimate and control algorithms and eventually on the physical world has not been taken into account by researchers. While the present information security tools can provide required mechanisms for the security of control systems, we contend that these mechanisms alone are insufficient for the security of control systems defence.

We think that by comprehending how the control system interacts with the physical world, we should be able

(1) To better comprehend the effects of an assault: It is important to note that there is currently no study on how an adversary would choose a strategy after gaining unauthorized access to some control network devices.
(2) To create original attack-detection algorithms: By figuring out how the physical process ought to react in response to our control instructions and sensor readings, we can tell whether a hacker is interfering with the control or sensor data.
(3) To create new algorithms and architectural designs that are resistant to attacks: If an attack is detected, we may be able to modify the control commands to improve the system's resilience.

2.6   How Artificial Intelligence Can Help Solve CPS Challenges

Data science with its machine learning and DL approaches is giving a significant contribution to the automatic analysis of CPSs to enable non-stop monitoring for increased security measures, as detailed in the antedating sections [52]. For the independent mitigation of cyberattacks on CPS, Kholidy presented a result with a foundational security frame. In order to ameliorate security, it incorporates standard IDS styles like SNORT and ML-grounded ways. In order to achieve the new Autonomous Response Controller (ARC) framework, it takes into account the security ways and threat assessment models formerly in use. They employed a probabilistic threat assessment model which evaluates possible pitfalls and takes the necessary conduct. Lv et al. delved into the use of AI to increase the security of CPS. A CPS-grounded inner terrain assessment and control system for smart structures was taken into consideration as a case study. They created a multi-agent system that consists of numerous rudiments including discovery control prosecution and communication. For classification and regression, they employed BPNN (Backpropagation Neural Network). The system is assessed using a variety of performance measures including Mean Absolute Error (MAE), NRMSE, and PSNR, and it was discovered to have excellent performance and the smallest false positives [26, 43].

Tantawy et al. suggested a model-based combined strategy for CPS security. They employed the Continuous Stirred Tank Reactor (CSTR) and its integration with cyberspace in a real-world company for their study on security. A business network and control network are also included in the CPS test bed employed for the exploration, in addition to data participation for machine-literacy-grounded monitoring of cyberattacks. Various corridors of the system including the factory control logic monitoring system commercial analysis and safety sense. To detect cyberattacks, a hybrid automation is created. To store data and use it efficiently for the discovery of assaults, a tree-grounded fashion is used. Regarding the physical subcaste of Cognitive Radio Networks, Salahdine and Kaabouch examined cybersecurity risks, weaknesses, and responses [55, 49]. They concentrated on several physical subcaste assaults related to vicious business injection belief manipulation wiretapping and dynamic diapason access. Compressive seeing, learning-grounded approaches, the IDS model pointgrounded approaches, data-supported approaches, localization-grounded approaches, belief propagation approaches, and diapason seeing ways have also been covered. The PU emulation attacks are divided into four orders: Hybrid attacks, point attacks, cryptography attacks, and game proposition attacks.

Sargolzaei et al. investigated the use of machine learning for defect discovery using a case study of a vehicular CPS. The False Data Injection (FDI) attack that leads to mishaps in vehicular networks was the main emphasis. In order to help the driver or controller determine the safe distance between his or her auto and the vehicle in front of it, they suggested a system that recognizes FDI attacks and takes remedial action in order to induce the right signal. For defect identification, it uses a neural network-grounded system. Goh et al. employed LSTM-RNN to model the system and an unsupervised literacy system of deep literacy to identify the anomalous gesture. A CPS can be told by one of two different time synchronization attacks that use GPS. Time Synchronization (TS) assault and Stealth Time Synchronization (STS) attack are the names of these types of attacks. ML approaches were delved into by Wang et al. to stop similar assaults. They proposed a discovery approach grounded on Artificial Neural Networks (ANNs) that took advantage of performance in identifying the threats [50, 16, 59].

2.6.1 Results

- **Intelligent transportation:** As the number of personal vehicles increases, related issues including air pollution traffic congestion and safety concerns are receiving greater attention. Next-generation transportation systems such air traffic railway and vehicle control will heavily rely on cutting-edge computer and sensor technologies to increase safety and throughput. The idea of an intelligent transportation system (ITS) is not new. It refers to a broad spectrum of integrated transportation management that must be precise, effective, and in real time. The old forms of transportation are evolving towards more intelligence on the basis of contemporary technology like electronics, computers, sensors, and networks. The United States AHS Consortium (NAHSC) whose members include General Motors and Carnegie Mellon University Bechtel Corporation and others was founded in 1994. In order to create more intelligent and safe traffic, NAHSC focuses on the research and development of autonomous high-speed systems or "AHS" (Automated Highway Systems) [57]. MIT created CarTel one of the American NSF-supported programmes. To meet these issues, the CarTel project integrates wireless networking, mobile computing, and sensors and data-intensive algorithms operating on cloud servers. CarTel enables applications to quickly gather, process, deliver, analyse, and display data from mobile unit sensors. CarTel has made contributions in the areas of danger identification, vehicular networking, traffic reduction, and monitoring of the road surface [20].
- **Precision agriculture:** In order to apply a full range of contemporary systems of agricultural management techniques and technologies, precision agriculture is a technology that was created in the early 1980s from conventional agriculture and backed by information technology [62]. Data management for production trials basic geographic data for farms information on microclimates and other data are all part of the precision agriculture concept. In the University of Nebraska-Lincoln Cyber-Physical Networking Lab, Agnelo R. Silva and Mehmet C. Vuran created the "underground wireless sensor network" project. They combined centre pivot systems with wireless underground sensor networks to create the CPS^2 for precision agriculture [53]. The Wireless Subterranean Sensor Networks (WUSNs) are made up of underground sensor nodes that are wirelessly linked and communicate via the soil. The outcomes of the experiment demonstrate the viability and high reliability of the CPS^2 idea when employing widely available wireless sensor motes. One of the usual CPS applications with promising futures combines CPS with precision agriculture.
- **Medical CPS:** In the future, medical CPS (MCPS) will take the place of conventional medical equipment that operate separately. Multiple medical devices collaborate with

sensors and networks to identify the patient's physical state in real time, particularly for essential patients like those with heart disease. The patient's portable terminal devices can constantly monitor the patient's status and deliver timely alerts or forecasts. Additionally, patients would find it much more convenient if medical equipment and real-time data distribution worked together. The highly reputable medical CPSs' development and problems were discussed by Insup Lee and Oleg Sokolsky [32] who also discussed the system's reliance on software for the creation of new functions, the need for network connections, and the need for ongoing patient monitoring to examine how medical CPS may develop in the future. Cheolgi Kim et al. established a general framework called the NASS (Network-Aware Supervisory Systems) to integrate medical devices into a clinical interoperability system that makes use of actual networks to address medical device interoperability concerns [8]. It offers a development environment where the supervisory logic for medical devices may be created using the presumption that operates a perfect, reliable network. The case study demonstrates how well the NASS architecture protects against actual network outages and is procedurally successful. More aspects including a higher need for security, real-time processing, and network latency will be taken into account while designing the medical CPS due to the uniqueness of medical applications [28].

2.7   Discussion

2.7.1   DDoS Attack in CPS

Hackers or cyber criminals use DDoS (Distributed Denial of Service) attacks, a sort of malicious cyberattack, to prevent intended users from accessing a host system network resource or online service. DDoS attacks flood the target system with thousands or millions of erroneous or undesired requests, overwhelming it and its supporting resources.

The goal of a DDoS attack is always to flood the system, although how this is accomplished can change. Three main types of DDoS attacks are listed below.

- **Application layer attacks:** Application layer DDoS attacks are designed to directly target the programme, focusing on specific bugs or issues and stop the application from providing content to the user. New attack tactics and vectors are developed by attackers for use in a new generation of attacks. When the defence gets good at blocking these new attacks, the attacker develops a new type of attack and continues this cycle.
- **Protocol attacks:** To deplete server resources, attackers target intermediate communication devices like load balancers, routing engines, servers, and firewalls. It can overpower mitigations such as firewalls at the edge, resulting in service interruption and DDoS.
- **Volumetric attacks:** Volumetric DDoS assaults are made to flood internal networks with a large amount of malicious traffic, including centralized DDoS mitigation scrubbing capabilities. These DDoS assaults try to use up all available bandwidth, either inside the target network or service or between it and the rest of the Internet. Numerous subtypes belong to one of the general categories mentioned above, yet have their distinctive traits. Here is a detailed list of modern DDoS attack techniques.
- Other attacks include SYN Flood, TCP Flood, SYN-ACK Flood, Fragmented ACK Flood, Session Flood, UDP Flood, UDP-LAG Flood, PORTMAP Flood, DNS Flood, NTP Flood, CHARGEN Flood, SSDP Flood, SNMP Flood, HTTP Flood, etc. We'll examine how these modern DDoS attacks affect r-physical systems in the next section.

2.7.1.1 How Does DDoS Attack Affect CPS?

CPSs are those that can act autonomously and make judgements on their own. Some examples of CPS are Smart Power Grid, Railway Systems Motion Sensors, etc. Let us see how a DDoS attack affects a CPS by looking into an example:

> According to the Department of Energy's Electric Emergency and Disturbance Report for March 2019 a DDoS attack had disrupted electrical grid operations in two sizable U.S. population centres: Salt Lake County in Utah and Los Angeles County in California. The Department of Energy acknowledged that the attack only resulted in "interruptions" in "electrical system operations" not any disruptions in electrical distribution or outages. In this context, "operations" refers to any computer systems used within the utility including those that run office applications or operational software and does not necessarily refer to the transmission of electricity to consumers.

This incident occurred in the year 2019 when there was a power outage in the cities mentioned above. The attackers tried to stop the power by flooding the system operations. This is a perfect example to tell that how a DDoS attack affects a CPS [10].

2.7.1.2 Understanding How to Prevent CPS from DDoS Attack Using AI

In this section, we are going to work on building a few ML models using Python for the CICDDoS2019 data set [48].

2.7.1.2.1 ABOUT THE DATA SET

CICDDoS2019 [48] is an evaluation data set that contains BENIGN and the most up-to-date Modern DDoS attacks. It is organized in a day-to-day format and includes the raw data of network traffic and event logs per machine, which are exported to CSV files for network analysis. This data set has more than 80 traffic features, some of which are listed Table 2.2. In this section, we are going to work on building a few ML models using Python.

2.7.1.2.2 IMPORTING-REQUIRED PYTHON LIBRARIES

Python libraries are blocks of code that contain built-in functions. These libraries provide access to the necessary packages or modules that can be installed to complete specific tasks. A quick explanation of the libraries used in the framework is provided in this section.

(1) **Pandas:** By providing data operations and data structures, the Pandas' library significantly aids in the study and manipulation of data, particularly time series and numerical tables. On top of NumPy, the Pandas' package was developed. It supports the effective implementation of a data frame. Series and data frames are the foundational data structures on which Pandas is built. Data frames are a two-dimensional structure in the form of a table with several columns, whereas series are a one-dimensional structure in the form of a list of items. Row and column labels for homogeneous and heterogeneous data types with or without missing data make up data frames. Pandas allow for the transformation of data structures into data frame objects, handling of missing data, and histogram or box plots.

Table 2.2 Features Present in the CICDDoS2019 with Description

No.	Feature Name	Description
1	Flow duration	Duration of the flow in microseconds
2	Flow duration	Total packets in the forward direction
3	total Fwd Packet	Total packets in the backward direction
4	total Length of Fwd Packet	Minimum size of the packet in the forward direction
5	total Length of Bwd Packet	Maximum size of the packet in the forward direction
6	Fwd Packet Length Min	Mean size of the packet in the forward direction
7	Fwd Packet Length Max	Standard deviation size of the packet in the forward direction
8	Fwd Packet Length Mean	Minimum size of the packet in the backward direction
9	Fwd Packet Length Std	Maximum size of the packet in the backward direction
10	Bwd Packet Length Min	Mean size of the packet in the backward direction
11	Bwd Packet Length Max	Standard deviation size of the packet in the backward direction
12	Active Min	The minimum time a flow was active before becoming idle
13	Active Mean	Meantime a flow was active before becoming idle
14	Active Max	The maximum time a flow was active before becoming idle
15	Active Std	Standard deviation time a flow was active before becoming idle
16	Idle Min	The minimum time a flow was idle before becoming active
17	Idle Mean	Meantime a flow was idle before becoming active
18	Idle Max	The maximum time a flow was idle before becoming active
19	Idle std	Standard deviation time a flow was idle before becoming active

(2) **Scikit-learn:** The collection of various classification, grouping, and regression algorithms in the Scikit-learn package helps learning and decision-making. It is a Python module built on top of the machine learning library SciPy. A widely used software for Python-based data science applications is called Scikit-learn. It allows the handling of data mining and machine learning tasks like classification, clustering, regression, dimensionality reduction, and model selection and offers a clear interface to a range of machine learning methods. Scikit-learn substantially aids data scientists in carrying out machine learning projects by incorporating efficient versions of several widely used algorithms.

(3) **Matplotlib:** The general-purpose Graphical User Interface (GUI) toolkits are the foundation of the object-oriented Application Programming Interface (API) which plays a crucial role in embedding the plots into programmes. Data can be represented as two-dimensional graphs and diagrams using Matplotlib, including scatter plots, histograms, and graphs using non-Cartesian coordinates. It substantially facilitates data science project visualization and works with a variety of operating systems and graphics backends. In the field of scientific Python, Matplotlib is also known as the cross-platform everything-to-everyone approach, with the benefit of supporting many output types and backends.

2.7.1.3   Methodology

Flood attacks include delivering massive volumes of empty or useless packets to the victim system to block communication between the two machines and clog up the entire network. The challenge with a DDoS attack is modelling or estimating the traffic on the target network.

The suggested approach seeks to produce two forms of network traffic:

(1) Regular packets
(2) Attacker packets

The suggested system for modelling has four phases:-

(1) Data Pre-processing
(2) Detection model
(3) Classification stage
(4) Performance evaluation stage

2.7.1.4   Data Pre-processing

The data should be preprocessed to achieve faster training and testing speeds, higher accuracy, etc., which typically involves steps such as encoding, normalization, and feature reduction using PCA.

Tag encoding is the process of metamorphosing tags into a numeric form so that they may be interpreted by machines. The operation of those tags can also be better determined by machine learning algorithms. It is a significant-supervised learning pre-processing step for the structured data set. The class "tag" is decoded for this case.

When the colourful characteristics (variables) are on a lower scale, machine learning algorithms generally perform more or meet more snappily, thus normalizing the data before training the machine learning models as its standard practice.

Also, normalization reduces the acuteness of the training process to the magnitude of the characteristics. As a result, following training, the coefficients improve. Point scaling is the process of rescaling features to make them more training-friendly.

Principal component analysis or PCA [3] is a fashion for reducing the number of confines in large data sets by condensing a large collection of variables into a lower set that retains the majority of the large set's information. Accuracy naturally suffers as a data set's variables are reduced, but the answer to dimensional reduction is to trade a little accuracy for simplicity. Machine learning algorithms can analyse data much more fast and easy with lower data sets because there are smaller gratuitous factors to reuse.

Figure 2.5 shows the correlation plot with all the features present in the data set. These features are sorted and named to prepare the predicted models.

2.7.1.5   Model

The two aspects of supervised classification are the development of the learning approach and the creation of predicted labels. These activities are performed using Scikit-learn [7].

2.7.1.5.1 MODELS SELECTION

In this work, four distinct classification techniques—Naive Bayes, Decision Tree, Random Forest, and Logistic Regression are tested and trained.

(1) **Naive Bayes**—This classifier is built on Bayes' Theorems on the presumption of event independence. Statistics refers to two events as independent if the chance of one happening does not affect an effect on the other. For example consider $P(B \mid A)$ to be the measure of the likelihood of any given occurrence. Let $P(B)$ represent the probability of B and $P(A)$ represent the probability of A; $P(A \mid B)$ represents the likelihood of A given B.

$$P(A \mid B) = \frac{P(B \mid A)P(A)}{P(B)} \tag{1}$$

(2) **Decision Tree**—Decision tree classifies data depending on the values of the pertinent attributes, starting at the root node. Every node represents a specific feature and all of its potential values. As it moves down from the root of the tree, the iterative method evaluates the information learned for each characteristic in the training set. Information gain is used to evaluate the level of selectivity imposed by the characteristics of the target classes. With more information gained, the attribute's value in categorizing each observation rises. The root node is replaced by the characteristic that provides the most information gain, and the programme continues to separate the data set by a selected feature either to construct or create a node.

(3) **Random Forest**—Random Forest and other guided machine learning algorithms are constantly used in retrogression and bracket problems. It builds decision trees from different samples using their normal for categorization and inviting votes for retrogression. One of the most important features of the Random Forest Algorithm is its capacity to handle data sets containing both statistical parameters as in vatication and categorical data as in bracket. It produces better results with regard to bracket problems.

(4) **Logistic Regression**—Analysis of scenarios where the dependent variable is binary makes use of predictive analytic techniques like logistic regression most successful. Logistic regression is used to describe the data and also shows why a binary dependent variable and various non-binary independent variables are correlated.

2.7.1.6 *Training*

During the training phase, training data is sent to the selected algorithms, so they can utilize it to create machine learning models. As a result, the training set is used. At this point in the process, the target attribute (class "Label") must exist in the incoming data source.

Table 2.3 Table of Scikit-Learn Libraries

Model	Scikit-Learn Classifiers
Naive Bayes	sklearn.naive_bayes.GaussianNB
Decision Tree	sklearn.tree.DecisionTreeClassifier
Random Forest	sklearn.ensemble.RandomForestClassifier
Logistic Regression	sklearn.linear_model.LogisticRegression

Pattern recognition is used during the training phase to link the input characteristics with the target property. Based on the patterns found, a model is developed. The DDoS attack on the application layer dataset is used as the input data source for the training of four algorithms in this study. The sort of network traffic such as an attack or about is the target attribute. These are directly imported from the Scikit-learn library as depicted in Table 2.3.

2.7.1.7 Testing

The models are evaluated using fictitious data after the conclusion of the modelling phase. The unobserved data utilized at this time is the test set that was produced by the data split (20%). Testing is carried out to assess a model's capacity to represent data and its future functionality. To ensure that the result of the test would only be used once, this study wanted to make sure that any model alterations were made before testing. Performance metrics used to evaluate the effectiveness of the DDoS datasets include accuracy precision recall and F-measure. These are covered in the paragraph that follows.

2.7.1.8 Evaluation

Creating performance metrics is an essential step in analysing a model's performance. Several metrics are generated for this investigation. Below is a description of these [41].

(1) **Confusion Matrix**— A confusion matrix contains data on the current classification and projected classification performed by a classification system or model. The matrix's data is regularly analysed to assess the effectiveness of these systems. Generally, every instance that relates to but isn't assigned to can have an entry in an x matrix. Ideally, all off-diagonals should be 0 to avoid categorization errors. It is feasible to pinpoint the precise faults caused using the class confusion matrix. Table 2.4 shows the prediction table. Where:

- TP denotes the True Positive
- TN denotes the True Negative
- FP denotes the False Positive
- FN denotes the False Negative

Table 2.4 Prediction Table

True Classification		Prediction	
		Yes	No
	Yes	TP	FN
	No	FP	TN

(2) **Accuracy**—The machine learning model's accuracy is a metric used to determine which model is best at identifying patterns and correlations between variables in a dataset using the input or training data. Depending on how well a model can generalize to "unseen" data, the better estimates and insights it can produce will in turn bring more commercial value.

$$AC = \frac{TP + TN}{TP + FN + FP + TP} \tag{2}$$

(3) **Precision**—Often accuracy alone is inadequate to determine a learning model's efficiency. Although accuracy can indicate whether a model has been trained correctly, it
does not offer detailed information about the specific application. Other performance
standards such as accuracy are consequently applied. Precision is the percentage of correctly classified positives also known as true positives.

$$P = \frac{TP}{TP + FP} \tag{3}$$

(4) **Recall**—Recall is another performance metric. Recall measures the number of true positives found or remembered. The fact that false positives or false negatives could have
harmful impacts in some situations makes it an extremely important statistic. For instance, the potential of disruption to the network and its users will increase if a DDoS
attack type is used, since harmful network traffic will probably go unnoticed.

$$P = \frac{TP}{TP + FN} \tag{4}$$

(5) **F-measure**—F-measure is a statistic indicator that combines recall and precision to calculate
the overall accuracy of a model. A model with a higher F-measure score detects threats effectively with few false alarms produced due to its low false positive and false negative rates.

$$F - measure = 2 \times \frac{TP}{TP + FP} \tag{5}$$

These evaluation measures are loaded directly from Scikit-learn Metrics Library as shown
in Table 2.5.

2.7.2 Model Analysis

The examination used the following metrics: Recall accuracy precision and F-measure as
shown in Table 2.6. The experiment's findings indicate that 99% and 99% respectively of

Table 2.5 Measures

S. No	Measure	Scikit-learn Metrics
1	Confusion Matrix	Sklearn.metrics.confusion_matrix
2	Accuracy	Sklearn.metrics.accuracy_score
3	Precision	Sklearn.metrics.precision_score
4	Recall	Sklearn.metrics.recall_score
5	F—Measure	Sklearn.metrics.f1_score

Table 2.6 Model Analysis

No	Method	Dataset	Data Availability	Accuracy	Precision	Recall	F-Measures
1	Random Forest	CICDDoS2019	Yes	0.98	0.99	0.99	0.99
2	Naive Bayes	CICDDoS2019	Yes	0.45	0.66	0.54	0.38
3	Logistic Regression	CICDDoS2019	Yes	0.98	0.99	0.98	0.99
4	Decision Tree	CICDDoS2019	Yes	0.99	0.99	0.99	0.99

the DT and RF algorithms provide the highest levels of accuracy. The precision recall and F-measure results were all 99% for both DT and RF.

2.7.2.1   Pros and Cons of the Above ML Methods

(1) **Naive Bayes:**

   (a) Pros: The forecasting of the test dataset's class is rapid and easy. Additionally, it excels at multi-class prediction. An NB classifier outperforms traditional algorithms like logistic regression while requiring less training data when the premise of independence is true.
   (b) Cons: Conditional independence of the premise class may reduce accuracy. The independence assumption may not hold for all attributes. Variables are practically interdependent.

(2) **Random Forest:**

   (a) Pros: It produces an output that is more accurate, dependable, and resistant to overfitting. It requires far fewer inputs and does not call for the feature selection process.
   (b) Cons: Real-time applications requiring large datasets may find it impractical to use RF because it generates many DTs.

(3) **Logistic Regression:**

   (a) Pros: It is simpler to use and to interpret and practice with When the dataset is linearly separated, it performs well and has good accuracy for many straightforward datasets.
   (b) Cons: LR should be avoided if the number of observations is less than the number of features, as overfitting may result. A significant flaw in LR is the assumption that the dependent and independent variables are linearly related.

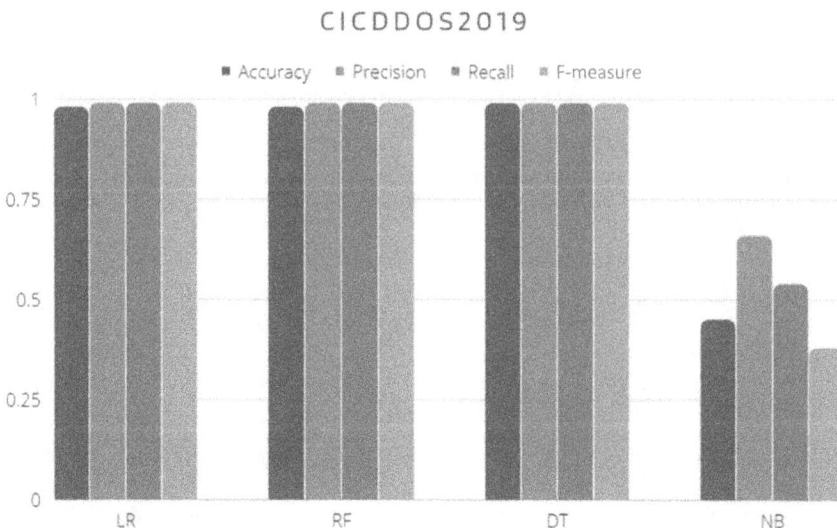

Figure 2.5 Model analysis results.

(4) **Decision Tree:**

   (a) Pros: Easy to use and straightforward.
   (b) Cons: Larger storage; is needed. It requires complicated computation; only when few DTs are utilized, it is simple to use.

2.8 Conclusion

The mutual penetration and integration of information space and physical space have directly aided in the development of CPS-related theories and applications as science and technology have advanced, and social production has improved. The CPS philosophy of "cooperative design and efficient operation" separates it from the Internet of Things even though it shares many characteristics with networked control systems, the latter being highly integrated and intricately coordinated control systems. As a result of its distinctive technical benefits, CPS has been extensively employed in important economic and social sectors, including smart manufacturing, smart transportation, telemedicine, power grids, and robotic systems. However, several issues have also been found as CPS application technology developed. The addition of the network layer forces CPS to address the physical layer's random instability elements in addition to the threat of hostile network layer assaults, which puts the system's security in grave jeopardy.

This chapter mainly investigates the security control and cross-layer architecture of CPS under DoS attacks and looks at the impact of network assaults on the system from the standpoints of control and communication. Active compensating methods with dynamic output feedback control or passive control methods with state feedback are suggested for continuous random DoS assaults. Additionally, a sliding mode control approach is recommended, and the network layer is designed using game theory to account for attackers with clever functions. Through the use of a cross-layer architecture, CPS's security and stability are effectively guaranteed against DoS assaults.

As the part of Discussion, we explicitly see that the model's performance in relation to the testing data is relatively high after applying machine learning techniques to the CICD-DOS2019 dataset. This suggests that the model can be employed to foretell DDOS attacks. As we can see, among the models Random Forest, Decision Tree, Naive Bayes, and Logistic Regression, accuracy is the highest for Decision Tree. Since the patterns of DDOS attacks can change over time, better pre-processing and techniques should be used to make better predictions that would actually aid in controlling them like ensemble learning and deep learning approaches, which would produce beneficial outcomes. Additional updating and maintaining of the trained AI models are also required.

References

[1] Ahmad, I., Khan, M., Qadir, J., and Salahuddin, M. A. A comparative study of deep learning architectures for detection of ddos attacks. *arXiv preprint arXiv:1811.12808* (2018).
[2] Alguliyev, R., Imamverdiyev, Y., and Sukhostat, L. Cyber-physical systems and their security issues. *Computers in Industry* 100 (2018), 212–223.
[3] Ali, M., Ahmed, E., ImRan, M., Yasin, M., and Iqbal, W. A DDoS attack detection using PCA dimensionality reduction and support vector machine. In *2017 13th International Conference on Emerging Technologies (ICET)*. IEEE, 2017, pp. 1–6.
[4] Amin, S., Cárdenas, A. A., and Sastry, S. S. Safe and secure networked control systems under denial-of-service attacks. In *International Workshop on Hybrid Systems: Computation and Control* (2009). Springer, pp. 31–45.

[5] Baheti, R., and Gill, H. Cyber-physical systems. *The Impact of Control Technology* 12, 1 (2011), 161–166.

[6] Celen, M., and Djurdjanovic, D. Operation-dependent maintenance scheduling in flexible manufacturing systems. *CIRP Journal of Manufacturing Science and Technology* 5, 4 (2012), 296–308.

[7] Chatterjee, K., Nandi, S., Chakraborty, T., Maiti, S., and Dasgupta, R. Feature selection approach to detect ddos attack using machine learning algorithms. *Journal of Information Security* 10, 4 (2019), 271–285.

[8] Cheolgi Kim, Mu Sun, Sibin Mohan, Heechul Yun, Lui Sha, and Tarek F. Abdelzaher. 2010. A framework for the safe interoperability of medical devices in the presence of network failures. In *Proceedings of the 1st ACM/IEEE International Conference on Cyber-Physical Systems (ICCPS '10)*. Association for Computing Machinery, New York, NY, USA, 149–158. https://doi.org/10.1145/1795194.1795215

[9] Choo, K.-K. R., Kermani, M. M., Azarderakhsh, R., and Govindarasu, M. Emerging embedded and cyber physical system security challenges and innovations. *IEEE Transactions on Dependable and Secure Computing* 14, 3 (2017), 235–236.

[10] CNBC. *DDoS Attack Caused Interruptions in Power System Operations*, 2019. https://www.cnbc.com/2019/05/02/ddos-attack-caused-interruptions-in-power-system-operations-doe.html.

[11] Ding, D., Han, Q.-L., Xiang, Y., Ge, X., and Zhang, X.-M. A survey on security control and attack detection for industrial cyber-physical systems. *Neurocomputing* 275 (2018), 1674–1683.

[12] Djouadi, S. M., Melin, A. M., Ferragut, E. M., Laska, J. A., Dong, J., and Drira, A. Finite energy and bounded actuator attacks on cyber-physical systems. In *2015 European Control Conference (ECC)* (2015). IEEE, pp. 3659–3664.

[13] Farivar, F., Haghighi, M. S., Jolfaei, A., and Alazab, M. Artificial intelligence for detection, estimation, and compensation of malicious attacks in nonlinear cyber-physical systems and industrial IoT. *IEEE Transactions on Industrial Informatics* 16, 4 (2019), 2716–2725.

[14] Garcia, S., and Uhlir, V. The CTU-13 dataset. A labeled dataset with botnet, normal and background traffic. (2011). https://www.impactcybertrust.org/dataset_view?idDataset=945

[15] Glasser, J., and Lindauer, B. Bridging the gap: A pragmatic approach to generating insider threat data. In *2013 IEEE Security and Privacy Workshops* (2013). IEEE, pp. 98–104.

[16] Goh, J., Adepu, S., Tan, M., and Lee, Z. S. Anomaly detection in cyber physical systems using recurrent neural networks. In *2017 IEEE 18th International Symposium on High Assurance Systems Engineering (HASE)* (2017). IEEE, pp. 140–145.

[17] Gupta, I., and Singh, A. K. Detection of DDoS attack using ensemble machine learning techniques. In *2018 Second International Conference on Computing Methodologies and Communication (ICCMC)*. IEEE, 2018, pp. 196–199.

[18] Hahn, A., Thomas, R. K., Lozano, I., and Cardenas, A. A multi-layered and kill-chain based security analysis framework for cyber-physical systems. *International Journal of Critical Infrastructure Protection* 11 (2015), 39–50.

[19] Hu, L., Xie, N., Kuang, Z., and Zhao, K. Review of cyber-physical system architecture. In *2012 IEEE 15th International Symposium on Object/Component/Service-Oriented Real-Time Distributed Computing Workshops* (2012). IEEE, pp. 25–30.

[20] Hull, B., Bychkovsky, V., Zhang, Y., Chen, K., Goraczko, M., Miu, A., Shih, E., Balakrishnan, H., and Madden. 2006. S. Cartel: A distributed mobile sensor computing system. In *Proceedings of the 4th International Conference on Embedded Networked Sensor Systems (SenSys '06)*. Association for Computing Machinery, New York, NY, USA, 125–138. https://doi.org/10.1145/1182807.1182821

[21] Hung, M.-H., Cheng, F.-T., and Yeh, S.-C. Development of a web-services-based e-diagnostics framework for semiconductor manufacturing industry. In *2003 IEEE International Conference on Robotics and Automation (Cat. No.03CH37422)*. IEEE, 2003, pp. 122–135.

[22] Jamal, A. A., Majid, A.-A. M., Konev, A., Kosachenko, T., and Shelupanov, A. A review on security analysis of cyber physical systems using machine learning. *Materialstoday: Proceedings* 80, 3 (2023), 2302–2306.

[23] Jazdi, N. Cyber physical systems in the context of industry 4.0. In *2014 IEEE International Conference on Automation, Quality and Testing, Robotics*. IEEE, 2014, pp. 1–4.

[24] Jin, C., Djurdjanovic, D., Ardakani, H. D., Wang, K., Buzza, M., Begheri, B., Brown, P., and Lee, J. A comprehensive framework of factoryto-factory dynamic fleet-level prognostics

and operation management for geographically distributed assets. In *2015 IEEE International Conference on Automation Science and Engineering (Case)* (2015). IEEE, pp. 225–230.

[25] Jing, X., Yan, Z., Jiang, X., and Pedrycz, W. Network traffic fusion and analysis against ddos flooding attacks with a novel reversible sketch. *Information Fusion* 51 (2019), 100–113.

[26] Kholidy, H. A. Autonomous mitigation of cyber risks in the cyber–physical systems. *Future Generation Computer Systems* 115 (2021), 171–187.

[27] Kim, C., Sun, M., Mohan, S., Yun, H., Sha, L., and Abdelzaher, T. F. A framework for the safe interoperability of medical devices in the presence of network failures. In *Proceedings of the 1st ACM/IEEE International Conference on Cyber-Physical Systems* (ICCPS '10). 2010. Association for Computing Machinery, New York, NY, USA, 149–158. https://doi.org/10.1145/1795194.1795215

[28] Koroniotis, N., Moustafa, N., Sitnikova, E., and Turnbull, B. Towards the development of realistic botnet dataset in the internet of things for network forensic analytics: Bot-IoT dataset. *Future Generation Computer Systems* 100 (2019), 779–796.

[29] Lee, Jay and Edzel Lapira. Fault detection in a network of similar machines using clustering approach. *Enginnering, computer science,* (2012). https://api.semanticscholar.org/CorpusID:113068586

[30] Lee, E. A. Time-critical networking-invited presentation. In *2009 IEEE/LEOS Summer Topical Meeting.* (2009) IEEE, pp. 149–150.

[31] Lee, E. A. CPS foundations. In *Proceedings of the 47th Design Automation Conference* (DAC '10). Association for Computing Machinery, New York, NY, USA, 737 -742. https://doi.org/10.1145/1837274.1837462

[32] Lee, I., Sokolsky, O., Chen, S., Hatcliff, J., Jee, E., Kim, B., King, A., Mullen-Fortino, M., Park, S., Roederer, A. and Venkatasubramanian, K.K., 2011. Challenges and research directions in medical cyber–physical systems. *Proceedings of the IEEE*, 100 (1), 75–90.

[33] Lee, J., Ardakani, H.D., Yang, S. and Bagheri, B., 2015. Industrial big data analytics and cyber-physical systems for future maintenance & service innovation. *Procedia cirp*, 38, pp. 3–7.

[34] Lee, J. Keynote presentation: Recent advances and transformation direction of PHM. In *Roadmapping Workshop on Measurement Science for Prognostics and Health Management of Smart Manufacturing Systems Agenda.*https://www.nist.gov/el/intelligent-systems-division-73500/roadmapping-workshop-measurement-science-prognostics-and, 2014

[35] Lee, J., Bagheri, B., and Jin, C. Introduction to cyber manufacturing. *Manufacturing Letters* 8 (2016), 11–15.

[36] Lee, J., Bagheri, B., and Kao, H.-A. A cyber-physical systems architecture for industry 4.0-based manufacturing systems. *Manufacturing Letters* 3 (2015), 18–23.

[37] Lee, J., Jin, C., and BagheRi, B. Cyber physical systems for predictive production systems. *Production Engineering* 11 (2017), 155–165.

[38] Lee, J., Jin, C., Liu, Z. Predictive Big Data Analytics and Cyber Physical Systems for TES Systems. In Redding, L., Roy, R., Shaw, A. (eds) *Advances in Through-life Engineering Services. Decision Engineering* (2017). Springer, Cham. https://doi.org/10.1007/978-3-319-49938-3_7

[39] Lee, J., Wu, F., Zhao, W., Ghaffari, M., Liao, L., and Siegel, D. Prognostics and health management design for rotary machinery systems—reviews, methodology and applications. *Mechanical Systems and Signal Processing* 42, 1–2 (2014), 314–334. https://doi.org/10.1016/j.ymssp.2013.06.004

[40] Li, L., and Ni, J. Short-term decision support system for maintenance task prioritization. *International Journal of Production Economics* 121, 1 (2009), 195–202.

[41] Lin, Y.-D., Chang, P.-K., Chen, J.-L., and Chen, C.-M. Information metrics for low-rate ddos attack detection: A comparative evaluation. *IEEE Transactions on Information Forensics and Security* 11, 10 (2016), 2229–2244.

[42] Lindauer, B., Glasser, J., Rosen, M., Wallnau, K. C., and Exactdata, L. Generating test data for insider threat detectors. *Journal of Wireless Mobile Networks, Ubiquitous Computing, and Dependable Applications* 5 (2014) 80–94.

[43] Lv, Z., Chen, D., Lou, R., and Alazab, A. Artificial intelligence for securing industrial-based cyber–physical systems. *Future Generation Computer Systems* 117 (2021), 291–298.

[44] Lv, Z., Chen, D., Lou, R., and Alazab, A. Artificial intelligence for securing industrial-based cyber–physical systems. *Future Generation Computer Systems* 117 (2021), 291–298.

[45] Mahmoud, M. S., Hamdan, M. M., and Baroudi, U. A. Modeling and control of cyber-physical systems subject to cyber attacks: A survey of recent advances and challenges. *Neurocomputing* 338 (2019), 101–115.

[46] Mitchell, R., and Chen, R. Effect of intrusion detection and response on reliability of cyber physical systems. *IEEE Transactions on Reliability* 62, 1 (2013), 199–210.

[47] Ni, J., and Jin, X. Decision support systems for effective maintenance operations. *CIRP Annals* 61, 1 (2012), 411–414.

[48] of New Brunswick, U. *DDoS 2019 Data Set*, 2019. https://www.unb.ca/cic/datasets/ddos-2019.html.

[49] Salahdine, F., and Kaabouch, N. Security threats, detection, and countermeasures for physical layer in cognitive radio networks: A survey. *Physical Communication* 39 (2020), 101001.

[50] Sargolzaei, A., Crane, C. D., Abbaspour, A., and Noei, S. A machine learning approach for fault detection in vehicular cyber-physical systems. 2016 15th *IEEE International Conference on Machine Learning and Applications (ICMLA)*, Anaheim, CA, USA, 2016, pp. 636-640, doi: 10.1109/ICMLA.2016.0112.

[51] Sarker, I. H., Kayes, A., Badsha, S., Alqahtani, H., Watters, P., and Ng, A. Cybersecurity data science: An overview from machine learning perspective. *Journal of Big Data* 7 (2020), 1–29.

[52] Shaukat, K., Luo, S., Varadharajan, V., Hameed, I. A., and Xu, M. A survey on machine learning techniques for cyber security in the last decade. *IEEE Access* 8 (2020), 222310–222354.

[53] Silva, A. R., and VuRan, M. C. (CPS) 2: Integration of center pivot systems with wireless underground sensor networks for autonomous precision agriculture. In *Proceedings of the 1st ACM/IEEE International Conference on Cyber-Physical Systems* (2010), pp. 79–88.

[54] Singhal, A. *Data warehousing and data mining techniques for cyber security* (2007), vol. 31. Springer Science & Business Media.

[55] Tantawy, A., Abdelwahed, S., Erradi, A., and Shaban, K. Model-based risk assessment for cyber physical systems security. *Computers & Security* 96 (2020), 101864.

[56] Tavallaee, M., Bagheri, E., Lu, W., and Ghorbani, A. A. A detailed analysis of the KDD cup 99 data set. In 2009 *IEEE Symposium on Computational Intelligence for Security and Defense Applications 2009*. IEEE, pp. 1–6.

[57] Tomizuka, M. Automated highway systems-an intelligent transportation system for the next century. In *ISIE'97 Proceeding of the IEEE International Symposium on Industrial Electronics* (1997), vol. 1. IEEE, pp. PS1–PS4.

[58] Wang, E. K., Ye, Y., Xu, X., Yiu, S.-M., Hui, L. C. K., and Chow, K.-P. Security issues and challenges for cyber physical system. In *2010 IEEE/ACM Int'l Conference on Green Computing and Communications & Int'l Conference on Cyber, Physical and Social Computing* (2010). IEEE, pp. 733–738.

[59] Wang, J., Tu, W., Hui, L. C., Yiu, S.-M., and Wang, E. K. Detecting time synchronization attacks in cyber-physical systems with machine learning techniques. In *2017 IEEE 37th International Conference on Distributed Computing Systems (ICDCS)* (2017). IEEE, pp. 2246–2251.

[60] Wang, T., Yu, J., Siegel, D., and Lee, J. A similarity-based prognostics approach for remaining useful life estimation of engineered systems. In *2008 International Conference on Prognostics and Health Management* (2008). IEEE, pp. 1–6.

[61] Yang, S., Bagheri, B., Kao, H.-A., and Lee, J. A unified framework and platform for designing of cloud-based machine health monitoring and manufacturing systems. *Journal of Manufacturing Science and Engineering* 137, 4 (2015), 040914

[62] Meng ZhiJun, Meng ZhiJun, Wang Xiu Wang Xiu, Zhao ChunJiang Zhao ChunJiang, and Xue XuZhang Xue XuZhang. "Development of field information collection system based on embedded COM-GIS and pocketPC for precision agriculture." *Transactions of the Chinese Society of Agricultural Engineering*, 2005, 21 (4), 91–96. https://www.cabidigitallibrary.org/doi/full/10.5555/20053130249

[63] Zhou, Y., and Jiang, X. Dissecting android malware: Characterization and evolution. In *2012 IEEE Symposium on Security and Privacy* (2012). IEEE, pp. 95–109.

Chapter 3

Toward Fast Reliable Intelligent Industry 5.0—A Comprehensive Study

Manash Kumar Mondal, Sourav Banerjee, and Yudong Zhang

Chapter Contents

3.1 Introduction

The first industrial revolution began in 1780 with the production of mechanical power from various sources, which was followed by the use of electrical energy for assembly lines.

DOI: 10.1201/9781003376712-3

The production industry has used information technology to automate tasks. For instance, the fourth industrial revolution, subsequently referred to as CPS [18,47], used IoT and the cloud to connect the virtual and physical space. Although the standard of Industry 4.0 transformed the manufacturing sector, process optimization overlooked human resources, leading to unemployment. Pioneers in the sector are therefore anticipating the next revolution in which machines and human intelligence will work together to create superior solutions. The fourth industrial revolution intended to transform manufacturing agents into cyber-physical systems (CPSs) from comprehensive physical systems through the effective integration of business operations and production.

This entails using IoT to integrate every component of the supply chain for the industrial sector, from suppliers to production lines to end users [24]. Industry 4.0 employs CPS through the IoT network to connect with every entity. Because of this, a cloud environment is used to store a large amount of data for efficient processing. Although Industry 4.0 transformed the manufacturing sector, process optimization overlooked human resources, distributed computing, Big data analytics, ambient intelligence, virtual reality, edge computing, and cybersecurity [18,25]. With increasing mass manufacturing, Industry 4.0 has decreased production, logistical, and quality control expenses. Industry 4.0 has reduced the cost of manufacturing, albeit at the expense of human costs through process optimization.

This unintentionally pushes employment backward and will increase labor union resistance, which will hinder Industry 4.0's full adoption [14]. This problem is anticipated to be resolved by Industry 5.0 with greater human involvement. The manufacturing sector is concentrating on the unfavorable effects of managing waste properly and lessening its effects on the ecosystem in response to the quickening rate of environmental damage caused by Industry 2.0. Nothing about Industry 4.0 protects the environment. The next industrial revolution has thus been sparked by the necessity for a technological answer to deliver manufacturing methods that are pollution-free [31,58]. By lowering waste production through the bio-economy, which creates a pollution-free surrounding, Industry 5.0 maintains the sustainability of civilization. The industrial revolution (1.0 to 5.0) has been depicted in Figure 3.1.

3.1.1 Motivation

By combining many technologies, including artificial intelligence (AI), the Internet of Things (IoT), cloud computing, CPSs, and cognitive computing, the Industry 4.0 standard has completely transformed the industrial industry. The basic idea behind Industry 4.0 is to build the manufacturing sector "smart" by connecting machines and creating equipment that can communicate with one another and control one another throughout their lives [42,55,56]. Process automation is given top attention in Industry 4.0 to minimize human involvement in the production process [27,43]. The aim of Industry 4.0 is to boost performance and productivity for all users by utilizing machine learning (ML) to provide intelligence across devices and apps [6,11,23,62].

Industry 5.0 is now being envisioned as a way to combine powerful, clever, and precise machinery with the distinctive creativity of human professionals. Many technical futurists think that Industry 5.0 will give production one more human touch [41]. Industry 5.0 is anticipated to bring together humans' critical, cognitive thinking and highly accurate technology. Another significant addition of Industry 5.0 is mass personalization, which allows clients to choose customized items based on their preferences and requirements. Industry 5.0 will enable adaptability between humans and robots, greatly boost industrial efficiency, and enable accountability for interaction and ongoing monitoring.

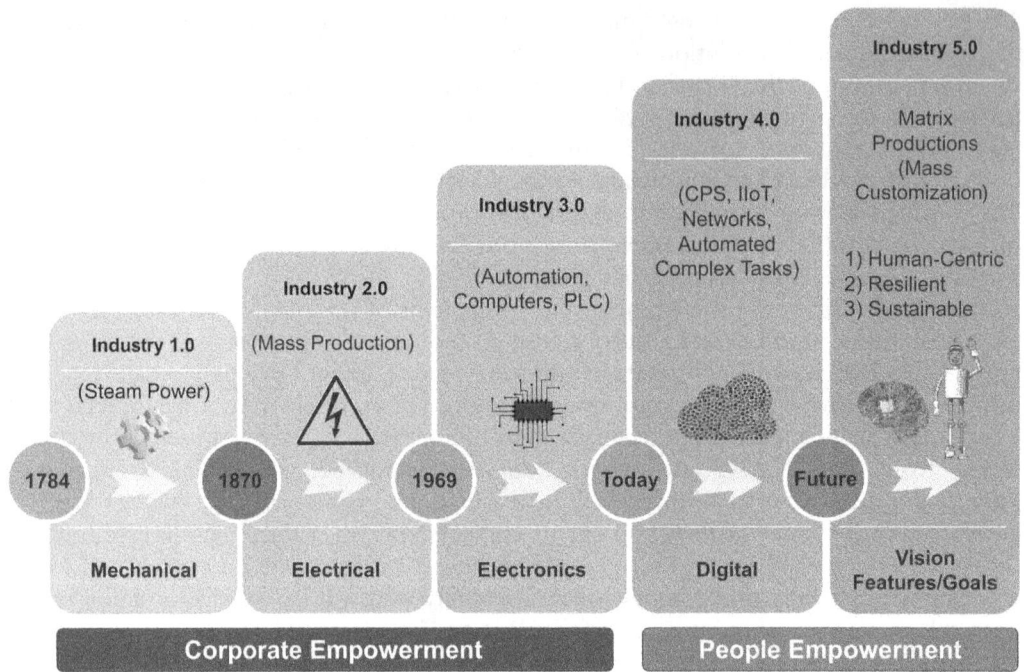

Figure 3.1 Industrial revolutions, up to Industry 5.0.

While Industry 4.0 focused on mass production with minimal waste and increased efficiency, Industry 5.0 aspires to achieve mass-customized production with zero waste, minimal cost, and ultimate precision [32,61]. The idea of Industry 5.0 has yet to fully develop, though. For instance, the discussion of the numerous definitions and views of top industry researchers and academicians about the Industry 5.0 perception is covered in Section 3.2. In order to build an intelligent society where humans conduct innovative, creative jobs and robots (cobots) handle the rest, all key Industry 5.0 concepts emphasize the coexistence of humans and robots. In this regard, the European Commission (EC) developed a futuristic prototype to supplement the industry 5.0 vision with resilience, human-centeredness, and sustainable approaches [1,21,31].

3.2   Industry 5.0

3.2.1   Industry 5.0

Industry 5.0 is a concept that transcends the term "industry." Every industry and organization that come to mind are covered. This indicates that it has a far broader application than Industry 4.0. As a result, we must adopt a comprehensive and universal viewpoint that applies to all industries when considering the strategic implications of Industry 5.0. Industry 5.0 is a proposed new phase in the evolution of industry that builds upon the foundation of Industry 4.0. While Industry 4.0 is focused on using digital technologies to automate and optimize industrial processes, Industry 5.0 takes a more holistic approach that emphasizes human-centricity, collaboration, and sustainability.

Industry 5.0 envisions a future in which advanced technologies such as robotics, artificial intelligence, and the Internet of Things are used not just to improve efficiency and productivity but also to enhance human creativity, problem-solving, and decision-making. This means integrating the best of both human and machine intelligence to create more innovative and sustainable solutions.

One key aspect of Industry 5.0 is the use of "cobots," or collaborative robots, that work alongside humans to perform tasks that require human-like dexterity, flexibility, and problem-solving ability. Cobots can help to reduce physical strain on workers, increase efficiency, and enable new levels of customization and personalization in industrial production.

Industry 5.0 also emphasizes the importance of sustainability and social responsibility, with a focus on reducing waste, increasing resource efficiency, and promoting ethical and equitable practices. This means taking a more holistic approach to industrial design and production and considering the broader social and environmental impacts of industrial activities.

Human-centrism, resilience, and sustainability are the three main pillars of Industry 5.0, as the European Commission explains in Figure 3.2. All three have a big impact on corporate planning.

3.2.1.1   Human-Centric

According to the image, a human-centric strategy "promotes abilities, diversity, and empowerment." The most significant change this implies is moving away from using people as means (such as in the case of human resources) and moving toward using them as ends. Or, to put it another way, there is a change in emphasis from organizations serving people to individuals serving organizations. This is more extreme than it might initially appear to be. Additionally, it fits in well with recent changes in the labor market.

Finding, serving, and retaining talent have become a considerably bigger task than discovering, offering, and retaining customers in several businesses and nations. If this trend persists, the business plan must provide it with a legitimate home, and that's what Industry 5.0 aims to do. Today's strategy is mostly focused on acquiring a competitive edge and utilizing it to produce distinctive additional value for clients. The work of Michael Porter, the most well-known design expert to date, is fundamentally influenced by this mindset.

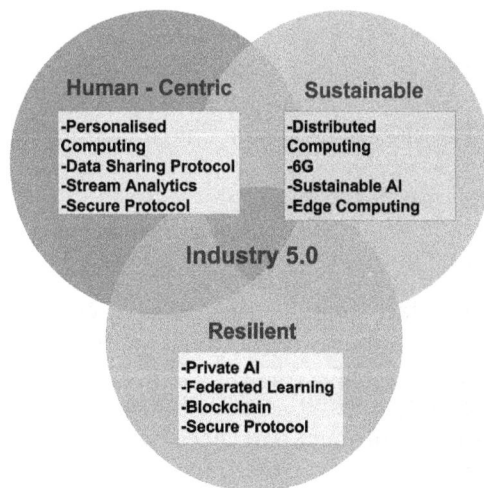

Figure 3.2 Three pillars of Industry 5.0.

3.2.1.2 Resilient

A robust strategy is, in the words of the European Commission, "agile and resilient using versatile and adaptable solutions." Few would disagree that resilience is essential—today and in the future in light of COVID-19, global supply constraints, and the Ukraine conflict. However, this transformation is more profound than it first appears. While flexibility and agility are currently higher on the corporate agenda, this does not automatically translate into greater resilience. If resilience is to truly be one of the three main pillars of Industry 5.0, then strategy's main emphasis must shift from growth, profitability, and effectiveness to building "anti-fragile" organizations that can anticipate, respond to, and learn the systematic way from any crisis order to ensure stable and long-term results.

3.2.1.3 Sustainable

With the many concerns we currently have over climate change, the concept of sustainability hardly needs an introduction. A sustainable strategy "leads to action on sustainability and respects planetary boundaries," according to the European Commission. This suggests, for instance, that organizations should focus on all 17 Sustainable Development Goals and the 3 pillars of the Triple Bottom Line. The third pillar is also a major change, just like the first two. Corporate sustainability initiatives have thus far mostly been concentrated on mitigating harm or engaging in green-washing but let's leave that out of the conversation. So it's business as usual but with more responsibility.

3.2.2 Needs of Industry 5.0

Future industries must play a crucial role in providing solutions to pressing societal issues, such as:

1) Protecting natural resources and the environment and combating climate change.
2) Adopting circular production models, developing and empowering ICT technologies, and revising energy consumption regulations to ensure the effective use of natural assets in the incident of external shocks, such as the COVID-19 pandemic (resiliency) [46,48,60].
3) Hyper-connectivity in the digital age and developing digital skills for empowering individuals and promoting social stability (human-centric value). It is crucial that the elements of the Industry 5.0 vision achieve and make the 17 Sustainable Development Goals (SDG) or Global Goals included in the United Nations' Agenda 2030 possible [4].

3.2.3 Features of Industry 5.0

3.2.3.1 Smart Manufacturing

Sustainable manufacturing, which supports producers in carrying out growth strategies, reduces pollution, and optimizes resource use throughout the development life cycle, is currently the most well-liked and cost-effective manufacturing strategy [45].

The environmentally friendly method used for industrial production, known as additive manufacturing, creates the product part layer by layer rather than as a monolithic block, resulting in the development of lighter but more durable parts one at a time. On the 3D objects, the material is added layer by layer. Intelligent additive manufacturing uses computer vision and AI algorithms to increase the accuracy and improve the graphical description of product designs in 3D printing. For improved compositions, 5D printing, a recent branch of additive manufacturing, is now used. The development of smart manufacturing has been significantly bolstered by the current leading technologies like AI, CPS, cloud computing, Big Data, 5G, IoT, DT, Fog computing, and EC. Intelligent empowering technologies are also gaining popularity. The key benefits of the smart manufacturing sector are productivity, profitability, and sustainability. Smart additive manufacturing (SAM) has emerged as a technique in the smart manufacturing space over the past ten years.

3.2.3.2 Predictive Maintenance

Industries are dealing with numerous difficulties as the globalization of the economy progresses. This is causing the production facilities to transition to impending changes like predictive maintenance (PdM). Manufacturers have been leveraging developing technology, such as CPS techniques and sophisticated analytical methodologies, to increase productivity and efficiency [52]. Industry transparency refers to the capacity to identify and evaluate ambiguities in order to estimate industrial capacity and opportunity. In essence, the majority of industrial plans presumptively anticipate constant equipment availability. However, in the actual industry, it never actually occurs. Industries are dealing with numerous difficulties as the globalization of the economy progresses. This is causing the production facilities to transition to impending changes like predictive maintenance.

In order to gain transparency, manufacturing facilities should switch to predictive maintenance. Modern prediction tools must be used in this transformation so that the workforce can make wise decisions. These tools convert data into information in a systematic manner and indicate uncertainties. Utilizing intelligent equipment and intelligent sensor networks, the Internet of Things (IoT) implementation offers the fundamental framework for preventive maintenance. Predictive maintenance's fundamental objective is to provide equipment and systems with the ability to be aware of themselves.

3.2.3.3 Hyper Customization

Aiming to link machines, building smart supply chains, encouraging the manufacture of smart products, and separating labor from automated sectors are all aspects of Industry 4.0. However, Industry 4.0 has been unable to keep up with the rising need for customization, whereas Industry 5.0 uses hyper customization to do so. Hyper customization is a personalized marketing strategy that uses real-time data and cutting-edge technology like artificial intelligence, machine learning, cognitive systems, and computer vision to offer each customer more specialized goods, services, and content. Robotics and human intelligence combined allow for the mass customization of items by producers. To accomplish this, various functional material variations are communicated with other staff members with the goal of customizing the product with various variations for consumer choice.

Industry 4.0 is intended for massive production with minimal waste and the highest efficiency, but Industry 5.0 wants to achieve mass customization at the lowest possible cost with the highest level of precision. Industries can organize manufacturing processes to incorporate client requests and market changes thanks to the interaction of humans, robots, and cognitive systems. The transformation to an agile supply chain and manufacturing process is the first stage in hyper-personalizing. Additionally, the production team, consumer preferences, and human intervention are required. Additionally, the viability of hyper customization is heavily reliant on how economically viable the generated items are [61].

3.2.3.4   Cyber-Physical Cognitive Systems

CPS has gained popularity in recent years as a result of the development of technologies like smart wearables, IoTs, cloud computing, fog and edge computing, and Big data analytics. The manufacturing process has changed from using fully manual systems to using CPS as a result of the fourth industrial revolution [30,57]. The IoT-enabled connection between CPSs serves as the foundation for Industry 4.0. Huge amounts of effective, secure data sharing and storage are made possible by cloud technology [12]. Additionally, cognitive techniques are used to improve the performance of the system in a variety of applications, including intelligent surveillance, industrial automation, smart grid, vehicular networks, and environment monitoring. These applications are referred to as cyber-physical cognitive systems (CPCSs) [51,54]. The nodes of the CPCSs contain cognitive capacities like the ability to observe or study the surroundings and take appropriate action.

In CPCSs, decision-making is mostly based on learning and knowledge. For human–robot collaborative (HRC) production, the CPCS has been introduced. The HRC works with a human and a robot to complete the assemblage of components in the manufacturing sector. For this real-time collaboration project, machine–human cognitive integration is modeled and used. The fourth industrial revolution's advantages were constrained by the fifth industrial revolution, which reinstates the use of human labor in production. Industry 5.0 is made possible by the fifth revolution, which makes it easier for skilled workers and robots to collaborate to generate personalized goods and services [9].

3.2.4   Applications of Industry 5.0

3.2.4.1   Intelligent Healthcare Development

These days, medical professionals use machine learning (ML) models to aid in the diagnosis of patients' illnesses. This aids in increasing the accuracy of disease diagnosis and thus helps patients save a great deal of time and money [5,13]. However, given the circumstances, this is insufficient. Technology that can guarantee individualized patient care, such as monitoring measurements of blood pressure and sugar levels, and provide patients with individualized treatment with support from doctors is urgently needed. The advent of Industry 5.0 can enable this. Smart watches, intelligent sensors, and other intelligent wearables can continuously capture a patient's healthcare data in real time and save this data in the cloud.

The medical state of the patients can then be determined using ML methods. These smart gadgets are able to connect with one another, and in the event that a doctor is needed, they can provide the doctors with the information of the patient's present condition and alert them to treat the patient. Doctors can use cobots to do surgery on patients by using robots that can communicate with one another. These are just a few instances of how Industry 5.0 can transform the healthcare sector. This revolution facilitates the production of implants, customized gadgets, etc. Corobots could take care of regular tasks through Industry 5.0, such as routine exams currently carried out by doctors. Doctors can focus on higher-level of employment in this fashion for the prescription of drugs to the patients.

3.2.4.2   Cloud Manufacturing

By incorporating cutting-edge technologies like cloud and EC, IoT, virtualization, and service-oriented technologies, cloud manufacturing is an innovative technique to transform the conventional manufacturing paradigm into an advanced manufacturing technique. Multinational stakeholders will work together in a cloud manufacturing process to run an effective and affordable manufacturing process. Reliability, excellent quality, cost-effectiveness, and on-demand capabilities are some of the characteristics that set cloud manufacturing apart. Additionally, cloud manufacturing benefits the environment by removing the need for lengthy raw material deliveries during the manufacturing process.

Cloud manufacturing enables designers to use manufacturing resources scattered across many geographical zones while protecting their intellectual property, such as design files for manufacturing goods, by storing them in the cloud with strict access controls [3]. By doing this, the designers are given the freedom to locate their manufacturing facilities nearer to the raw materials as well as in nations where manufacturing is more affordable. Here, the cloud is in charge of managing the operations of the manufacturing life cycle, including service composition [29] and scheduling [19]. IoT sensors can be used to gather and evaluate in the cloud the operating condition data of the manufacturing process [50]. The use of cloud manufacturing as a service-oriented manufacturing framework was demonstrated by Li et al. [26] and Tao et al. [49].

The potential business policies for cloud manufacturing, including the pay-as-you-go business strategy, were described by Xu et al. [59]. Several energy-awareness-related researches were done by researchers in cloud computing environments [33–36]. The upcoming generation of cloud manufacturing systems, known as Industry 5.0, is anticipated to support various and complicated requirements in the contexts of engineering, production, and logistics. The development of EC features, 5G-based telecommunications networks, and AI/ML technologies opens up new opportunities to significantly increase the capabilities of future cloud manufacturing systems.

3.2.4.3 Supply Chain Managements

In order to fulfill demand and produce individualized and customized products more quickly, companies can benefit from disruptive technologies that enable Industry 5.0, such as DT, cobots, 5G and beyond, ML, IoT, and EC [28]. As mass customization is a core idea in Industry 5.0, this aids supply chain management (SCM) in incorporating it into their manufacturing processes. The SCM, which consists of warehouses, inventory positions, assets, and logistics, can be recreated digitally using DT. The DT includes all manufacturing sites, suppliers, contract manufacturers, shipping channels, distribution centers, and client locations. From the design stage through construction and commissioning and on to operations, DT provides assistance for the SCM during its full life cycle [37,22]. DT can sense real-world data via IoT sensors by imitating real-time SCM systems. These data can be used by ML, Big data, etc., to forecast the challenges encountered at various stages of SCM. Table 3.1 illustrates the paradigm shift from Industry 4.0 to Industry 5.0.

Table 3.1 Industry 4.0 to Industry 5.0 Paradigm Shift

Industry 4.0	Industry 5.0
Focus on equipment connectivity	Focus on customer experience
Mass personalization	Hyper customization
Smart supply chain	Interactive product (experience-activated)
Smart products	Interactive products
Remote workforce	On-site workforce

3.3   5G Technology

5G is the next generation of wireless connectivity, offering a truly connected future. Industry 5.0 has been heralded as being one of the most important innovations in the industry. With 5G, manufacturers can create more intelligent products that are online, offline, mobile, or stationary.

The numerous access approaches in the existing networks are almost at a standstill and require immediate improvement when the current 5G network is taken into account. At least for the next 50 years, current technology like OFDMA will be functional. Furthermore, the wireless configuration that was changed from 1G to 4G does not need to be changed. As an alternative, the user requirements could be satisfied by simply adding an application or improving the basic network. The package providers will be prompted by this to start planning for a 5G network as soon as 4G is put in place commercially [53]. A significant shift in the approach to building the 5G wireless cellular architecture is required to satisfy user requests and address the issues raised by the 5G system. Researchers' broad observations have revealed in [8] that the majority of wireless users spend about 80% of their time indoors and approximately 20% of the time is spent outside.

In the recent wireless cellular framework, an outside base station (BS) located in the center of a cell facilitates communication between mobile users whether they are inside or outside. Signals must therefore pass through the walls of the inside in order for the inside users to exchange information with the outside base station. This will result in massive penetration loss, which decreases the spectrum effectiveness, data rate, and energy efficiency of wireless communications. To overcome this obstacle, a fresh idea or design approach for the 5G cellular architecture is to distinguish between external and internal configurations [53]. The signal loss through the building's walls will be marginally lessened using this method of designing. Massive MIMO technology will be used to support this concept [44], where a widely dispersed array of antennas with tens or hundreds of antenna units is deployed. Massive MIMO systems have the advantage of using the benefits of a large array of antenna devices in terms of enormous capacity improvements, whereas current MIMO systems only use two or four antennas.

In order to develop or create a big massive MIMO network, the outer base stations will first be outfitted with substantial antenna arrays, some of which are spread across the hexagonal cell and connected to the base station via optical fiber cables, supported by massive MIMO technology. Outside, mobile users are often equipped with a specific number of antenna units, but with collaboration, a sizable virtual antenna array may be built, which, combined with base station antenna arrays, creates huge virtual MIMO linkages. To communicate with outdoor base stations using line-of-sight components, every building will also have a vast array of external antennas built. For communication with inside users, the massive antenna arrays are connected to the wireless access points inside the building through cables. This will result in significant improvements to the cellular system's energy efficiency, cell average throughput, high data rate, and spectrum efficiency, albeit at the expense of higher infrastructure establishment costs. With the implementation of such a design, users inside buildings will only need to connect to or communicate with internal wireless access points, while external buildings would continue to have larger antenna arrays built [53]. Certain technologies, such as Wi-Fi, Small-Cell, Ultra-Wideband, and Millimeter Wave Communications, are used for indoor communication [7]. However, higher frequencies are being used by technologies like visible light communication and millimeter wave communication frequently for cellular communications. Cloud and fog computing get the benefit of 5g directly due to high data rate and lower latency time [38].

However, it is not a good idea to employ these high-frequency waves for outdoor and long-distance applications since they are quickly scattered by raindrops, gases, and the atmosphere. They also do not penetrate dense materials well. However, due to their huge bandwidth, visible light and millimeter wave communications technologies can increase the transmission data rate for indoor settings. Another strategy to address the issue of spectrum scarcity is the introduction of additional spectrum that is not typically used for wireless communication. This strategy is known as Cognitive Radio (CR) networks, which increase the spectrum utilization of existing radio spectra [20]. The heterogeneous 5G cellular architecture necessitates the use of macrocells, microcells, tiny cells, and relays. The idea of a mobile tiny cell is crucial to 5G wireless. Mobile relay and tiny cell concepts are used to some extent in cellular networks [16].

It is being adopted to accommodate those with high levels of mobility who are riding in cars and high-speed trains. While gigantic MIMO units made up of enormous antenna arrays are deployed outside moving vehicles to interact with the base station, mobile small cells are placed within moving vehicles to communicate with users inside those vehicles. The idea of separating indoor and outdoor installations is supported by the fact that a mobile small cell appears to users as a standard base station and that all of its associated users are seen as a single unit to the base station. In order to take advantage of data rate services with high-frequency signaling overhead, mobile small-cell users have a high data rate, as illustrated in [53]. The radio network and the network cloud are the only two logical levels in the architecture of the 5G wireless cellular network. The radio network is made up of various types of parts with various functions. The User plane entity (UPE) and Control plane entity (CPE) that perform higher-layer functions linked to the User and Control planes, respectively, make up the network function virtualization (NFV) cloud. Resource pooling is one example of the special network functionality as a service (XaaS) that will offer services based on demand. The radio network and network cloud are connected by XaaS. In [53] and [2], the architecture of the 5G cellular network is described. It is equally crucial for the front-end and backhaul networks. The general 5G cellular network design is presented in this research according to Figure 3.3. It discusses the connections between many cutting-edge technologies, including massive MIMO networks, cognitive radio networks, and mobile and static small-cell networks. The function of the network function virtualization (NFV) cloud in the proposed architecture for the 5G cellular network is also explained. This suggested 5G cellular network architecture also takes into account the ideas of Device to Device (D2D) communication, small cell access points, and the Internet of Things (IoT). Overall, the suggested 5G cellular network design might offer a solid foundation for the next generation of 5G standardized networks.

3.3.1 Advantages of 5G

High speeds: In comparison to 4G and 4G LTE, 5G operates more quickly on mobile phones and other devices. Instead of taking minutes, it enables customers to download movies, videos, and music in seconds. Organizations can use the network's 20 Gbps speed for services like automation and enhanced web conferencing. According to a recent poll, customers who used 5G downloaded content in less than 23 hours per day.

Low latency: In comparison to 4G, 5G offers lower latency, which will effectively enable emerging applications like AI, IoT, and virtual reality. Additionally, it makes it simple for mobile phone users to browse the web and open websites. Another benefit is that it provides a means to access the Internet whenever you need to find some crucial information.

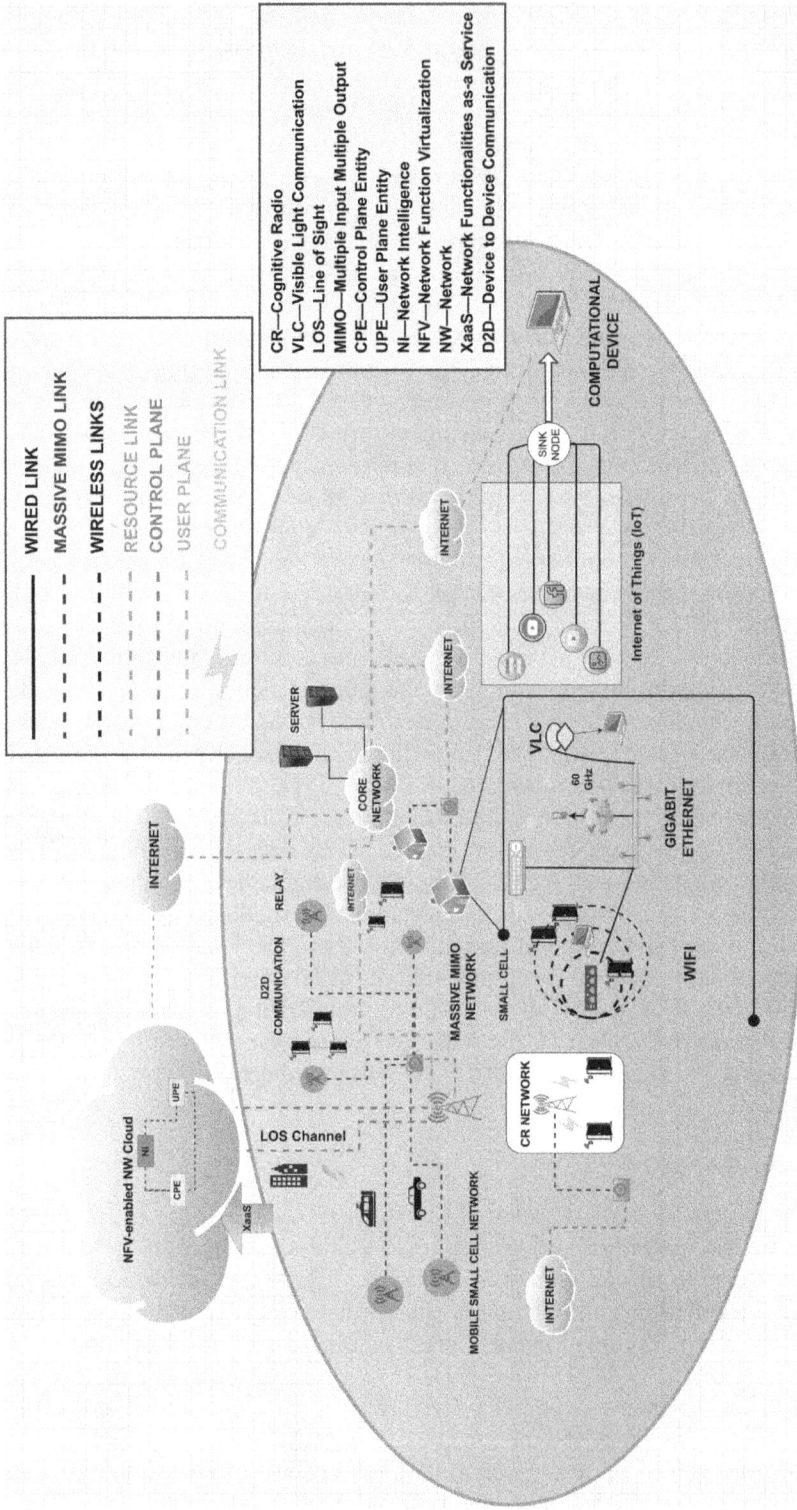

WIRED LINK
MASSIVE MIMO LINK
WIRELESS LINKS
RESOURCE LINK
CONTROL PLANE
USER PLANE
COMMUNICATION LINK

CR—Cognitive Radio
VLC—Visible Light Communication
LOS—Line of Sight
MIMO—Multiple Input Multiple Output
CPE—Control Plane Entity
UPE—User Plane Entity
NI—Network Intelligence
NFV—Network Function Virtualization
NW—Network
XaaS—Network Functionalities as-a Service
D2D—Device to Device Communication

COMPUTATIONAL DEVICE

SINK NODE

INTERNET

Internet of Things (IoT)

INTERNET

VLC

60 GHz

GIGABIT ETHERNET

WIFI

SMALL CELL

MASSIVE MIMO NETWORK

SERVER

CORE NETWORK

INTERNET

RELAY

D2D COMMUNICATION

CR NETWORK

LOS Channel

NFV-enabled NW Cloud

UPE

NI

CPE

XaaS

MOBILE SMALL CELL NETWORK

INTERNET

Figure 3.3 Architectural framework of 5G.

Increased capacity: Up to 100 times more capacity than 4G is possible with 5G. It enables businesses to transition between cellular and Wi-Fi wireless solutions, greatly enhancing performance. In addition, it offers highly effective ways to access the Internet.

More bandwidth: One of the key benefits of 5G is that it increases bandwidth, which will aid in quick data transfer. Additionally, by choosing a 5G network, mobile phone consumers can avail of a faster connection with greater bandwidth.

Powering innovation: Connecting with a wide variety of devices, such as drones and sensors, requires the use of 5G technology. It offers strategies for accelerating IoT adoption, enabling businesses to increase productivity and do other things.

Less tower congestion: Accessing critical information may become difficult due to 4G cellphone networks' frequent congestion. However, the improved speed and increased capacity of 5G networks enable consumers to bypass them.

3.3.2 Disadvantages of 5G

Expensive initial rollout costs: Because network operators would have to both upgrade their current network infrastructures and construct new ones to fulfill the requirements of the 3GPP standard, developing and implementing 5G capabilities would be costly. In order to upgrade and construct infrastructures, it is necessary to buy new hardware, obtain new licenses, integrate and advance complementary technologies like multiuser MIMO and massive MIMO as well as beamforming, and lease both public and private areas. Building and expanding their fifth-generation network capabilities would take some time for developing and undeveloped nations, as well as for rural areas like provinces because the entrance barrier for local network operators is still high due to cost, expertise, and resource needs.

Not compatible with older devices: Newer generations of cellular network technologies have various hardware requirements, just like the older versions did. To connect to a 5G network, devices must have the required hardware. Fifth-generation networks will not be supported by the majority of flagship smartphones from 2019 and earlier, as well as midrange to entry-level devices from 2020 and entry-level devices from 2021. For users to benefit from 5G technology, they must purchase equipment that is capable of doing so.

Variations between mmWave and sub-6: The fact that the entire 5G standard is built on two separate standards is another issue. In contrast to mmWave, sub-6 employs various technologies and operates on different principles. Additionally, while some devices only support the mmWave specification, others only support the sub-6 specification. The sub-6 5G standard covers the so-called C-Band 5G. To determine whether the gadgets they intend to purchase are compliant with both or just one of the criteria, consumers would need to read and comprehend the tiny print of those devices. There are fundamental differences between various 5G networks and 5G gadgets. Given the variations in these two specifications, it is preferable to buy a device that supports both.

mmWave specification restrictions: In terms of bandwidth, latency, and data transmission speed, mmWave 5G is superior to sub-6 5G. There is one problem, though. The mmWave specification's range is constrained because it calls for employing higher frequencies. To completely cover a certain area, network operators would need to construct and place hundreds to thousands of smaller cells. Users must also be within a block of a mmWave cell site and along a line of sight in order to join a mmWave network. This specification's restrictions make it perfect for congested urban regions or specified target locations like stadiums and airports.

Cybersecurity: One of the issues with 5G is cybersecurity since hacking will happen. The increase in bandwidth makes it simple for thieves to grab the database. Additionally, the software it employs makes it subject to attacks. Attacks are quite likely when 5G connects to more devices. As a result, organizations and corporations should invest in a security operations center to safeguard their infrastructure.

3.4 6G Technology

6G technology is the next step in mobile communications for the Industry to upgrade its systems and operations. To develop a fully functional 5G ecosystem, it is important that the entire industry has a common standard that supports 5G standards, from the core network to WANs and mobile operators. The first-generation service node of 6G ensures that all operators can start deploying 5G in real time. 6G technology will have a disruptive effect on entire supply chains as it enables high-definition, real-time, interactive, and virtual experiences in an increasingly digital world through advanced communication technologies such as Machine-to-Machine (M2M) and Remote Cellular Networks (RCNs).

In conjunction with a secure and automated orchestration architecture, 6G architecture includes building blocks spanning important architectural domains of a communication network, starting at the physical layer and moving up to the service layer. As shown in Figure 3.4, we define and formulate architectural 6G building blocks. The Nokia Bell Labs 6G architectural breakdown into building pieces consists of four key interconnected components that offer an open and distributed reference architecture. The "het-cloud" component of the 6G architectural cloud transition effectively serves as the architecture's infrastructure platform and contains features such as an open, scalable, and agnostic run-time environment; data flow centricity; and hardware acceleration.

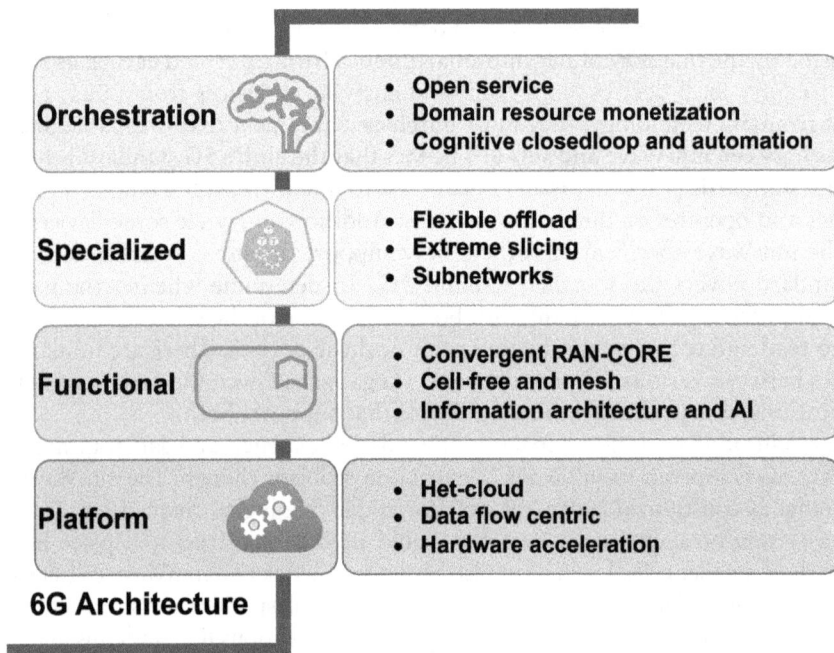

Figure 3.4 6G architectural framework.

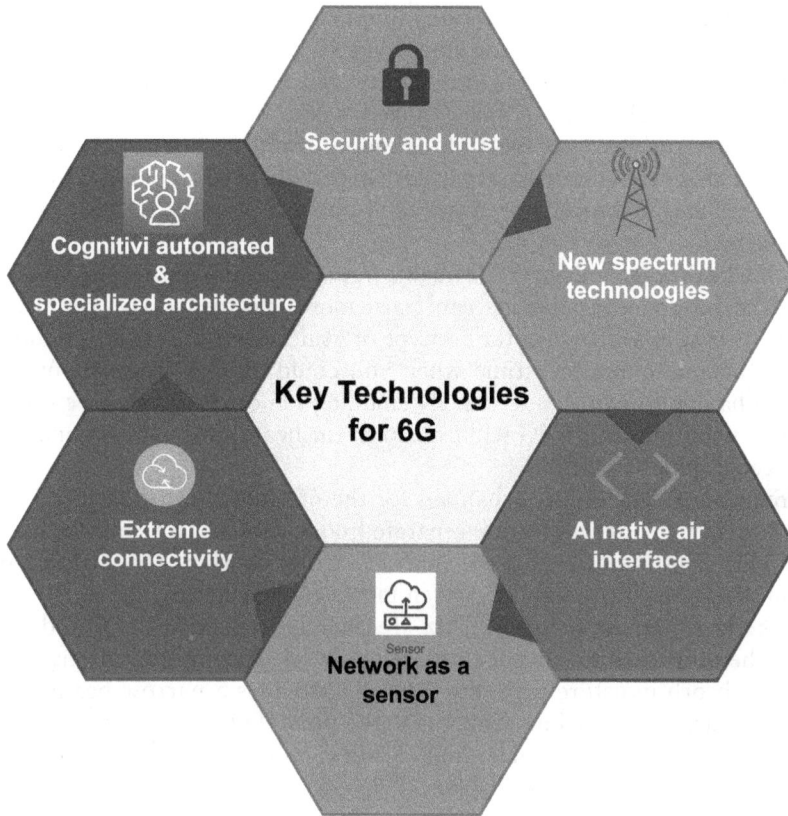

Figure 3.5 Six key technologies of 6G.

Information architecture, AI, and the subjects of RAN-CORE convergence, cell-free, and mesh connectivity, as well as functional architecture, are all included in the "functions" component. The advent of specialized networks and related performance characteristics are a key transformational theme of the 6G era; architectural enablers such as flexible off-load, extreme slicing, and subnetworks are demonstrated as parts of the "specialized" building block. The "orchestration" element of the 6G architectural transformation, which will ensure open service enabling an ecosystem play, domain resource monetization as well as cognitive closed loop and automation, is important in terms of its financial impact. Figure 3.5 shows six key technologies used in 6G

3.4.1 Advantages of 6G

6G technology has several advantages, and the advantages are discussed below.

Provides higher data rates: The 6G network's ability to sustain larger data speeds is an additional advantage. Remember that this form of connection will only be accessible in frequencies utilizing the mmWave spectrum, just like 5G. Such high-frequency waves are currently incompatible with the available devices. Very fast data rate (Tb/sec) and very low latency (sub-ms) are features of 6G.

Increases the number of mobile connections supported: The fact that 6G technology is intended to handle more mobile connections than 5G capacity is one of its key advantages. As a result, there will be less device interference, which will result in an improved service. It is intended to handle more mobile connections than the approximately 10 × 105 capacity of 5G per km. Because, most of the indoor traffic is produced by mobile devices. Furthermore, effective interior coverage has never truly been a goal of cellular networks' design. By utilizing femtocells or distributed antenna systems (DASs), 6G overcomes these obstacles.

Transforms the healthcare industry: Through surgeries and simulations in a real-world setting, medical interns and students can learn more effectively. The key benefit of 6G technology is that it will change the concept of healthcare for patients and medical professionals. Imagine living in a time when you could quickly learn about your health rather than having to wait for weeks. By enabling remote surgery and ensuring healthcare workflow optimization, 6G will transform the healthcare industry and break down time and geographic limitations.

Separates frequencies: The control channels for the 6G standard are assigned a frequency range of 8 to 12 GHz. It will feature separate frequencies and a frequency bandwidth of up to 3.5 kHz. This indicates that the channels do not overlap. This makes it possible to give room to various transmissions, which increases data speeds.

Utilizes terahertz (THz) frequencies: THz (Terahertz) frequencies are used in 6G. THz waves are helpful for high-speed, short-range wireless communications because they can readily absorb moisture from the air. THz provides a narrow-beam, better directivity, and secure transmission, which is made possible by its potent anti-interference abilities. From 108 to 1,013 GHz, high wireless bandwidth (a few tens of GHz) can enable higher Tb/sec transmission rates. THz waves are utilized in space communication to enable lossless transmission between satellites. In order to meet the needs for urban coverage, beamforming and massive MIMO multiplexing gain assist in overcoming rain attenuation and fading propagation. The extremely low photon energy of THz waves (10^{-3} eV) allows for greater energy efficiency. THz waves can be employed for various unique communication methods since they can permeate materials with minimal attenuation.

Visible lights: Visible lights used by 6G wireless take advantage of LED's lighting and high-speed data connection capabilities. Electromagnetic radiation (EM) cannot be generated by visible light communication (VLC). As a result, it is immune to EM interference from outside sources. VLC aids in enhancing network security as well.

3.4.2 Disadvantages of 6G

Cell-less architecture and multi-connectivity: Multi-connectivity and cell-less architecture are used in 6G. Therefore, flawless scheduling is required for seamless mobility and integration of various networks (THz, VLC, mmWave, sub-6 GHz). The UE connects to the RAN and not to a specific cell in a cell-less architecture. Here, creating a new network architecture is a difficult task.

THz (Terahertz) frequencies: The limitations of THz (terahertz) frequencies might be thought of as limitations of 6G wireless technology since THz is used for some of its communications. The term "terahertz frequency" refers to the electromagnetic (EM) wave spectrum between 0.1 and 10 THz with a wavelength of 30 to 3,000 micrometers. THz waves are well suited for use between satellites and can be employed widely in space communications. The THz signal's sensitivity to shadows has a significant impact on coverage. Additionally, THz frequencies with a lower frequency experience have a higher free space fading. Ultra-large-scale antenna, which demands high bandwidth and vast quantitative high resolution, is a significant problem in THz. Designing 6G devices with low power and cheap cost involves significant processing power challenges.

Health-related consequences: Ongoing discussions surround the usage of 6G technology and its potential for harm to people. For instance, exposure to high-frequency radiation has been related to illnesses like autism, ADHD, PTSD, migraines, nausea, and blurred vision due to OCD. Additionally, RF exposure from using a cell phone may cause cancer, notwithstanding the findings of other studies. The usage of cell phones over an extended period of time, according to numerous scientists, may have an impact on some aspects of human cells.

Visible light frequencies: Since visible light wavelengths are used in some of 6G's communications, its disadvantages are also those of 6G wireless technology. The wavelength range for visible light is 390–700 nm.

Energy consumption and efficiency: A 6G system is required to manage a high number of terminals and networking devices in an energy-efficient and cost-effective manner. Circuitry in network and terminal equipment as well as the design of the communication protocol stack must be optimized to achieve this. To meet this need, energy-harvesting strategies are employed.

3.4.3 Comparison of 5G and 6G

Within the next ten years, it is anticipated to witness the birth of a new 6G technology due to the quick growth of fifth-generation (5G) applications and rising demand for even faster communication networks. According to numerous sources, the 6G wireless network standard might appear around 2030. The severe technological limitations of 5G wireless networks are therefore examined critically in this research, together with the projected difficulties of

Table 3.2 Comparing 5G and 6G Specifications

Key Points	5G	6G
Operating bandwidth	Up to 400 MHz for sub-6 bands Up to 3.25 GHz for mmWave bands	Up to 400 MHz for sub-6 GHz bands Up to 3.25 GHz for mmWave band Indicative value: 10–100 GHz for THz bands
Carrier bandwidth	400 MHz	To be defined
Peak data rate	200 Gbps	≥ 1 $Tbps$ (Holographic, VR/AR, and tactile)
User experience rate	100 Mbps	1 Gbps
User plane latency	4 ms (eMBB) and 1 ms (uRLLC)	25 μs to 1 ms
Mobility	500 km/h	1,000 km/h
Control plane latency	20 ms	20 ms

6G communication networks. Table 3.2 highlights the comparison between 5G and 6G technology.

3.5 Cyber-Physical System

3.5.1 Definition (NIST)

"CPS are complex systems of computational, physical, and human components integrated to achieve some function over one or more networks."

CPSs are intelligent systems with constructed networks of interconnected physical and computational elements. It is widely acknowledged that CPS and related systems, including the Industrial Internet and the Internet of Things (IoT), have enormous potential to enable novel applications and have an impact on a variety of economic sectors in the global economy [15].

A cyber-physical system is a combination of physical, computational, and information technologies to create new and dynamic systems. Industry 5.0 is a new paradigm that recognizes that people, technology, and processes are all intertwined. The scope of CPS has been explored in two articles: A case study on the city of Redondo Beach, California, where an obstacle detection system was used as a demonstration for a county-wide (electronic logging devices) ELD initiative; and a presentation on cooperative control concepts as applied in Industry 4.0 research for controlling automated machinery.

In order to perform time-sensitive tasks with variable levels of contact with the environment, including human involvement, CPSs integrate computation, communication, sensing, and actuation with physical systems. Figure 3.6 displays a conceptual representation of the CPS. This illustration highlights the possible interconnections between

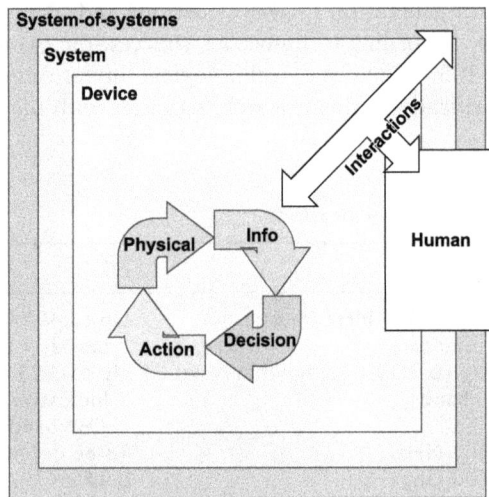

Figure 3.6 CPS conceptual model.

systems and devices in a system of systems (SoS) (such as a CPS infrastructure). A CPS can be as straightforward as a single device, or it can be an SoS made up of several systems made up of various devices or one or more cyber-physical devices that create a plan.

3.5.2 Different Applications of Smart CPS

For CPSs to function, the cyber and physical must be combined, and they must be interconnected. Typically, a CPS comprises sensing, processing, and actuation. Traditional information technology (IT) is used in CPSs to process data as it moves from sensors to computation. Traditional operational technology (OT) is also used in CPSs for actuation and control functions. A particularly novel aspect of CPSs is the fusion of these OT and IT worlds, along with the temporal limitations that go along with it.

An SoS could be a CPS. As a result, it may span several objectives as well as temporal and data domains, necessitating techniques for translating between or accommodating various domains. For instance, several time domains may refer to various time scales or have various levels of granularity or accuracy.

CPSs can be used for purposes other than that for which they were originally intended. For instance, a mobile traffic sensor might be created from a cell phone in a car, and equipment failures could be identified using data on energy use. The smart city is another application of CPSs [40, 10].

Due to the open nature of the CPS composition, emergent behaviors are to be expected. One of the main analysis issues is to comprehend a behavior that cannot be explained by a single CPS component but rather results from the interplay of potentially numerous CPS subsystems. For instance, a traffic jam is a negative emergent behavior, whereas the smart grid's optimal energy distribution, in which power suppliers and consumers collaborate, is a desirable positive emergent impact [39].

CPSs' potential impact on the physical world and its interconnectedness raise questions regarding its reliability. Stressing security, privacy, safety, reliability, and resilience is more urgently needed, along with corresponding assurance for widely used networked infrastructures and devices. For instance, publish and subscribe messages, certificate authorities, type and object registries, and other infrastructure-based components and aggregators in CPS networks may be owned and operated by outside parties. This could lead to possible trust problems.

To guarantee interoperability, manage evolution, and deal with emergent effects, CPSs need a methodology. Many of the subsystems, particularly those in large-scale CPSs like smart grid and smart city, are the responsibility of various manufacturers.

Smart CPSs have a wide range of applications in various domains. Here are some examples of different applications of smart CPSs:

Smart grid: CPSs can be used to optimize the generation, distribution, and consumption of energy in smart grids. CPSs can monitor and control energy generation and consumption in real time, helping to balance supply and demand.

Smart transportation: CPSs can be used to optimize traffic flow, improve road safety, and reduce congestion. CPSs can monitor traffic flow, control traffic signals, and provide real-time information to drivers to avoid accidents and reduce travel time.

Smart healthcare: CPSs can be used to improve patient care by monitoring patients in real time and providing timely interventions. CPSs can monitor vital signs, administer medication, and alert healthcare professionals in case of emergencies.

Smart manufacturing: CPSs can be used to optimize manufacturing processes by monitoring and controlling production lines in real time. CPSs can track inventory levels, monitor equipment performance, and predict maintenance needs to reduce downtime and improve efficiency.

Smart buildings: CPSs can be used to optimize energy consumption in buildings by monitoring and controlling heating, cooling, and lighting systems. CPSs can adjust the temperature and lighting based on occupancy and other factors to reduce energy waste.

Smart agriculture: CPSs can be used to optimize crop production by monitoring and controlling irrigation, fertilization, and pest control. CPSs can monitor soil moisture, temperature, and nutrient levels to optimize crop growth and yield.

Smart City: CPSs can be used to optimize city services such as waste management, public transportation, and emergency services. CPSs can monitor and control traffic flow, track waste disposal, and provide real-time information to emergency services.

3.6   Digital Trust

The term "digital trust" refers to the combination of technical and organizational characteristics of digital systems that make them less likely to cause physical harm or allow attacks on the integrity, confidentiality, or availability of data or services.

As per the World Economic Forum: "*Digital trust is individuals' expectation that digital technologies and services—and the organizations providing them—will protect all stakeholders' interests and uphold societal expectations and values.*"

Digital trust, also known as digital reputation, is the confidence that people have in the reliability and value of a website. It's both an objective and subjective notion many websites face every day. Digital trust is about a lack of provenance or the lack of certain attributes that would signal to others whether something is trustworthy. Trust is the flip side of risk. The more people are distrustful of a product, service, or company, the less likely they are to engage with it. Today, we live in an increasingly digital landscape. Digital technology delivers on its promise of efficiency, convenience, and convenience as well as growing concerns about cybersecurity. But how widespread are these concerns? Does digital trust really exist? We find that while most consumers do use online services in their daily lives, they are generally wary of them due to security and privacy concerns. The majority of these concerns relate back to their own data privacy.

Trust is a complex phenomenon with many social, psychological, and economic dimensions. The term "trust" is used to capture a variety of related feelings, thoughts, and judgments about the moral, social, and political role that people attach to other people and entities. For example, trust can be both an individual's subjective assessment of another person's reliability or performance as well as an objective measure of the extent to which the recipient of the trust is responsive to the sender in terms of providing or not providing desired outcomes. It could also be distinguished from a belief based on evidence or logical reasoning. The concept of trust is thus seen at different levels such as individual ("I trust you"), group ("we all trust each other"), and culture ("everyone trusts their government").

Management and consumers have grown to rely on readily available, quick, and secure mobile connectivity as a result of the growing digital delivery of government services. However, as 5G networks link millions of users, Internet-of-things (IoT) gadgets, driverless vehicles, and smart city services, the advantages of ubiquitous connectivity must be weighed against the expanded attack surface that would be made possible. Governments must keep up with the escalating security concerns if they want to build trust in these digitally delivered services.

Not simply over smartphones, 5G will enable new use cases, applications, and services. Large-scale IoT sensor installations are supported, as are connections with extremely low latency for automated systems, better coverage, and more bandwidth for several latest applications in transportation, healthcare, and public security, to mention a few.

Cyberattacks are also becoming more complex every day. Government networks are frequently attacked by criminal gangs, disgruntled citizens, past and present employees, nation-states, and actors with governmental sponsorship. Thankfully, the foundation of software-defined networking and virtualization is one of the distinguishing characteristics of 5G network design. This indicates that 5G networks are very flexible, scalable, and integrated throughout. Most importantly, a lot of their tasks can be automated [52].

These networks are capable of changing to counter extremely sophisticated cyberattacks against targets in the government. In comparison to earlier generations of mobile standards,

Figure 3.7 Digital trust framework.

they offer much higher levels of security, thanks to their greater data encryption and stringent user authentication.

This is significant because hacker vulnerabilities usually use automation and dynamic network response adaptation. Government networks need to be able to respond even more quickly because "dwell time," or how long a hacker or an attack remains unnoticed, must be reduced. It has been demonstrated that the use of software analytics, machine learning, orchestration, and automation may cut dwell time by 80%. Figure 3.7 represents frameworks of digital trust.

Goal of digital trust: The digital trust framework defines shared goals or values that inform the concept of digital trust, including:

- Security and reliability
- Accountability and oversight
- Inclusive, ethical, and responsible use

Dimensions: The framework also defines dimensions against which the trustworthiness of digital technologies can be operationalized and evaluated:

- Privacy
- Fairness
- Redressability
- Auditability
- Interoperability
- Transparency
- Safety
- Cybersecurity

3.7   Conclusions

In conclusion, the study provides valuable insights into the current state and future direction of Industry 5.0. The survey results suggest that Industry 5.0 is expected to bring significant benefits in terms of productivity, efficiency, and quality, driven by advanced technologies such as AI, IoT, and cloud computing. The survey highlights the importance of fast and reliable connectivity, as well as the need for intelligent automation and data analytics to enable real-time decision-making. The results also indicate that there is a strong demand for Industry 5.0 solutions in various sectors, including manufacturing, healthcare, transportation, and energy. However, the chapter also identifies several challenges that need to be addressed for the successful implementation of Industry 5.0, including cyber-security, data privacy, and the need for new skill sets and workforce training. The results of the study suggest that the success of Industry 5.0 will depend on a collaborative effort by industry, academia, and government.

Industry 5.0 will evolve into a leading industrial revolution in a few years. 5G and 6G technologies recreate a crucial role in Industry 5.0. These technologies accomplish the communication demands of Industry 5.0. The cyber-physical system was introduced in Industrial 4.0, whereas it makes a major impact on Industry 5.0. In this chapter, we highlight all the aspects of Industry 5.0. Additionally, the 5G and 6G technologies and the cyber-physical system are conferred in the subsequent sections. This chapter furnishes a brief idea regarding Industry 5.0 and how 5G and 6G technologies are utilized in CPSs. Further, this chapter highlights digital trust in brief. This chapter will assist the researchers by introducing these emerging topics in the future research direction.

Overall, the chapter provides a comprehensive overview of the current trends and future outlook for Industry 5.0, highlighting the need for continued research and development to overcome the challenges and realize the full potential of this emerging paradigm.

Bibliography

[1] Giuseppe Aceto, Valerio Persico, and Antonio Pescapé. Industry 4.0 and health: Internet of things, big data, and cloud computing for healthcare 4.0. *Journal of Industrial Information Integration*, 18:100129, 2020.

[2] Patrick Kwadwo Agyapong, Mikio Iwamura, Dirk Staehle, Wolfgang Kiess, and Anass Benjebbour. Design considerations for a 5g network architecture. *IEEE Communications Magazine*, 52(11):65–75, 2014.

[3] Hossein Akbaripour, Mahmoud Houshmand, Tom Van Woensel, and Nevin Mutlu. Cloud manufacturing service selection optimization and scheduling with transportation considerations: Mixed-integer programming models. *The International Journal of Advanced Manufacturing Technology*, 95(1):43–70, 2018.

[4] UN General Assembly. *Work of the statistical commission pertaining to the 2030 agenda for sustainable development*. United Nations: New York, NY, 2017.

[5] Sweta Bhattacharya, Praveen Kumar Reddy Maddikunta, Saqib Hakak, Wazir Zada Khan, Ali Kashif Bashir, Alireza Jolfaei, Usman Tariq, et al. Antlion re-sampling based deep neural network model for classification of imbalanced multimodal stroke dataset. *Multimedia Tools and Applications*, 1–25, 2020.

[6] Nirmal Kr Biswas, Sourav Banerjee, Utpal Biswas, and Uttam Ghosh. An approach towards development of new linear regression prediction model for reduced energy consumption and sla violation in the domain of green cloud computing. *Sustainable Energy Technologies and Assessments*, 45:101087, 2021.

[7] Ariel Bleicher et al. Millimeter waves may be the future of 5g phones. *IEEE Spectrum*, 8, 2013.

[8] Vikram Chandrasekhar, Jeffrey G. Andrews, and Alan Gatherer. Femtocell networks: A survey. *IEEE Communications Magazine*, 46(9):59–67, 2008.

[9] Xiao Chen, Martin A. Eder, and A.S.M. Shihavuddin. A concept for human-cyberphysical systems of future wind turbines towards industry 5.0. *DTU Library*, 2020.10.36227/techrxiv.13106108.v1,

[10] Debashis Das, Sourav Banerjee, Kousik Dasgupta, Pushpita Chatterjee, Uttam Ghosh, and Utpal Biswas. Blockchain enabled sdn framework for security management in 5g applications. In *24th International Conference on Distributed Computing and Networking*, Association for Computing Machinery, 414–419, 2023.

[11] Ignacio de la Peña Zarzuelo, María Jesús Freire Soeane, and Beatriz López Bermúdez. Industry 4.0 in the port and maritime industry: A literature review. *Journal of Industrial Information Integration*, 20:100173, 2020.

[12] Caterine Silva De Oliveira, Cesar Sanin, and Edward Szczerbicki. Visual content representation and retrieval for cognitive cyber physical systems. *Procedia Computer Science*, 159:2249–2257, 2019.

[13] Natarajan Deepa, B. Prabadevi, Praveen Kumar Maddikunta, Thippa Reddy Gadekallu, Thar Baker, M. Ajmal Khan, and Usman Tariq. An ai-based intelligent system for healthcare analysis using ridge-adaline stochastic gradient descent classifier. *The Journal of Supercomputing*, 77(2):1998–2017, 2021.

[14] Kadir Alpaslan Demir, Gõlzde Dõlven, and Bũllent Sezen. Industry 5.0 and human-robot co-working. *Procedia Computer Science*, 158:688–695, 2019.

[15] Edward R. Griffor, Christopher Greer, David A. Wollman, Martin J. Burns, et al. Framework for cyber-physical systems: Volume 1, overview. *National Institute of Standards and Technology, U.S. Department of Commerce*, 2017.

[16] Fourat Haider, Cheng-Xiang Wang, Harald Haas, Dongfeng Yuan, Haiming Wang, Xiqi Gao, Xiao-Hu You, and Erol Hepsaydir. Spectral efficiency analysis of mobile femtocell based cellular systems. In *2011 IEEE 13th international conference on communication technology*, 347–351. IEEE, 2011.

[17] Abid Haleem and Mohd Javaid. Industry 5.0 and its applications in orthopaedics. *Journal of Clinical Orthopaedics & Trauma*, 10(4):807–808, 2019.

[18] Debiao He, Mimi Ma, Sherali Zeadally, Neeraj Kumar, and Kaitai Liang. Certificateless public key authenticated encryption with keyword search for industrial internet of things. *IEEE Transactions on Industrial Informatics*, 14(8):3618–3627, 2017.

[19] Petri Helo, Duy Phuong, and Yuqiuge Hao. Cloud manufacturing—scheduling as a service for sheet metal manufacturing. *Computers & Operations Research*, 110:208–219, 2019.

[20] Xuemin Hong, Cheng-xiang Wang, Hsiao-hwa Chen, and Yan Zhang. Secondary spectrum access networks. *IEEE Vehicular Technology Magazine*, 4(2):36–43, 2009.

[21] Miao Hu, Xianzhuo Luo, Jiawen Chen, Young Choon Lee, Yipeng Zhou, and Di Wu. Virtual reality: A survey of enabling technologies and its applications in IoT. *Journal of Network and Computer Applications*, 178:102970, 2021.

[22] Dmitry Ivanov and Alexandre Dolgui. New disruption risk management perspectives in supply chains: Digital twins, the ripple effect, and resileanness. *IFAC-PapersOnLine*, 52(13):337–342, 2019.

[23] Mohd Javaid, Ibrahim Haleem Khan, Ravi Pratap Singh, Shanay Rab, and Rajiv Suman. Exploring contributions of drones towards industry 4.0. *Industrial Robot: The International Journal of Robotics Research and Application*, 49(3):476-490, 2021.

[24] Ibrahim Haleem Khan and Mohd Javaid. Role of internet of things (IoT) in adoption of industry 4.0. *Journal of Industrial Integration and Management*, 2150006, 2021.

[25] Jin Ho Kim. A review of cyber-physical system research relevant to the emerging it trends: Industry 4.0, IoT, big data, and cloud computing. *Journal of Industrial Integration and Management*, 2(3):1750011, 2017.

[26] Bo-Hu Li, Lin Zhang, Shi-Long Wang, Fei Tao, Jun Wei Cao, Xiao Dan Jiang, Xiao Song, and Xu Dong Chai. Cloud manufacturing: A new service-oriented networked manufacturing model. *Computer Integrated Manufacturing Systems*, 16(1):1–7, 2010.

[27] Ling Li. China's manufacturing locus in 2025: With a comparison of "made-in-China 2025" and "industry 4.0". *Technological Forecasting and Social Change*, 135:66–74, 2018.

[28] Ling Li. Education supply chain in the era of industry 4.0. *Systems Research and Behavioral Science*, 37(4):579–592, 2020.

[29] Yongkui Liu, Xun Xu, Lin Zhang, and Fei Tao. An extensible model for multitask-oriented service composition and scheduling in cloud manufacturing. *Journal of Computing and Information Science in Engineering*, 16(4), 2016.

[30] Yang Lu. Cyber physical system (cps)-based industry 4.0: A survey. *Journal of Industrial Integration and Management*, 2(3):1750014, 2017.

[31] Yang Lu. Industry 4.0: A survey on technologies, applications and open research issues. *Journal of Industrial Information Integration*, 6:1–10, 2017.

[32] Praveen Kumar Reddy Maddikunta, Quoc-Viet Pham, B. Prabadevi, Natarajan Deepa, Kapal Dev, Thippa Reddy Gadekallu, Rukhsana Ruby, and Madhusanka Liyanage. Industry 5.0: A survey on enabling technologies and potential applications. *Journal of Industrial Information Integration*, 26:100257, 2022.

[33] Riman Mandal, Manash Kumar Mondal, Sourav Banerjee, and Utpal Biswas. An approach toward design and development of an energy-aware vm selection policy with improved SLA violation in the domain of green cloud computing. *The Journal of Supercomputing*, 76(9):7374–7393, 2020.

[34] Riman Mandal, Manash Kumar Mondal, Sourav Banerjee, Pushpita Chatterjee, Wathiq Mansoor, and Utpal Biswas. Design and implementation of an sla and energy-aware vm placement policy in green cloud computing. In *2022 IEEE Globecom Workshops (GC Wkshps)*, IEEE, Rio de Janeiro, Brazil, 777–782, 2022.

[35] Riman Mandal, Manash Kumar Mondal, Sourav Banerjee, Pushpita Chatterjee, Wathiq Mansoor, and Utpal Biswas. PBV MSP: A priority-based vm selection policy for vm consolidation in

green cloud computing. In *2022 5th International Conference on Signal Processing and Information Security (ICSPIS)*, Dubai, IEEE, United Arab Emirates, 32–37, 2022.

[36] Riman Mandal, Manash Kumar Mondal, Sourav Banerjee, Gautam Srivastava, Waleed Alnumay, Uttam Ghosh, and Utpal Biswas. Mecpvms: An SLA aware energy-efficient virtual machine selection policy for green cloud computing. *Cluster Computing*, 1–15, 2022.

[37] Jose Antonio Marmolejo-Saucedo, Margarita Hurtado-Hernandez, and Ricardo Suarez-Valdes. Digital twins in supply chain management: A brief literature review. In *International Conference on Intelligent Computing & Optimization*, 653–661. Springer, 2019.

[38] Manash Kumar Mondal and Madhab Bandyopadhyay. A comparative study between cloud computing and fog computing. *Brainwave: A Multidisciplinary Journal*, 2(1):36–42, 2021.

[39] Manash Kumar Mondal, Riman Mandal, Sourav Banerjee, Utpal Biswas, Pushpita Chatterjee, and Waleed Alnumay. A CPS based social distancing measuring model using edge and fog computing. *Computer Communications*, 194:378–386, 2022.

[40] Manash Kumar Mondal, Riman Mandal, Sourav Banerjee, Utpal Biswas, Jerry Chun-Wei Lin, Osama Alfarraj, and Amr Tolba. Design and development of a fog-assisted elephant corridor over a railway track. *Sustainability*, 15(7):5944, 2023.

[41] Saeid Nahavandi. Industry 5.0—a human-centric solution. *Sustainability*, 11(16):4371, 2019.

[42] Quoc-Viet Pham, Kapal Dev, Praveen Kumar Reddy Maddikunta, Thippa Reddy Gadekallu, Thien Huynh-The, et al. Fusion of federated learning and industrial internet of things: A survey. *arXiv preprint arXiv:2101.00798*, 2021.

[43] V. Priya, I. Sumaiya Thaseen, Thippa Reddy Gadekallu, Mohamed K. Aboudaif, and Emad Abouel Nasr. Robust attack detection approach for IIoT using ensemble classifier. *arXiv preprint arXiv:2102.01515*, 2021.

[44] Fredrik Rusek, Daniel Persson, Buon Kiong Lau, Erik G. Larsson, Thomas L. Marzetta, Ove Edfors, and Fredrik Tufvesson. Scaling up mimo: Opportunities and challenges with very large arrays. *IEEE Signal Processing Magazine*, 30(1):40–60, 2012.

[45] Manuel Sanchez, Ernesto Exposito, and Jose Aguilar. Autonomic computing in manufacturing process coordination in industry 4.0 context. *Journal of Industrial Information Integration*, 19:100159, 2020.

[46] Mahak Sharma, Rajat Sehrawat, Sunil Luthra, Tugrul Daim, and Dana Bakry. Moving towards industry 5.0 in the pharmaceutical manufacturing sector: Challenges and solutions for Germany. *IEEE Transactions on Engineering Management*, pp. 1–18 2022.

[47] Pradip Kumar Sharma, Neeraj Kumar, and Jong Hyuk Park. Blockchain-based distributed framework for automotive industry in a smart city. *IEEE Transactions on Industrial Informatics*, 15(7):4197–4205, 2018.

[48] Rahul Sindhwani, Shayan Afridi, Anil Kumar, Audrius Banaitis, Sunil Luthra, and Punj Lata Singh. Can industry 5.0 revolutionize the wave of resilience and social value creation? A multi-criteria framework to analyze enablers. *Technology in Society*, 68:101887, 2022.

[49] Fei Tao, Lin Zhang, V.C. Venkatesh, Y. Luo, and Ying Cheng. Cloud manufacturing: A computing and service-oriented manufacturing model. *Proceedings of the Institution of Mechanical Engineers, Part B: Journal of Engineering Manufacture*, 225(10):1969–1976, 2011.

[50] Fei Tao, Ying Zuo, Li Da Xu, and Lin Zhang. IoT-based intelligent perception and access of manufacturing resource toward cloud manufacturing. *IEEE Transactions on Industrial Informatics*, 10(2):1547–1557, 2014.

[51] Ozan Alp Topal, Mehmet Ozgun Demir, Zekai Liang, Ali Emre Pusane, Guido Dartmann, Gerd Ascheid, and Gunes Karabulut Kur. A physical layer security framework for cognitive cyber-physical systems. *IEEE Wireless Communications*, 27(4):32–39, 2020.

[52] Jeff Verrant. *Building digital trust with 5G*, 2020. https://gcn.com/cybersecurity/2020/11/building-digital-trust-with-5g/315413/.

[53] Cheng-Xiang Wang, Fourat Haider, Xiqi Gao, Xiao-Hu You, Yang Yang, Dongfeng Yuan, Hadi M. Aggoune, Harald Haas, Simon Fletcher, and Erol Hepsaydir. Cellular architecture and key technologies for 5g wireless communication networks. *IEEE Communications Magazine*, 52(2):122–130, 2014.

[54] Shulan Wang, Haiyan Wang, Jianqiang Li, Huihui Wang, Junaid Chaudhry, Mamoun Alazab, and Houbing Song. A fast cp-abe system for cyber-physical security and privacy in mobile healthcare network. *IEEE Transactions on Industry Applications*, 56(4):4467–4477, 2020.

[55] Li Da Xu. The contribution of systems science to industry 4.0. *Systems Research and Behavioral Science*, 37(4):618–631, 2020.

[56] Li Da Xu. Industry 4.0—frontiers of fourth industrial revolution. *Systems Research and Behavioral Science*, 37(4):531–534, 2020.

[57] Li Da Xu and Lian Duan. Big data for cyber physical systems in industry 4.0: A survey. *Enterprise Information Systems*, 13(2):148–169, 2019.

[58] Li Da Xu, Eric L Xu, and Ling Li. Industry 4.0: State of the art and future trends. *International Journal of Production Research*, 56(8):2941–2962, 2018.

[59] Xun Xu. From cloud computing to cloud manufacturing. *Robotics and Computer-Integrated Manufacturing*, 28(1):75–86, 2012.

[60] Xun Xu, Yuqian Lu, Birgit Vogel-Heuser, and Lihui Wang. Industry 4.0 and industry 5.0—inception, conception and perception. *Journal of Manufacturing Systems*, 61:530–535, 2021.

[61] Hasan Yetjffs and Mehmet Karakoĺse. Optimization of mass customization process using quantum-inspired evolutionary algorithm in industry 4.0. In *2020 IEEE International Symposium on Systems Engineering (ISSE)*, 1–5. IEEE, 2020.

[62] Caiming Zhang and Yong Chen. A review of research relevant to the emerging industry trends: Industry 4.0, IoT, blockchain, and business analytics. *Journal of Industrial Integration and Management*, 5(1):165–180, 2020.

Chapter 4

Software Development for Medical Devices

A Comprehensive Guide

Pravin Mundra, Uttam Ghosh, Sudip Barik, and Sourav Banerjee

Chapter Contents

4.1 Introduction

The Internet of Medical Things (IoMT) refers to the network of interconnected medical devices and systems that collect and exchange health data in real time [9, 46]. The proliferation of IoMT devices has led to a surge in sensitive health data, including patient electronic health records, diagnostics, treatment plans, and more. IoMT is evolving rapidly, driven by emerging technologies that are transforming healthcare and medical practices. Artificial Intelligence (AI) and Machine Learning (ML) are revolutionizing IoMT by enabling predictive analytics, disease diagnosis, and treatment optimization [8]. AI/ML algorithms can analyze large datasets from the medical devices to identify trends, predict patient outcomes, and personalize treatment plans [4, 39]. Edge computing in medical system can play an important role for processing the vast amount of critical medical data locally in real time and reducing the latency. Blockchain technology [25] can be applied to enhance the data integrity, security, and patient privacy in medical systems. Moreover, 5G networks enable

DOI: 10.1201/9781003376712-4

real-time monitoring, high-quality video consultations, and the rapid exchange of medical data among devices, physicians, and patients [10].

In contemporary times, the integration of software has emerged as a fundamental component within medical devices, ushering in a transformative era for the healthcare sector. This influence extends across a spectrum of applications, ranging from diagnostic instruments to implantable devices, where software assumes a pivotal role in elevating both the functionality and effectiveness of these medical tools.

Software as a Medical Device (SaMD) [17] is a software that performs a medical function independently, without being an integral part of a physical medical device. SaMD encompasses a wide range of applications, including software that aids in the display, processing, analysis, or evaluation of medical images. Additionally, it includes software that manages the functionality of a connected medical device, like a pacemaker, and software that provides essential parameters for the operation of different medical devices or software. SaMD is versatile and can operate on various platforms, such as smartphones, computers, or cloud servers.

In Figure 4.1, we can observe various examples of medical devices commonly employed within healthcare settings. These devices play pivotal roles in tasks such as diagnosing, monitoring, and treating a variety of medical conditions [23].

Software-Driven Medical Devices (SdMDs) include hardware medical devices with significant software components such as Software in Medical Devices (SiMD). SiMD includes software that helps to run a hardware medical device by powering its mechanics or producing a graphical interface. Some examples of SiMD [54] include:

- Software that controls the inflation or deflation of a blood pressure cuff
- Software that controls the delivery of insulin on an insulin pump
- Software that monitors the heart rate on an electrocardiogram (ECG) machine

This chapter will explore the various aspects of software development in medical devices, including the challenges, regulations, and best practices involved in this specialized field.

Figure 4.1 Examples of medical devices with integrated software.

4.1.1 Types of Software for Medical Devices

Before initiating a new project, it is necessary to properly identify the needs for your medical device software development, as the scope and technology can vary greatly. Embedded coding and Software as a Medical Device (SaMD) are examples of this. The bulk of real-world medical equipment startups and initiatives employ a mix of technologies. Let's look at the important areas of medical device software development [29].

Embedded Medical Systems and the Development of Embedded Medical Software: This domain encompasses the low-level programming of micro-components such as microcontrollers and microchips that are combined with microprocessors and embedded memory. These components are typically found within the inner workings of various healthcare devices [20]. Medical equipment that use embedded systems driven or configured by embedded code include pulse oximeters, electronic defibrillators, smart (bio)sensors, automated infusion pumps, glucometers, electronic thermometers, electronic blood pressure sensors, a wide range of laboratory equipment, and medical imaging equipment such as X-ray, ECG, EEG, MRI, and CT.

Embedded programming plays an important role in healthcare equipment and biomedical applications because it regulates the operation of numerous electronic components and allows medical devices to be integrated with nonspecific or general-purpose software and hardware. This integration may encompass PCs, EHRs, Wi-Fi, and other systems [5].

While some medical devices just require basic programming abilities for embedded system development, some projects necessitate considerable knowledge of healthcare device engineering. Programming and calibrating all of the embedded electronics in a large and complex machine, such as a modern MRI tomograph, require significant skill.

4.1.1.1 Software as a Medical Device (SaMD)

Software as a Medical Device (SaMD) represents a transformative paradigm shift in the field of healthcare. SaMD refers to software applications intended for medical purposes, encompassing an extensive range of functions from diagnostics to treatment planning. What distinguishes SaMD from conventional medical devices is its ability to operate independently, without the need for additional hardware. This groundbreaking technology has the potential to revolutionize patient care, offering innovative solutions that enhance accuracy, efficiency, and accessibility within the healthcare ecosystem.

SaMD applications can perform diverse tasks, such as analyzing medical images, monitoring vital signs, and providing decision support to clinicians. These software solutions can significantly expedite diagnoses, allowing healthcare professionals to make more informed decisions swiftly. Moreover, SaMD's adaptability and potential for remote operation have facilitated the democratization of healthcare, enabling patients in remote or underserved areas to access expert medical guidance and monitoring.

So, this burgeoning field also presents challenges. Ensuring the safety, efficacy, and data security of SaMD is paramount. Regulatory bodies worldwide are actively engaged in developing guidelines and standards to govern the development, validation, and deployment of SaMD. As technology continues to advance, the synergy between software and medicine will undoubtedly shape the future of healthcare, offering new horizons for patient care and disease management.

4.1.2 Scope of Medical Device Software Development

The process of planning, producing, and deploying software solutions specifically for medical devices used in healthcare settings is referred to as medical device software development.

These gadgets cover a broad spectrum of technology, such as monitoring tools, treatment systems, diagnostic equipment, and therapeutic software.

The main goal of developing software for medical devices is to improve patient care by utilizing software skills to offer advanced functions, increase accuracy, and facilitate better healthcare results. It entails the creation of software programs that interact with the hardware elements of medical apparatus to form an integrated system that aids medical practitioners in the diagnostic, observation, and therapeutic processes.

The development of medical device software goes beyond conventional software development techniques since it requires a thorough knowledge of the medical industry, regulatory compliance, and the particular difficulties and factors present in healthcare contexts. It covers the whole process of developing software, from gathering requirements to creating the software architecture through coding, testing, validation, obtaining regulatory approval, and continuous maintenance [15].

The scope of medical device software development covers a number of areas, including the following:

Diagnostic Devices: Software applications that aid in the detection and diagnosis of medical conditions, such as imaging software for radiology or pathology.

Monitoring Devices: Software solutions are used for continuous monitoring of vital signs, patient data, and physiological parameters, such as remote patient monitoring systems or wearable devices.

Treatment Systems: Software integrated with medical devices to deliver specific treatments or therapies, such as software-controlled infusion pumps or radiation-therapy-planning software.

Data Analysis and Decision Support: Software applications that analyze and interpret medical data and support healthcare professionals in making informed decisions and treatment planning.

User Interfaces and Medical Visualization: Software components that provide intuitive and user-friendly interfaces for healthcare professionals and patients, facilitating efficient interaction with medical devices.

Connectivity and Interoperability: Software solutions that enable seamless integration and communication among medical devices, hospital information systems, electronic health records, and other healthcare infrastructure.

Due to the crucial nature of medical device software, it must be developed by regulatory requirements, quality assurance procedures, and industry-specific best practices. To ensure the efficient and secure operation of medical devices and the provision of high-quality patient care, it is imperative to address the specific difficulties related to safety, dependability, regulatory compliance, and data security [12].

The significance, difficulties, life cycle, regulatory requirements, best practices, and data privacy and security issues related to medical device software development will all be discussed in the parts that follow. To promote innovation, guarantee patient safety, and optimize healthcare results in the quickly developing field of medical technology, developers, healthcare practitioners, and regulatory agencies need to comprehend these factors.

4.1.3 Role and Significance of Medical Device Software in Healthcare

Modern healthcare systems depend heavily on medical device software, which has revolutionized patient care, diagnosis, and treatment outcomes [16]. Its importance can be seen in several ways:

Enhanced functionality and accuracy: Software plays a vital role in enhancing the functionality and accuracy of medical devices. Through software, medical devices can perform complex calculations, automate processes, and provide advanced features that were previously not possible. For example, the software can enable a precise control of medical devices, allowing for accurate dosage delivery in drug administration or precise surgical interventions. Software-driven algorithms can also enhance diagnostic device accuracy, improving test results' reliability and aiding in early detection and treatment [14].

Remote monitoring and connectivity: The software enables medical devices to establish connectivity and facilitate remote monitoring of patients. This capability is especially valuable in telemedicine and remote patient-monitoring scenarios. Through software, medical devices can transmit data in real time to healthcare professionals or centralized systems, enabling remote monitoring of vital signs, tracking patient progress, and facilitating timely interventions. Remote monitoring and connectivity enhance patient convenience, reduce the need for hospital visits, and improve access to healthcare in remote or underserved areas [21].

Real-time data analysis and decision support: Software in medical devices allows for real-time data analysis and decision support. By processing and analyzing data collected by medical devices, the software can provide valuable insights and support healthcare professionals in making informed decisions [51]. Real-time data analysis helps detect patterns, trends, and anomalies, enabling early intervention and personalized treatment plans. Software algorithms can also provide decision support, such as suggesting optimal treatment options or alerting healthcare professionals to critical situations, leading to more efficient and effective healthcare delivery [30].

Improved patient outcomes and safety: The integration of software in medical devices contributes to improved patient outcomes and safety. Software can aid in the prevention, early detection, and management of medical conditions. For example, software-enabled monitoring devices can detect irregularities in vital signs and alert healthcare professionals, enabling timely intervention and reducing the risk of adverse events. Additionally, software-driven automation minimizes human errors in tasks like medication administration or data recording. By enhancing patient outcomes and safety, software in medical devices positively impacts healthcare quality and patient satisfaction [19, 22].

Operational efficiency and workflow streamlining: Medical device software improves operational efficiency in healthcare environments by automating procedures and streamlining workflows. Data entry, documentation, and reporting tasks can all be automated, which eases the administrative burden on medical staff. This increase in productivity enables them to concentrate more on patient care, spend less time on mundane duties, and reduce errors associated with manual data management [42].

Reduction of errors and adverse events: By minimizing errors and unfavorable outcomes, medical device software improves patient safety. Automated systems can carry out error checks, identify potential problems, and notify healthcare practitioners of any issues. The danger of pharmaceutical errors is reduced by software-integrated devices that assure precise medicine delivery, dosage computations, and protocol adherence. In addition, it is possible to include software-driven safety systems to stop injury in dangerous circumstances [6].

Fostering innovation and advancements: Medical device software fosters innovation in healthcare by enabling the development of new diagnostic and therapeutic techniques. It allows for the integration of emerging technologies like artificial intelligence (AI), machine learning, the Internet of Things (IoT), and cloud computing. These advancements facilitate the creation of smart devices, personalized medicine approaches, and real-time data analysis, leading to improved patient outcomes and advancements in medical research [18].

In summary, medical device software is an essential part of contemporary healthcare systems, offering important advantages such as enhanced patient care, precise diagnoses, increased operational efficiency, and innovation stimulation. It is essential to achieving improved health outcomes for people and communities because of its involvement in changing healthcare practices and advancing technology.

4.2   Background

With the introduction of digital technology in recent years, the healthcare sector has undergone substantial development. Medical device software, for example, is critical in aiding diagnoses, therapy, and patient care. Medical device software has become an essential component of modern healthcare delivery, ranging from embedded software in medical equipment to mobile applications and cloud-based platforms [34].

The vital role of medical device software resides in its ability to boost clinical results, increase patient safety, expedite healthcare operations, and enable remote monitoring and telemedicine. These software solutions help healthcare practitioners to collect, and analyze patient data, giving them vital insights for accurate diagnosis, personalized therapies, and continuous monitoring of patient's health states. Furthermore, medical device software allows for the seamless integration and interoperability of various healthcare systems, enabling efficient information sharing and collaboration among healthcare practitioners [50].

Yet, developing medical device software is not without its difficulties. The healthcare industry's unique characteristics, tight regulatory standards, the necessity for thorough testing and validation, and the complexity of combining software with hardware components all provide significant challenges. Furthermore, due to the sensitive nature of patient health information and the increasing possibility of cybersecurity breaches, protecting data privacy and security become critical.

To solve these issues, healthcare organizations and software developers must navigate a complicated terrain that includes regulatory regulations, best practices, and data privacy and security concerns. Compliance with regulatory standards, such as those established by regulatory agencies such as the United States Food and Drug Administration (FDA) [38] and international standards such as IEC 62304, guarantees that medical device software meets quality and safety requirements. Best practices in software development, such as adhering to certain processes, documentation, and traceability, aid in ensuring the stability and effectiveness of medical device software. Furthermore, to protect against potential breaches and unauthorized access, rigorous data protection measures and the incorporation of strong security systems become necessary.

4.3   Challenges in Software Development for Medical Devices

Developing software for medical devices poses unique challenges that must be addressed to ensure the safety, effectiveness, and quality of the devices. This section explores some of the key challenges faced by software developers in the medical device industry [36].

4.3.1   Regulatory Compliance and Approval Processes

Adhering to regulatory standards and obtaining necessary approvals are critical aspects of medical device software development. Regulatory bodies, such as the US Food and Drug

Administration (FDA), impose stringent requirements to ensure the safety, effectiveness, and quality of medical devices. Navigating the complex regulatory landscape presents significant challenges that require dedicated efforts and expertise.

Medical device software developers must thoroughly understand the regulatory framework and stay updated with evolving guidelines. The FDA, for example, classifies medical device software into different categories based on risk levels, such as Class A, B, or C. Each classification has specific requirements for approval, including premarket notifications, premarket approvals, or exemptions [37].

IEC 62304 categorizes medical device software based on the level of risk it poses to patients or users. The classification system is as follows:

- Class A: No potential for injury or harm to health.
- Class B: Possibility of nonserious injury.
- Class C: Possibility of serious injury or death.

This classification scheme is similar to ISO 14971 [28] Clauses 4.4, 5, and 6.1. In the case of safety-critical software systems, they can be divided into individual items, with each item running a different software element, each having its own safety classification. These items can be further subdivided into additional software elements. The overall classification of the software system is determined by the highest classification among all the software elements it contains. For instance, if a software system comprises five software elements, with four classified as Class A and one as Class C, the overall device would be classified as Class C. This concept is illustrated in Figure 4.2.

However, IEC 62304 [31] allows for the segregation of a specific software item from the overall software system. This means that the segregated software item can independently receive a lower safety classification.

The process of regulatory compliance involves several stages throughout the software development life cycle. From the initial design phase, developers need to consider regulatory requirements, including risk management, documentation, and quality management systems. Robust documentation, including design specifications, hazard analysis, and software validation plans, is essential to demonstrate compliance.

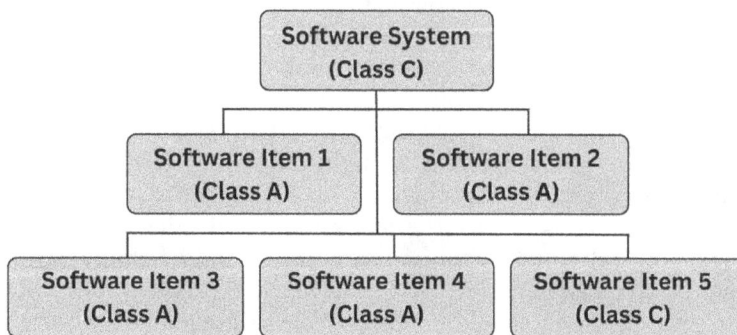

Figure 4.2 Classification of software components within a comprehensive software system.

Testing and validation are crucial steps in the approval process. Developers must conduct comprehensive testing, including software verification and validation, to ensure that the software meets predefined requirements and functions as intended. Validation involves verifying that the software performs safely and effectively within its intended use.

Post-market surveillance is equally important. Developers must establish processes to monitor and address any potential risks or issues that may arise after the software is released to the market. This includes collecting feedback, conducting post-market studies, and promptly addressing any reported adverse events or complaints [11].

Achieving compliance requires collaboration among cross-functional teams, including software engineers, regulatory experts, quality assurance personnel, and legal advisors. Establishing a clear regulatory strategy early in the development process is essential to streamline the approval process and avoid unnecessary delays.

Additionally, it is crucial to stay informed about changes in regulatory requirements and evolving best practices. Regularly monitoring updates from regulatory bodies and participating in industry conferences, workshops, and forums can help developers stay ahead of regulatory changes and ensure ongoing compliance.

Overcoming the challenges of regulatory compliance and approval processes requires a comprehensive understanding of the regulations, meticulous planning, and proactive engagement with regulatory authorities. By investing the necessary time, resources, and expertise, developers can navigate the regulatory landscape and bring safe and compliant medical device software to market.

4.3.2   Ensuring the Safety and Reliability of Software

Ensuring the safety and reliability of medical device software is of paramount importance in healthcare. Patient safety heavily relies on the accurate and error-free functioning of software that drives medical devices. A single software glitch or malfunction could have serious consequences for patient health and well-being. Therefore, developers must take comprehensive measures to mitigate risks and ensure the safety and reliability of medical device software [2].

One crucial aspect of achieving safety and reliability is implementing robust risk management processes throughout the software development life cycle. This involves identifying and assessing potential risks associated with the software, its interactions with hardware components, and its impact on patient care. By conducting thorough risk assessments, developers can identify potential hazards and take necessary steps to mitigate or eliminate them. Risk management also involves considering factors such as software complexity, potential failure modes, and the severity of potential harm to patients [52].

Rigorous testing and quality assurance play a vital role in validating the safety and reliability of medical device software. Various testing methodologies, including functional testing, performance testing, and interoperability testing, should be employed to assess the software's behavior under different scenarios. Rigorous testing helps identify and rectify software defects, ensures compliance with functional requirements, and verifies that the software performs as intended. Additionally, conducting validation testing with simulated or real-world data helps evaluate the accuracy and reliability of the software in delivering desired outcomes [33].

Quality assurance measures should be implemented throughout the software development process to maintain high standards of safety and reliability. This includes adhering to industry best practices, following established software development methodologies, and conducting comprehensive code reviews. Quality assurance also involves documenting software development processes, maintaining traceability between requirements and design, and establishing robust change control procedures [41].

To address vulnerabilities and potential security risks, medical device software development should follow secure design and coding practices. This includes incorporating encryption techniques, secure authentication mechanisms, and secure communication protocols to protect patient data and prevent unauthorized access. Ongoing vulnerability management is essential to identify and address potential security vulnerabilities promptly, including regular software updates, patch management, and security audits.

Collaboration with healthcare professionals, regulatory experts, and end users is crucial in ensuring the safety and reliability of medical device software. By involving these stakeholders throughout the development process, developers can gain valuable insights into the specific requirements, use cases, and potential risks associated with the software. Regular communication, feedback loops, and user testing can help identify and address usability issues, optimize workflows, and improve overall software performance and safety.

4.3.3 Integration with Hardware and Existing Healthcare Systems

Integration with hardware and existing healthcare systems is a critical aspect of medical device software development. Medical devices often rely on the integration of software with various hardware components, sensors, and actuators to perform their intended functions. Additionally, seamless interoperability with existing healthcare systems, such as EHRs or HIS, is essential for data exchange, communication, and streamlined workflows. However, achieving smooth integration poses several challenges that need to be addressed [53].

One of the main challenges is the heterogeneity of hardware devices and systems used in healthcare settings. Different medical devices may employ different communication protocols, data formats, and interfaces. Integrating software with such diverse hardware components requires careful consideration and adaptation. The software must be designed to handle the specific communication requirements of each device and ensure data compatibility and integrity [43].

Another challenge arises from the complex nature of the existing healthcare systems. These systems may have their own unique data structures, workflow processes, and security protocols. Medical device software needs to align with these systems to exchange data accurately and securely. Achieving interoperability often requires collaboration with healthcare IT professionals who possess the knowledge of the existing infrastructure and can provide guidance on integrating the software effectively [13].

To overcome integration challenges, collaboration with hardware engineers and system integrators is crucial. Hardware engineers can provide insights into the technical specifications, interfaces, and communication protocols of the devices. They can assist in designing software interfaces that facilitate seamless communication with the hardware components.

System integrators, on the other hand, play a vital role in bridging the gap between the software and the existing healthcare systems. They possess expertise in integrating different software systems, managing data exchange, and ensuring compatibility across platforms. Collaborating with system integrators helps ensure that the medical device software aligns with the established standards and workflows of the healthcare environment [1].

A multidisciplinary approach involving software developers, hardware engineers, and healthcare IT professionals is necessary to address the challenges of integration. It is essential to establish clear communication channels, define data exchange protocols, and conduct rigorous testing to verify the compatibility and functionality of the integrated system [57].

Moreover, keeping abreast of emerging standards and interoperability frameworks, such as HL7 (Health Level Seven) or FHIR (Fast Healthcare Interoperability Resources), can provide valuable guidance in designing software that seamlessly integrates with existing healthcare systems.

By addressing the integration challenges, medical device software can effectively leverage hardware capabilities and seamlessly communicate with healthcare systems, contributing to improved patient care, streamlined workflows, and enhanced data exchange in the healthcare ecosystem.

Risk management: Managing risks associated with medical device software is crucial to ensure patient safety. Identifying potential hazards, conducting risk assessments, implementing risk mitigation strategies, and documenting the risk management process require expertise and rigorous analysis. Balancing the need for innovative functionalities with the need for risk reduction presents an ongoing challenge for developers.

Verification and validation: Validating the performance and safety of medical device software is a complex task. Rigorous testing, including unit testing, integration testing, and system testing, must be conducted to verify that the software meets the specified requirements and functions as intended. Validating the software's effectiveness in real-world clinical scenarios through clinical evaluations and user acceptance testing adds an additional layer of complexity [47].

Interoperability and integration: Medical devices often need to interact and exchange data with other devices and systems within the healthcare ecosystem. Ensuring seamless interoperability and integration is a challenge, as different devices may use different communication protocols or data formats. Developers must address compatibility issues, design standardized interfaces, and ensure secure and reliable data exchange [3, 45].

Software complexity: Medical device software can be highly complex, incorporating advanced algorithms, machine learning, and artificial intelligence capabilities. Managing the complexity of such software, while ensuring its reliability, maintainability, and usability, requires specialized skills and expertise. Developers must implement robust software engineering practices, maintain clean and modular code, and ensure effective documentation to handle the complexity effectively [6].

Usability and user-centered design: Medical device software must be designed with the end users in mind, including healthcare professionals, patients, and caregivers. Ensuring usability and user-centered design is a challenge due to the diverse user base, varying levels of technical expertise, and specific user requirements. Incorporating user feedback, conducting usability testing, and iteratively improving the software's usability present ongoing challenges throughout the development process.

Data security and privacy: With the increasing connectivity of medical devices, data security and privacy are critical concerns. Protecting patient data from unauthorized access, ensuring secure data transmission, and complying with data privacy regulations require robust cybersecurity measures. Developers must implement encryption, authentication mechanisms, vulnerability management, and secure coding practices to safeguard patient information.

Software maintenance and updates: Medical device software requires ongoing maintenance and periodic updates to address bugs, vulnerabilities, and evolving regulatory requirements. Ensuring seamless software updates without disrupting device functionality, managing software configuration, and maintaining version control present challenges for developers. Balancing the need for updates with minimizing disruption to healthcare workflows and patient care is a continuous challenge.

Addressing these challenges requires collaboration among multidisciplinary teams, including software developers, regulatory experts, clinical professionals, and quality assurance personnel. Implementing best practices, leveraging industry standards, and staying updated with the evolving regulatory landscape are crucial for successful software development in the medical device industry.

4.4 Software Development Life Cycle (SDLC) for Medical Devices

The software development life cycle (SDLC) provides a structured framework for the development of software in medical devices. It consists of a series of phases and activities that guide the development process, as shown in Figure 4.3, ensuring adherence to regulatory requirements, quality standards, and patient safety. The SDLC for medical devices typically includes the following key stages:

Requirements gathering: In this initial phase, the requirements for the software are identified and documented. This involves understanding the intended use of the device,

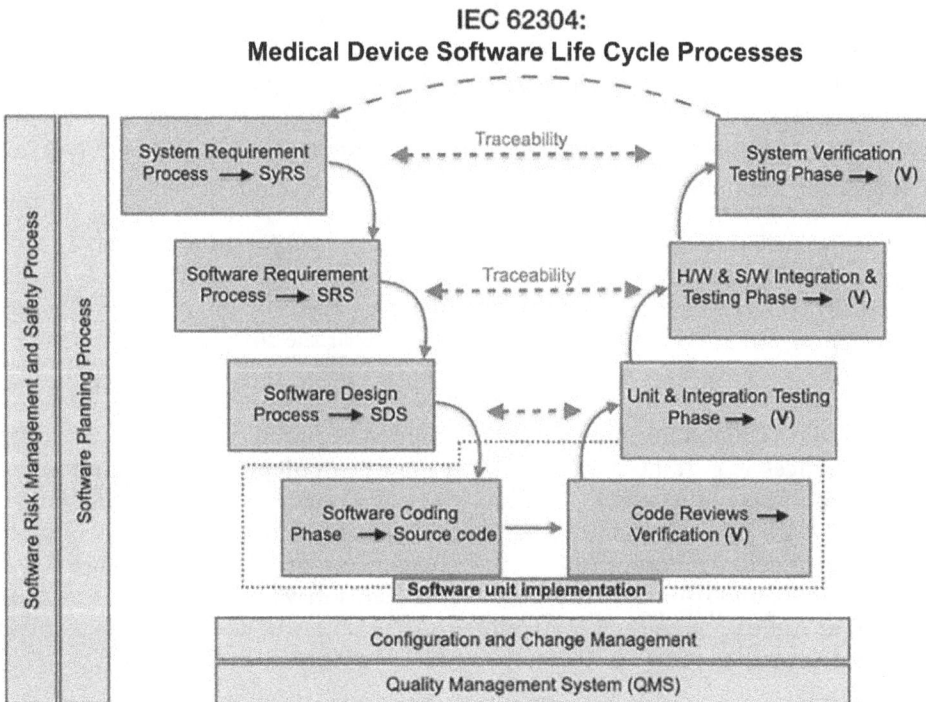

Figure 4.3 IEC 62304: Summary of medical device software development life cycle.

user needs, functional requirements, and any regulatory or quality requirements that must be met. Requirements gathering may involve interactions with stakeholders, including healthcare professionals, patients, regulatory bodies, and quality assurance teams.

Design and architecture: Once the requirements are established, the software design and architecture are developed. This phase involves defining the overall system architecture, including hardware–software interfaces, data flows, and module interactions. Design decisions are made to ensure that the software meets the intended use, is scalable, maintainable, and compliant with relevant standards and regulations.

Implementation: In this phase, the software is developed on the basis of the design specifications. It involves coding, unit testing, and integration of software modules. Best practices for coding, such as following coding standards, using appropriate programming languages, and applying secure coding practices, are crucial to ensure the quality and reliability of the software.

Verification and validation: The verification and validation phase aims to ensure that the developed software meets the specified requirements and performs reliably. Verification involves activities such as unit testing, integration testing, and system testing to identify and fix defects. Validation involves evaluating the software's performance in real-world scenarios, often through clinical evaluations and user acceptance testing, to ensure that it functions safely and effectively.

Risk management: Throughout the SDLC, risk management is a critical component. Risk assessment and analysis are performed to identify potential hazards associated with the software and to develop risk mitigation strategies. Manufacturers must conduct a thorough risk analysis, implement risk control measures, and document the risk management process to demonstrate the safe operation of the software.

Documentation and traceability: Comprehensive documentation is essential at each stage of the SDLC. This includes documenting the software requirements, design specifications, test plans, verification and validation activities, and risk management processes. Traceability matrices are used to establish and maintain traceability among requirements, design elements, test cases, and risk controls.

Release and maintenance: Once the software is deemed ready for deployment, it undergoes a release process that includes configuration management, version control, and documentation of the software release. After deployment, ongoing maintenance is crucial to address any software updates, bug fixes, or enhancements. Post-market surveillance and vigilance processes ensure a continuous monitoring of the software's performance and safety.

Throughout the SDLC, adherence to regulatory requirements, quality standards, and best practices is paramount. Manufacturers must comply with regulations from regulatory bodies, such as the FDA, and adhere to international standards, such as IEC 62304, which provide guidance for the software development process in medical devices. Additionally, incorporating risk management practices, adhering to secure coding principles, and following industry-recognized software development methodologies, such as Agile or Waterfall, contribute to the successful development of software in medical devices.

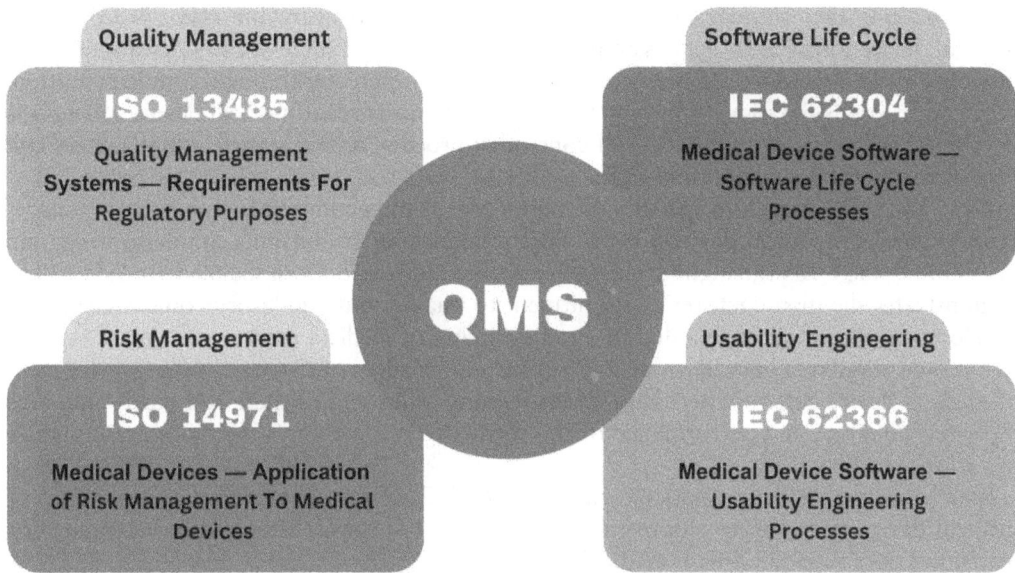

Figure 4.4 A simple breakdown of ISO and IEC Standards for SaMD.

4.5   Regulatory Compliance and Standards

4.5.1   FDA Regulations for Medical Device Software (e.g., 21 CFR Part 820)

The U.S. Food and Drug Administration (FDA) provides regulatory oversight for medical devices including software. Compliance with FDA regulations is crucial for the development, testing, and marketing of medical device software in the United States. One important regulation is 21 CFR Part 820, which outlines the Quality System Regulation (QSR) for medical devices.

21 CFR Part 820 specifies requirements for the establishment and maintenance of a quality management system (QMS) by medical device manufacturers as shown in Figure 4.4. It covers various aspects of the development process, including design controls, document controls, device history records, complaint handling, and corrective and preventive actions. Compliance with these regulations ensures that medical device software is developed and manufactured in a controlled and quality-oriented manner.

The FDA also provides guidance documents specific to medical device software such as the "General Principles of Software Validation" and the "Content of Premarket Submissions for Software Contained in Medical Devices." These guidance documents provide recommendations and expectations for the validation, documentation, and risk management of medical device software [44].

International Standards (e.g., IEC 62304, ISO 13485): International standards play a significant role in guiding the development of medical device software and ensuring global regulatory compliance. Two notable standards in this context are IEC 62304 and ISO 13485 [48].

IEC 62304: This standard provides guidance on the software life cycle processes for medical device software. It defines activities and tasks at each stage of the software development life cycle, including requirements specification, architectural design, implementation, verification, and validation. Adhering to IEC 62304 ensures that medical device software is developed following a systematic and controlled process, with a specific emphasis on risk management and documentation [35].

ISO 13485: This standard specifies the requirements for a comprehensive quality management system for medical devices. It provides guidance on establishing, implementing, and maintaining the QMS throughout the entire product life cycle. Compliance with ISO 13485 demonstrates the manufacturer's commitment to quality and regulatory compliance in the development, production, installation, and servicing of medical device software. These international standards help harmonize practices and ensure consistency in the development of medical device software across different regions. Adhering to these standards enhances the safety, quality, and performance of the software and facilitates regulatory compliance in various markets [55].

Risk Management Frameworks (e.g., ISO 14971): Risk management is a crucial aspect of medical device software development, and ISO 14971 is a widely recognized standard for managing risks associated with medical devices.

ISO 14971 provides guidance on establishing a risk management process throughout the entire life cycle of a medical device, including its software components. The standard outlines a systematic approach to identify, analyze, evaluate, and control risks associated with the device and its software. It emphasizes the importance of risk mitigation and ongoing risk assessment to ensure the safety and effectiveness of the software.

Complying with ISO 14971 involves conducting a comprehensive risk analysis, assessing the probability and severity of potential hazards, implementing risk control measures, and monitoring the effectiveness of these measures. The standard also emphasizes the importance of maintaining a risk management file that documents the risk management activities and decisions throughout the development process.

By adopting risk management frameworks like ISO 14971, medical device software developers can systematically identify and address potential risks, enhance the safety and reliability of the software, and meet regulatory requirements related to risk management. These frameworks help ensure that potential risks are proactively managed, reducing the likelihood of adverse events and promoting patient safety [55].

4.6 Best Practices in Software Development for Medical Devices

Developing software for medical devices requires adherence to best practices to ensure the safety, efficacy, and quality of the devices. This section highlights some key best practices in software development for medical devices:

Agile development methodology: Adopting an agile development methodology is beneficial in the context of medical device software development. Agile methodologies, such as Scrum or Kanban, promote iterative development, collaboration, and flexibility. They allow for frequent feedback and adjustments, leading to faster delivery of high-quality software. Agile practices enable teams to respond to changing requirements, incorporate user feedback, and identify and resolve issues in a timely manner [40].

Software testing and validation: Robust testing and validation processes are essential to ensure the reliability and safety of medical device software. This includes comprehensive unit testing, integration testing, system testing, and validation in real-world clinical settings. Rigorous testing helps identify and fix defects, ensure the software meets the specified requirements, and validate its performance and accuracy. The use of automated testing frameworks and tools can expedite the testing process and improve efficiency [40, 49].

Documentation and traceability: Comprehensive documentation is crucial throughout the software development life cycle. This includes documenting software requirements, design specifications, test plans, risk management processes, and any changes made during the development process. Documentation facilitates transparency, reproducibility, and compliance with regulatory requirements. Additionally, maintaining traceability matrices helps establish and maintain traceability between requirements, design elements, test cases, and risk controls [24].

Collaboration and cross-functional teams: Promoting collaboration among cross-functional teams is essential for successful software development in medical devices. This includes close collaboration between software developers, regulatory experts, clinical professionals, quality assurance personnel, and other stakeholders. Collaborative environments foster effective communication, knowledge sharing, and problem-solving. By involving all relevant stakeholders throughout the development process, the resulting software aligns with regulatory requirements, user needs, and quality standards [56].

Risk management: Risk management is a critical aspect of software development for medical devices. Manufacturers must conduct a thorough risk analysis, identify potential hazards associated with the software, and implement risk control measures. It is important to document the risk management process and maintain traceability between identified risks and implemented risk controls. Regularly reviewing and updating the risk management documentation ensure that potential risks are addressed throughout the device's life cycle [26].

Usability and user-centered design: Considering usability and adopting a user-centered design approach are essential for medical device software. Understanding the needs, capabilities, and workflows of the end users, including healthcare professionals, patients, and caregivers, helps design intuitive and user-friendly interfaces. Conducting usability studies, gathering user feedback, and incorporating user-centered design principles contribute to the development of software that is efficient, and effective, and enhances user satisfaction [27].

Cybersecurity and data privacy: With the increasing connectivity of medical devices, robust cybersecurity measures and data privacy practices are crucial. Implementing security controls, encryption mechanisms, access controls, and vulnerability management processes helps protect against unauthorized access, data breaches, and cyber threats. Adhering to data privacy regulations, such as the General Data Protection Regulation (GDPR), ensures the proper handling and protection of patient data throughout the software development process [32].

Post-market surveillance and maintenance: Post-market surveillance and ongoing maintenance are important aspects of software development for medical devices. Establishing processes for monitoring the performance, safety, and effectiveness of the software in real-world settings allows for the timely identification and mitigation of any issues that may arise. Regular software updates, bug fixes, and enhancements should be carried out to address potential vulnerabilities and ensure the software remains up-to-date [7].

By following these best practices, manufacturers can enhance the quality, reliability, and safety of medical device software. Incorporating agile methodologies, rigorous testing, effective documentation, and ongoing validation processes are essential steps in achieving these goals.

4.7   Software Verification and Validation

Software testing and validation play a crucial role in ensuring the reliability, safety, and effectiveness of medical device software. Thorough testing and validation processes help identify defects, assess compliance with regulatory requirements, and verify that the software functions as intended. This section provides more details on the key aspects of software testing and validation in the context of medical device software development [49].

Unit testing: Unit testing involves testing individual components or units of the software to ensure they function correctly. Developers write test cases to verify the behavior of specific functions, modules, or classes. By isolating and testing individual units, defects and errors can be identified early in the development process. Unit testing ensures that the building blocks of the software are reliable, enhancing the overall quality of the final product.

Integration testing: Integration testing verifies the interaction and interoperability between different components or modules of the software. It ensures that the integrated system functions as expected and that data flows correctly between various modules. Integration testing helps detect issues related to communication, data integrity, and dependencies between different software components. By validating the integration, the overall performance and functionality of the software can be ensured.

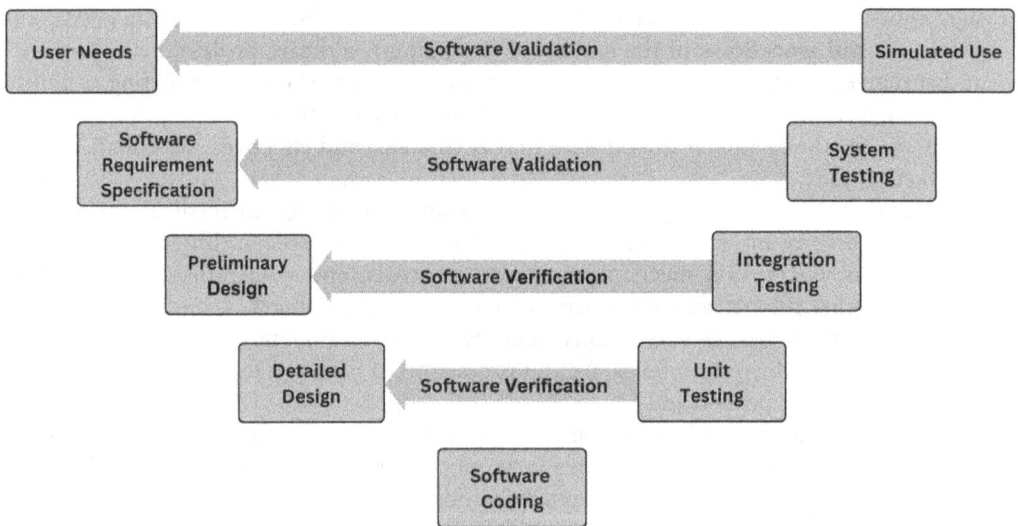

Figure 4.5 Software Verification and Validation (V&V) for medical devices.

System testing: System testing evaluates the entire medical device software system as a whole. It focuses on validating the system's compliance with functional and non-functional requirements, including user interfaces, system behavior, and performance. System testing involves executing test scenarios that simulate real-world usage conditions, ensuring that the software operates as intended in different scenarios. By testing the system in a comprehensive manner, potential defects or shortcomings can be identified and addressed.

Verification and validation testing: Verification and validation (V&V) testing is a comprehensive testing process that ensures that the software meets the intended purpose and satisfies regulatory requirements, as shown in Figure 5.4. Verification testing involves assessing whether the software has been developed according to the specified requirements and design. Validation testing, on the other hand, evaluates the software's performance in real-world clinical settings and verifies that it achieves the desired clinical outcomes. V&V testing provides evidence that the software is fit for its intended purpose and meets the necessary quality standards.

Usability testing: Usability testing focuses on evaluating the user experience and the ease of use of medical device software. It involves gathering feedback from users, such as healthcare professionals and patients, to assess the software's intuitiveness, navigation, and overall user satisfaction. Usability testing helps identify areas for improvement in terms of user interface design, workflow efficiency, and user satisfaction. Incorporating user feedback through usability testing ensures that the software is user-friendly and meets the needs of its intended users.

4.7.1 Validation in Real-World Clinical Settings

Validating the software in real-world clinical settings is a crucial step in ensuring its safety and effectiveness. This involves testing the software with actual users, such as healthcare professionals, in realistic healthcare environments. Real-world validation helps identify any issues or challenges that may arise during actual usages, such as interoperability issues, performance limitations, or usability concerns. By validating the software in real-world scenarios, developers can refine and optimize its performance for practical clinical use.

By employing comprehensive software testing and validation processes, medical device software developers can identify and address defects; ensure compliance with regulatory requirements; and validate the software's safety, reliability, and effectiveness. These testing and validation efforts contribute to the overall quality and trustworthiness of medical device software, ultimately enhancing patient safety and improving healthcare outcomes.

4.8 Cybersecurity and Data Privacy

Cybersecurity and data privacy are critical considerations in the development of medical device software. As the healthcare industry becomes increasingly interconnected and reliant on digital systems, protecting sensitive patient data and ensuring the security of medical devices are of utmost importance. This section explores the key aspects of cybersecurity and data privacy in medical device software development.

Security by design: Security should be an integral part of the software development process, starting from the initial design phase. Adopting a security-by-design approach involves identifying potential vulnerabilities, threat modeling, and incorporating security controls into the software architecture. Developers need to follow industry best practices and coding standards that prioritize security. This includes using secure coding frameworks; adhering to secure coding guidelines; and avoiding common vulnerabilities such as buffer overflows, injection attacks, or insecure authentication mechanisms. By adopting secure coding practices, developers can minimize the likelihood of software vulnerabilities that could be exploited by malicious actors. This proactive approach helps minimize security risks and ensures that security measures are implemented throughout the development life cycle.

Threat modeling and risk assessment: Threat modeling involves identifying potential threats and vulnerabilities that could compromise the security of medical device software. This process helps prioritize security measures and allocate resources effectively. Conducting a comprehensive risk assessment assists in identifying and assessing potential risks, their impact, and the likelihood of occurrence. By understanding the specific threats and risks associated with the software, developers can implement targeted security measures to mitigate vulnerabilities.

Secure communication and encryption: Medical device software often involves the transfer of sensitive patient data, such as health records and diagnostic information. Ensuring secure communication channels and encryption of data during transit is crucial. Implementing robust encryption protocols, such as Transport Layer Security (TLS), helps protect data from unauthorized access or interception. Additionally, mechanisms like Public Key Infrastructure (PKI) and digital signatures can enhance the integrity and authenticity of transmitted data.

Access controls and user authentication: Implementing appropriate access controls and user authentication mechanisms is vital to prevent unauthorized access to medical device software. Strong authentication methods, such as two-factor authentication or biometric authentication, can enhance user verification and limit access to authorized personnel. Role-based access control (RBAC) can be employed to assign specific privileges to different user roles, ensuring that only authorized individuals can perform certain actions or access sensitive data.

Vulnerability management and patching: Regular vulnerability assessments and patch management are crucial to address security vulnerabilities and protect against emerging threats. Medical device software developers should actively monitor security vulnerabilities, stay updated on security patches released by software and hardware vendors, and promptly apply patches to address known vulnerabilities. Establishing a process for ongoing vulnerability management and timely patch deployment helps mitigate the risk of potential exploits.

Incident response and recovery: Developing an incident response plan is essential to address security incidents effectively. This plan outlines the steps to be taken in the event of a security breach, including containment, investigation, notification, and recovery. Timely response and recovery measures help minimize the impact of security incidents and facilitate the restoration of normal operations. Regular testing of the incident response plan ensures its effectiveness and enables continuous improvement.

Compliance with regulatory requirements: Medical device software must comply with applicable regulatory requirements and standards, such as the FDA's guidance on cybersecurity in medical devices or the European Union's Medical Device Regulation (MDR). Adhering to these regulations ensures that the software meets the necessary security and privacy standards. Compliance with standards like ISO 27001 for information security management systems and ISO 27799 for healthcare information security further enhances the security posture of medical device software.

By incorporating robust cybersecurity and data privacy measures, medical device software developers can protect sensitive patient data, prevent unauthorized access, and ensure the safe and secure operation of medical devices. These measures not only safeguard patient privacy but also contribute to maintaining the trust of healthcare providers and patients in the digital healthcare ecosystem.

4.9 Future Trends in Software Development for Medical Devices

The field of software development for medical devices is constantly evolving, driven by advancements in technology, regulatory requirements, and the need for improved patient care. This section explores some of the future trends that are expected to shape the landscape of software development for medical devices.

Artificial intelligence and machine learning: Artificial intelligence (AI) and machine learning (ML) have the potential to revolutionize medical device software development. These technologies can enable devices to analyze complex data, identify patterns, and make intelligent decisions in real time. AI and ML algorithms can be used for tasks such as image recognition, diagnostics, predictive analytics, and personalized medicine. Integrating AI and ML into medical device software can enhance accuracy, efficiency, and patient outcomes.

Internet of Medical Things (IoMT): The Internet of Medical Things (IoMT) refers to the network of medical devices, sensors, and healthcare systems connected through the Internet. IoMT enables real-time monitoring, data collection, and remote healthcare management. In the future, medical device software will need to support seamless integration with IoMT ecosystems, ensuring secure and reliable data transfer, interoperability, and remote device management capabilities.

User experience and human-centered design: User experience (UX) and human-centered design (HCD) will play an increasingly significant role in the development of medical device software. Emphasizing the ease of use, intuitive interfaces, and workflows tailored to healthcare professionals and patients will enhance adoption and usability. Incorporating user feedback through iterative design processes and usability testing will be crucial for creating software that meets the needs of end users.

Cloud computing and Big data analytics: Cloud computing and Big data analytics offer opportunities for scalable storage, processing power, and advanced analytics capabilities. Medical device software developers can leverage the cloud for data storage, data analysis, and remote access to software and updates. Big data analytics can help derive meaningful insights from large datasets, enabling personalized medicine, population health management, and real-time monitoring.

These future trends will shape the landscape of software development for medical devices, influencing the design, functionality, and regulatory requirements of medical device software. By embracing these trends, developers can create innovative solutions that improve patient outcomes, enhance healthcare delivery, and drive advancements in the field.

4.10 Conclusion

Throughout this chapter, we have explored various aspects of software development in medical devices, including the importance of software in enhancing functionality, remote monitoring, real-time data analysis, and improving patient outcomes and safety. We have also discussed the critical considerations such as regulatory compliance, risk management, cybersecurity, and data privacy. Software development in the context of medical devices demands adherence to stringent regulatory requirements and international standards such as FDA regulations, IEC 62304, ISO 13485, and ISO 14971. Compliance with these standards ensures that the software is developed following systematic and controlled processes, with a focus on risk management, quality assurance, and patient safety. We have also delved into the challenges faced in software development for medical devices, including the complexities of regulatory compliance, interoperability, usability, and cybersecurity. Addressing these challenges requires a multidisciplinary approach, involving collaboration between software engineers, healthcare professionals, regulatory experts, and cybersecurity specialists. Furthermore, we discussed best practices in software development for medical devices, emphasizing the importance of software testing and validation, regulatory compliance, risk management, and cybersecurity. These best practices help ensure the development of high-quality, safe, and reliable software that meets regulatory requirements, mitigates risks, and maintains the privacy and security of patient data. Looking toward the future, we explored the emerging trends in software development for medical devices, including advancements in artificial intelligence, machine learning, telemedicine, and interoperability. These trends have the potential to revolutionize healthcare delivery, enabling more personalized and efficient patient care.

In conclusion, software development in medical devices continues to drive innovation and improvement in the healthcare industry. With careful considerations of regulatory compliance, risk management, cybersecurity, and data privacy, developers can create software solutions that enhance patient care, improve outcomes, and contribute to the advancement of healthcare technology. As technology continues to evolve, it is crucial to stay abreast of new developments, standards, and regulations to ensure the continued success of software development in the field of medical devices.

Bibliography

[1] Amelie Abadie, Melanie Roux, Soumyadeb Chowdhury, and Prasanta Dey. Inter-linking organisational resources, AI adoption and omnichannel integration quality in Ghana's healthcare supply chain. *Journal of Business Research*, 162:113866, 2023.

[2] Noorul Husna Abd Rahman, Ayman Khallel Ibrahim, Khairunnisa Hasikin, Nasrul Anuar Abd Razak, et al. Critical device reliability assessment in healthcare services. *Journal of Healthcare Engineering*, 2023, 2023.

[3] David Arney, Yi Zhang, Lauren R. Kennedy-Metz, Roger D. Dias, Julian M. Goldman, and Marco A. Zenati. An open-source, interoperable architecture for generating real-time surgical team cognitive alerts from heart-rate variability monitoring. *Sensors*, 23(8):3890, 2023.

[4] M. Aruna, S. Ananda Kumar, B. Arthi, and Uttam Ghosh. Smart security for industrial and healthcare IoT applications. In *Intelligent Internet of Things for Healthcare and Industry*, pages 353–371. Springer, 2022.

[5] Katie L. Ayers, Stefanie Eggers, Ben N. Rollo, Katherine R. Smith, Nadia M. Davidson, Nicole A. Siddall, Liang Zhao, Josephine Bowles, Karin Weiss, Ginevra Zanni, et al. Variants in sart3 cause a spliceosomopathy characterised by failure of testis development and neuronal defects. *Nature Communications*, 14(1):3403, 2023.

[6] Almir Badnjevic. Evidence-based maintenance of medical devices: Current shortage and pathway towards solution. *Technology and Health Care*, 31:293–305, 2023.

[7] Almir Badnjevic, Amar Deumic, Zijad Džemic, and Lejla Gurbeta Pokvic. A novel method for conformity assessment testing of anaesthesia machines for post-market surveillance purposes. *Technology and Health Care* (Preprint):1–11, 2023.

[8] Sourav Banerjee, Sudip Barik, Debashis Das, Uttam Ghosh, and Narayan C. Debnath. Federated learning assisted Covid-19 detection model. In AboulElla Hassanien, Rawya Y. Rizk, Dragan Pamucar, Ashraf Darwish, and Kuo-Chi Chang, editors, *Proceedings of the 9th International Conference on Advanced Intelligent Systems and Informatics 2023*, pages 392–399. Springer Nature Switzerland, 2023.

[9] Siddharth Banyal, Deepanjali Mehra, Amartya, Siddhant Banyal, Deepak Kumar Sharma, and Uttam Ghosh. Computational intelligence in healthcare with special emphasis on bioinformatics and internet of medical things. In *Intelligent Internet of Things for Healthcare and Industry*, pages 145–170. Springer, 2022.

[10] Deborsi Basu, Vikram Krishnakumar, Uttam Ghosh, and Raja Datta. Deep-care: Deep learning-based smart healthcare framework using 5g assisted network slicing. In *2022 IEEE International Conference on Advanced Networks and Telecommunications Systems (ANTS)*, Gandhinagar, Gujarat, India, 2022, pp. 201-206, doi: 10.1109/ANTS56424.2022.10227802.

[11] R. Beckers, Z. Kwade, and F. Zanca. The EU medical device regulation: Implications for artificial intelligence-based medical device software in medical physics. *Physica Medica*, 83:1–8, 2021.

[12] Shaurya Bhatt, Deepak Joshi, Pawan Kumar Rakesh, and Anoop Kant Godiyal. Advances in additive manufacturing processes and their use for the fabrication of lower limb prosthetic devices. *Expert Review of Medical Devices*, 20(1):17–27, 2023.

[13] Bharat Bhushan, Avinash Kumar, Ambuj Kumar Agarwal, Amit Kumar, Pronaya Bhattacharya, and Arun Kumar. Towards a secure and sustainable internet of medical things (IoMT): Requirements, design challenges, security techniques, and future trends. *Sustainability*, 15(7):6177, 2023.

[14] Adam Bohr and Kaveh Memarzadeh. The rise of artificial intelligence in healthcare applications. *Artificial Intelligence in Healthcare*, 25–60, 2020.

[15] Michael Bretthauer, Sara Gerke, Cesare Hassan, Omer F. Ahmad, and Yuichi Mori. The New European medical device regulation: Balancing innovation and patient safety. *Annals of Internal Medicine*, 176(6):844–848, 2023

[16] Steven Brown and Apurva Desai. Legal and regulatory issues related to the use of clinical software in healthcare delivery. In *Clinical Decision Support and Beyond*, pages 651–692. Elsevier, 2023.

[17] Richard J. Chen, Judy J. Wang, Drew F.K. Williamson, Tiffany Y. Chen, Jana Lipkova, Ming Y. Lu, Sharifa Sahai, and Faisal Mahmood. Algorithmic fairness in artificial intelligence for medicine and healthcare. *Nature Biomedical Engineering*, 7(6):719–742, 2023.

[18] Oriana Ciani, Patrizio Armeni, Paola Roberta Boscolo, Marianna Cavazza, Claudio Jommi, and Rosanna Tarricone. De innovatione: The concept of innovation for medical technologies and its implications for healthcare policy-making. *Health Policy and Technology*, 5(1):47–64, 2016.

[19] Linda Connor, Jennifer Dean, Molly McNett, Donna M. Tydings, Amanda Shrout, Penelope F. Gorsuch, Ashley Hole, Laura Moore, Roy Brown, Bernadette Mazurek Melnyk, et al. Evidence-based practice improves patient outcomes and healthcare system return on investment: Findings from a scoping review. *Worldviews on Evidence-Based Nursing*, 20(1):6–15, 2023.

[20] Genevieve Dammery, Louise A. Ellis, Kate Churruca, Janani Mahadeva, Francisco Lopez, Ann Carrigan, Nicole Halim, Simon Willcock, and Jeffrey Braithwaite. The journey to a learning health system in primary care: A qualitative case study utilising an embedded research approach. *BMC Primary Care*, 24(1):22, 2023.

[21] Delshi Howsalya Devi, Kumutha Duraisamy, Ammar Armghan, Meshari Alsharari, Khaled Aliqab, Vishal Sorathiya, Sudipta Das, and Nasr Rashid. 5g technology in healthcare and wearable devices: A review. *Sensors*, 23(5):2519, 2023.

[22] Margaret Hardt DiCuccio. The relationship between patient safety culture and patient outcomes. *Journal of Patient Safety*, 11(3):135–142, 2015.

[23] Jean Feng, Scott Emerson, and Noah Simon. Approval policies for modifications to machine learning-based software as a medical device: A study of bio-creep. *Biometrics*, 77(1):31–44, 2021.

[24] Senay A. Gebreab, Khaled Salah, Raja Jayaraman, and Jamal Zemerly. Trusted traceability and certification of refurbished medical devices using dynamic composable NFTs. *IEEE Access*, 11:30373–30389, 2023.

[25] Uttam Ghosh, Debashis Das, Pushpita Chatterjee and Sachin Shetty, "Quantum-Enabled Blockchain for Data Processing and Management in Smart Cities," *2023 IEEE 24th International*

Symposium on a World of Wireless, Mobile and Multimedia Networks (WoWMoM), Boston, MA, USA, 2023, pp. 425-430, doi: 10.1109/WoWMoM57956.2023.00075.

[26] William J. Gordon and Ariel D. Stern. Challenges and opportunities in software-driven medical devices. *Nature Biomedical Engineering*, 3(7):493–497, 2019.

[27] Thomas J. Hagedorn, Sundar Krishnamurty, and Ian R. Grosse. An information model to support user-centered design of medical devices. *Journal of Biomedical Informatics*, 62:181–194, 2016.

[28] Vaishali Hegde. Case study—risk management for medical devices (based on ISO 14971). In *2011 Proceedings-Annual Reliability and Maintainability Symposium*, pages 1–6. IEEE, 2011.

[29] Juhamatti Huusko, Ulla-Mari Kinnunen, and Kaija Saranto. Medical device regulation (MDR) in health technology enterprises—perspectives of managers and regulatory professionals. *BMC Health Services Research*, 23(1):1–12, 2023.

[30] Mohd Javaid, Abid Haleem, Ravi Pratap Singh, Rajiv Suman, and Shanay Rab. Significance of machine learning in healthcare: Features, pillars and applications. *International Journal of Intelligent Networks*, 3:58–73, 2022.

[31] Peter Jordan, "Standard IEC 62304 - Medical Device Software - Software Lifecycle Processes," *2006 IET Seminar on Software for Medical devices*, London, UK, pp. 41-47, 2006.

[32] Ilhan Firat Kilincer, Fatih Ertam, Abdulkadir Sengur, Ru-San Tan, and U. Rajendra Acharya. Automated detection of cybersecurity attacks in healthcare systems with recursive feature elimination and multilayer perceptron optimization. *Biocybernetics and Biomedical Engineering*, 43(1):30–41, 2023.

[33] Siqi Li, Zheng Guo, and Xuehui Zang. Advancing the production of clinical medical devices through chatgpt. *Annals of Biomedical Engineering*, 1–5, 2023.

[34] Zhenyu Luo, Sihui Liu, Linhe Yang, Shuyan Zhong, and Lihua Bai. Ambulance referral of more than 2 hours could result in a high prevalence of medical-device-related pressure injuries (MDRPIS) with characteristics different from some inpatient settings: A descriptive observational study. *BMC Emergency Medicine*, 23(1):1–7, 2023.

[35] Maria Raffaella Martina, Elisabetta Bianchini, Sara Sinceri, Martina Francesconi, and Vincenzo Gemignani. Software medical device maintenance: Devops based approach for problem and modification management. *Journal of Software: Evolution and Process*, e2570, 2023.

[36] Martin McHugh, Oisin Cawley, Fergal McCaffcry, Ita Richardson, and Xiaofeng Wang. An agile v-model for medical device software development to overcome the challenges with plan-driven software development lifecycles. In *2013 5th International Workshop on Software Engineering in Health Care (SEHC)*, pages 12–19. IEEE, 2013.

[37] Martin McHugh, Fergal McCaffery, and Valentine Casey. Software process improvement to assist medical device software development organisations to comply with the amendments to the medical device directive. *IET Software*, 6(5):431–437, 2012.

[38] Sanjana Mukherjee. The united states food and drug administration (FDA) regulatory response to combat neglected tropical diseases (NTDS): A review. *PLOS Neglected Tropical Diseases*, 17(1):e0011010, 2023.

[39] Senthil Murugan Nagarajan, Ganesh Gopal Deverajan, Puspita Chatterjee, Waleed Alnumay, and Uttam Ghosh. Effective task scheduling algorithm with deep learning for internet of health things (IoHT) in sustainable smart cities. *Sustainable Cities and Society*, 71:102945, 2021.

[40] Saroj Kumar Nanda, Sandeep Kumar Panda, and Madhabananda Dash. Medical supply chain integrated with blockchain and IoT to track the logistics of medical products. *Multimedia Tools and Applications*, 1–23, 2023.

[41] Charles Odilichukwu R. Okpala and MaYgorzata Korzeniowska. Understanding the relevance of quality management in agro-food product industry: From ethical considerations to assuring food hygiene quality safety standards and its associated processes. *Food Reviews International*, 39(4):1879–1952, 2023.

[42] Kevin Pierre, Adam G. Haneberg, Sean Kwak, Keith R. Peters, Bruno Hochhegger, Thiparom Sananmuang, Padcha Tunlayadechanont, Patrick J. Tighe, Anthony Mancuso, and Reza Forghani. Applications of artificial intelligence in the radiology roundtrip: Process streamlining, workflow optimization, and beyond. In *Seminars in Roentgenology*, volume 58, pages 158–169. Elsevier, 2023.

[43] Janet Pigueiras-del Real, Lionel C. Gontard, Isabel Benavente-Fernández, Simón P. Lubián-López, Enrique Gallero-Rebollo, and Angel Ruiz-Zafra, "NRP: A Multi-Source, Heterogeneous,

Automatic Data Collection System for Infants in Neonatal Intensive Care Units," in *IEEE Journal of Biomedical and Health Informatics*, 28(2):678–689, 2024.

[44] Vernessa T. Pollard. FDA issues draft predetermined change control plan for machine-learning-enabled device software functions. *Mondaq Business Briefing*, 2023. https://www.duanemorris.com/alerts/fda_issues_draft_guidance_predetermined_change_control_plans_artificial_intelligence_0623.html

[45] Andreas Puder, Jacqueline Henle, and Eric Sax. Threat assessment and risk analysis (TARA) for interoperable medical devices in the operating room inspired by the automotive industry. In *Healthcare*, volume 11, page 872. MDPI, 2023.

[46] Dukka Karun Kumar Reddy, H.S. Behera, Janmenjoy Nayak, Ashanta Ranjan Routray, Pemmada Suresh Kumar, and Uttam Ghosh. A fog-based intelligent secured iomt framework for early diabetes prediction. In *Intelligent Internet of Things for Healthcare and Industry*, pages 199–218. Springer, 2022.

[47] Matthias Seibold, José Miguel Spirig, Hooman Esfandiari, Mazda Farshad, and Philipp Fürnstahl. Translation of medical AR research into clinical practice. *Journal of Imaging*, 9(2):44, 2023.

[48] Kaapo Seppällal. New business creation in health technology. In *Design Thinking in Healthcare: From Problem to Innovative Solutions*, pages 101–111. Springer, 2023.

[49] Veenu Singh, Vijay Kumar, and V.B. Singh. A hybrid novel fuzzy AHP-topsis technique for selecting parameter-influencing testing in software development. *Decision Analytics Journal*, 6:100159, 2023.

[50] M. Srivani, Abirami Murugappan, and T. Mala. Cognitive computing technological trends and future research directions in healthcare—a systematic literature review. *Artificial Intelligence in Medicine*, 102513, 2023.

[51] Muhammad Turab and Sonain Jamil. A comprehensive survey of digital twins in healthcare in the era of Metaverse. *BioMedInformatics*, 3(3):563–584, 2023.

[52] Tomaso Vairo, Margherita Pettinato, Andrea P. Reverberi, Maria Francesca Milazzo, and Bruno Fabiano. An approach towards the implementation of a reliable resilience model based on machine learning. *Process Safety and Environmental Protection*, 172:632–641, 2023.

[53] Sai Srinivas Vellela, B. Venkateswara Reddy, Kancharla K. Chaitanya, and M. Venkateswara Rao. An integrated approach to improve e-healthcare system using dynamic cloud computing platform. In *2023 5th International Conference on Smart Systems and Inventive Technology (ICSSIT)*, pages 776–782. IEEE, 2023.

[54] Chunhu Xie, Huachun Wu, and Jian Zhou. Vectorization programming based on HR DSP using SIMD. *Electronics*, 12(13):2922, 2023.

[55] Nataliya Yakymets, Mihai Adrian Ionescu, and David Atienza Alonso . Metamodel for safety risk management of medical devices based on ISO 14971. In *The ACM/IEEE 26th International Conference on Model-Driven Engineering Languages and Systems*, EPFL Scientific Publications, 2023.

[56] Zhe Yin, Carlos Caldas, Daniel de Oliveira, Sharareh Kermanshachi, and Apurva Pamidimukkala. Cross-functional collaboration in the early phases of capital projects: Barriers and contributing factors. *Project Leadership and Society*, 100092, 2023.

[57] Youshan Yu, Nicolette Lakemond, and Gunnar Holmberg, "AI in the Context of Complex Intelligent Systems: Engineering Management Consequences," in *IEEE Transactions on Engineering Management*, vol. 71, pp. 6512-6525, 2024.

Chapter 5

6G Communication Technology for Industry 5.0

Prospect, Opportunities, Security Issues, and Future Directions

Priyanka Das, Sangram Ray, Dipanwita Sadhukhan, and Mahesh Chandra Govil

Chapter Contents

5.1 Introduction

Industry Internet of Things (IIoT) is a forethoughtful perception of the futuristic industrial paradigm that focuses on systems' resiliency and agility with the deployment of adaptable and flexible technologies. IIoT is expected to support various industrial verticals such as smart farming, healthcare, smart grids, and supply chain production ecosystems to leverage huge production with customer-centric customization [1].

IIoT eventually has a huge economic impact that brings entirely new infrastructures to our most vital and impactful societal systems. It puts emphasis on the communication between machines (M2M), the use of machine learning, and the handling of Big data, which leads to improved reliability and efficiency in industrial operations [2].

IIoT applications are a simple evolution of Internet of Things (IoT). Thus, IIoT shares some security weaknesses with the IoT, including the integrity of CPS, the protection of data, the establishment of keys for system pairing, and the management of systems.

5.1.1 Evolution of IIoT

Before the evolution era of IIoT, agriculture and simple handicrafts were the significant factors of the global economy and later as said "necessity is the mother of disruption," the transformation in the IIoT ecosystem took pace.

Industry 1.0 revolution started around 17th century from the period between 1760 to 1840 in which the industrial machines were powered by water, steam, etc. This stage was

DOI: 10.1201/9781003376712-5

referred to as mechanization [3]. Henry Ford invented the assembly line with electricity in 1870, which led to the evolution of Industry 2.0 termed electrification. The fundamental concept of Industry 2.0 was mass production [4]. Industry 3.0 marked a shift from traditional mechanical methods to modern digital production processes along with partial automation. In this generation, huge-sized computers with memory-programmable controllers were used to make industrial plans, which resulted in a reduction of the need of human efforts. In the year 2011, the world witnessed a drastic shift from Industry 3.0 toward Industry 4.0 termed as digitization. The digital era encompasses the use of information and communication technologies to support remote production processes that are managed through a network of connected devices [5, 6]. One of the major concerns of Industry 4.0 is IoT-driven CPS. The entire evolution of IIoT is well depicted in Figure 5.1.

The revolution of Industry 5.0 involves the blending of artificial intelligence with human knowledge which is the driving force behind the development of this era. Industry 5.0 is visualized as a way to improve the quality of production which allows the robot to handle boring and repetitive tasks while human handles intelligent and critical thinking tasks [11]. The key definitions of Industry 5.0 given by different authors are shown in Table 5.1.

5.2   State-of-the-Art of IIoT

In this section, the essential concepts—IIoT architecture, connectivity, and IIoT standardization are highlighted.

A.   *IIoT Architecture*
IIoT architecture focuses on scalability, extensibility, flexibility, and interoperability between diverse smart devices, which uses different technologies such as Artificial Intelligence and Blockchain [12] The essential components of IIoT architecture with its specifications are shown in Figure 5.2.

B.   *IIoT Connectivity*
The IIoT connectivity varies on the basis of the combination of edge architecture and the backbone that is appropriate for a specific situation. A primary goal is to enable sharing of data and interoperability between closed subsystems. Neither the five-layer Internet model nor the seven-layer OSI model is adequate to take into consideration the distributed traits of controllers, sensors, gateways, and other components of IIoT [12]. Figure 5.3 mirrors the hourglass of the IIoT protocol stack in which the upper layers allow "smart devices" to transmit "smart data." These layers are not only restricted to sharing relevant information but also provide awareness to the intended users. Similarly, the lower layers are responsible for scalability and flexibility requirements.

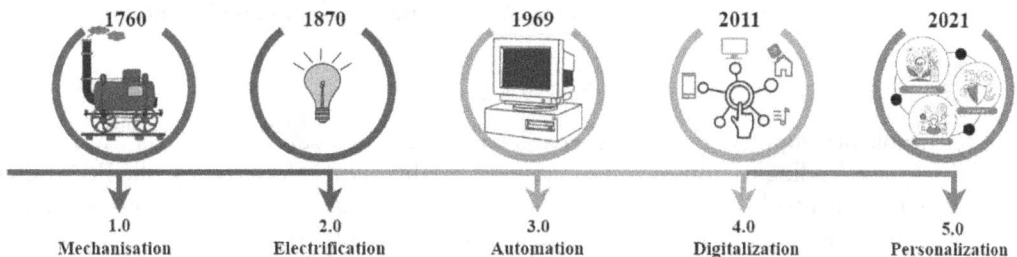

Figure 5.1 A prospective evolution of IIoT.

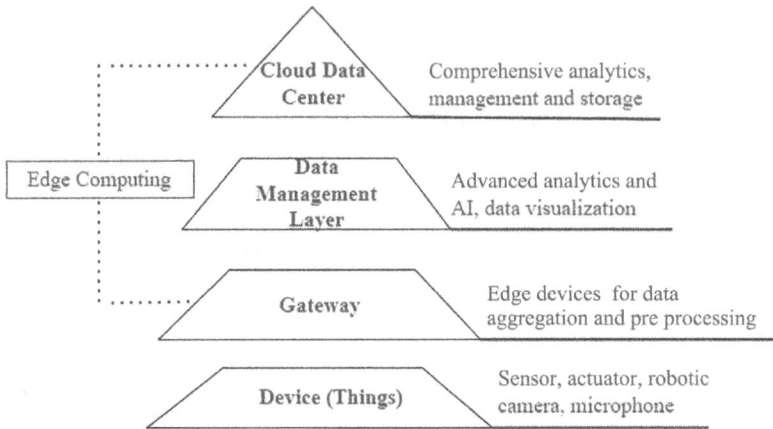

Figure 5.2 IIoT architecture.

Table 5.1 Fundamental Definitions of Industry 5.0 Stated by Experts

Authors	Definition
[7]	The very first human-initiated industrial evolution is Industry 5.0, based on 6R principles which are Recognize, Reconsider, Reuse, Realize, Reduce, and Recycle. It is a technique that aims to eliminate waste in a systematic manner and manufactures good-quality custom items.
[8]	Industry 5.0 is an amalgamation of human intelligence and the potentiality of CPSs aimed at resolving the problem of labor shortage faced in Industry 4.0.
[9]	Industry 5.0 is an amalgamation of human-intelligence and the potentiality of CPSs to build cooperate factories.
[10]	Industry 5.0 refers to a network of smaller factories where robots and humans have a direct, seamless interaction.

C. IIoT Standardization

Standardization is a crucial step for a technology to be widespread. The process of standardization faces numerous challenges. The standardization of 5G toward IIoT has a crucial significance in Industry 5.0 but has various shortcomings with the transformation of the industrial era. This comes under the umbrella of Ultra Reliable Low Latency Communications (URLLC) that aims to provide 1 *ms* latency to grant a valid transmission time interval (TTI) operation [13]. With the evolution of Industry 5.0, 6G communication came into the scenario which plays a crucial role to connect a variety of IoT devices. 6G, an upgradation of 5G, emphasizes its ample amount of benefits, for example—higher data transmission rate, high bandwidth and low latency.

5.2.1 Feasible Developments of Industry 5.0

The feasible development of IIoT has enabled new efficiencies and capabilities within industrial settings, leading to improved productivity, quality, and safety. Today, the IIoT system is being used in a wide range of industrial applications which help organizations to improve operational efficiency, reduce costs, enhance product quality, etc. A few of the aspects that are influenced by the evolution of Industry 5.0 are discussed below:

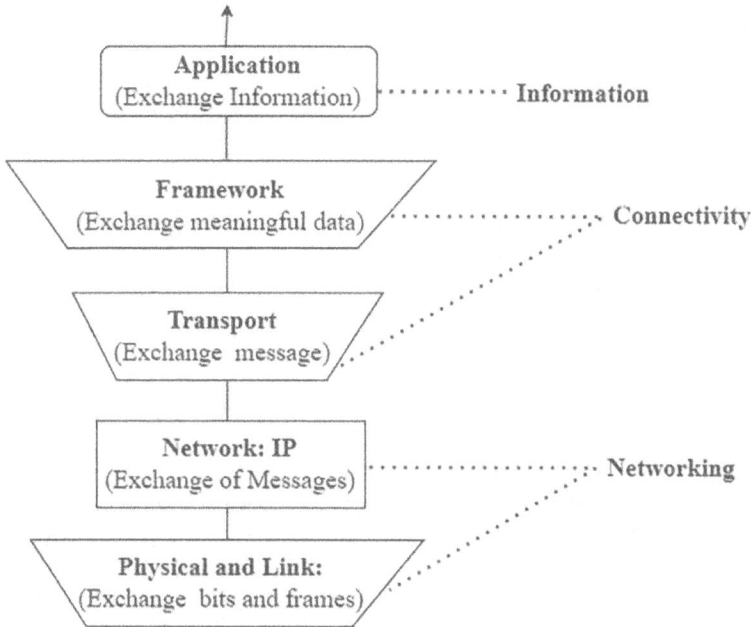

Figure 5.3 The hourglass-shaped IIoT protocol stack.

- *Human Capital Development*
  The significant human capital development influenced by Industry 5.0 focuses on characteristics such as—Information, Internet, Internet of Things, and Innovations. These characteristics contribute to human capital development in two ways, namely:
    i. Minimizing the quantity of employment assigned to human labor.
    ii. Raising the need for a skilled labor force equipped with the necessary abilities in emerging technologies.
  Thus, the dominance of evolving Industry 5.0 has brought a revolution in human capital development upon a wide range scope in the work environment with automation, digitization, technological aspects, etc.

- *Changes in Business Strategies/Models*
  Businesses operating in Industry 5.0 adopt new strategies and business models that can effectively leverage these new technologies while still retaining a human touch. One way to achieve this is through a focus on customization and personalization. With the rapid growth of technological change and increasing competition, companies must be able to adapt quickly to new trends and shifts in the market. Additionally, businesses influenced by Industry 5.0 also prioritize sustainability and social responsibility in their operations.

- *Education System*
  The integration of Industry 5.0 in the education system refers to the use of IIoT technologies in educational settings such as—classrooms, laboratories, and online learning environments. This involves Industry 5.0 technologies to enhance teaching

and learning experiences, improve student outcomes, and prepare students for careers in the field of industrial automation and digital transformation. Some potential applications of Industry 5.0 in the education system include smart classrooms, virtual laboratories, personalized learning experiences, and real-time monitoring of student progress and performance.

- *IoMT*
  Internet of Medical Things (IoMT) is a vital component of Industry 5.0, which focuses on combining the strengths of humans and machines to improve productivity and create new value for the healthcare system. The benefits of IoMT in Industry 5.0 include improved patient outcomes, personalized treatment plans, etc.

- *Financial Industry*
  Industry 5.0 has the potential to transform the financial industry with innovation, improving efficiency, and enhancing customer experiences. A significant aspect of Industry 5.0 in the financial industry is the need for a greater emphasis on sustainability and social responsibility. Further, Industry 5.0 can be applied in several ways in the financial industry. For example, financial institutions can use advanced technologies such as artificial intelligence and machine learning to improve decision-making processes and risk management [14].

5.3   Transition From 1G to 6G for IIoT

The paradigms of a different generation of communication technology are listed below. Additionally, a detailed comparison of 1G to 6G communication technology is tabulated in Table 5.2.

- *1G: Voice Calls*
  During the age of 1G communication technology, phones were heavy and bulky in size. They had big antennas with huge batteries, and the network reception was sketchy. The analog-system-based 1G wireless communication enabled the exchange

Table 5.2 Comparison between Different Generations of Communication Technology

Specifications	1G	2G	3G	4G	5G	6G
Year	1980–1990	1990–2000	2000–2010	2010–2020	2020–2030	2030–2040
Core network	PSTN	PSTN	Packet N/W	Internet	IoT	IoE
Services	Voice	Text	Photos	Video	3D VR/AR	Tactil
Multiplexing	FDMA	FDMA, TDMA	CDMA	OFDMA	OFDMA	OFDMA Smart OFDMA plus IM
Architecture	SISO	SISO	SISO	MIMO	Massive MIMO	Intelligent Surface
Data transfer rate	2.4 Kb/sec	144 Kb/sec	2 Mb/sec	1 Gb/sec	34.6 Gb/sec	100 Gb/sec
Maximum frequency	894 MHz	1900 MHz	2100 MHz	6 GHz	90 GHz	10 THz

of information between two devices which were supported only with poor-quality voice call features. One of the key challenges of 1G communication technology was its fixed geographical area since it lacked roaming support by the network.

- *2G: Text*
  2G communication technology facilitated better quality of voice calls as compared to 1G. The analog system based on 1G of wireless transmission was substituted by a more sophisticated digital technology known as Global System for Mobile Communication (GSM). In addition, it also supported new services such as—Short/Multimedia Message Service which is abbreviated as (SMS) and (MMS), respectively.

- *3G: Era of Applications*
  The 3G communication technology was introduced with high-speed Internet services, laying the foundation for the development of smartphones equipped with a diverse range of applications. It also fabricated the concept of online radio services, mobile television, emails on phones, etc.

- *4G: Internet Calling*
  The incorporation of the LTE (Long-Term Evolution) system significantly improved the data rate allowing synchronous transmission of both data and voice. VoLTE (Voice over LTE) or Internet calling is one of the fundamental improvements of the 4G communication network. It also facilitates Voice over Wi-Fi (VoWi-Fi) feature which allowed one to make voice calls in either low-network areas or even in no-network-coverage area [15].

- *5G: IoT*
  5G technology is still in the process of being fully rolled out globally, and it will likely take a few years for it to be widely available and adopted. As compared to all the existing communication technologies, the advantage of recently launched 5G technology is low latency with higher throughput features, which makes the network ideal with automation features and connected ecosystem.

- *6G: IoE*
  The transition from 5G to 6G is not an immediate concern. It is expected that by the year 2030, 6G will bring a drastic evolution in communication mediums as it will enable the newly introduced intelligent Internet of Everything (IoE) concept. Future 6G is forecasted to bring high and sophisticated Quality of Service (QoS), for example, holographic communication and virtual reality [16]. The concept of the IoE is making progress in enhancing people's lives by enhancing the IoT by creating connections among individuals, processes, information, and devices. The advent of 6G technology will make communication instantaneous with low latency, thereby creating a seamless connection between the digital and physical world.

Although 5G communication can transform more data as compared to earlier communication technologies, one of the major flaws of 5G communication is the higher frequency bands of 5G which have shorter ranges; therefore, more cell towers are needed to be built closer together. Thus, there is the necessity of upgradation from existing 5G to advance 6G communication technology. The key factors for moving from 5G to futuristic 6G are mentioned below:

- *Wider Network Coverage*
  The network coverage is directly correlated to usability. Prevailing communication technology (5G) provides a fair network coverage accessibility on the ground, but the network infrastructure of existing communication technologies is not well-built enough to provide network connectivity at high altitudes and under the deep sea. Therefore, the futuristic 6G communication is expected to raise both—the need for wider coverage area as well as usability. Moreover, 6G communication is presumed to provide network connectivity at high altitudes as well as under the deep sea.

- *Network Capability*
  The futuristic 6G communication technology is expected to be with a better network capability than the existing 5G communication technology to support the need for modernized services. The vital features from which the capability of a network can be determined are resource management, throughput, effective usage of underlying network architecture for modernized applications, etc.

- *Service Provisioning*
  The process of setting up a specific service for an end user is referred to as service provisioning. 6G communication is expected to offer a wide range of advanced services to end users.

5.3.1 Key Areas of 6G Communication

The key areas in which 6G communication technology plays a vital role are described below.

- Real-Time Intelligent Edge
  The existing 5G communication technology supports autonomous driving, but services such as self-awareness and self-adaption are not supported yet. To overcome these shortcomings of 5G, 6G communication technology is introduced which enables AI-assisted services.

- Distributed Artificial Intelligence
  The futuristic 6G network is expected to support the revolutionary Internet of Everything (IoE), which will make the 6G communication technology capable enough to take smart decisions of its own [17].

- 3D Intercoms
  6G is predicted to be with an optimized network, and the design will make a smooth shift from two dimension (2D) to three dimension (3D). 6G is expected to support 3D communication which will enable communication in heterogeneous scenarios such as—Unmanned Aerial Vehicles (UAVs), satellites, and undersea communication. A 3D intercom could facilitate these features with the exact time as well as location. Additionally, resource management and characteristics such as mobility and routing also need network optimization in 3D intercom [18].

- Intelligent Radio
  The transceiver system could be separated in futuristic 6G while it was designed together in existing generations. Thus, it has the ability to self-update. Figure 5.4 illustrates the new logical state of 6G communication technology [19].

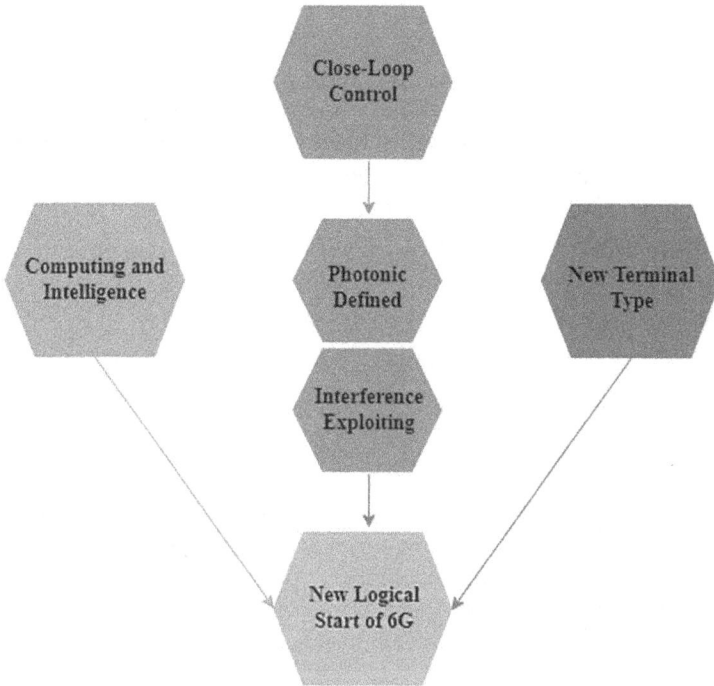

Figure 5.4 New logical start of 6G.

5.3.2 Key Technologies of 6G Communication

In this subsection, the key technologies of 6G communication are discussed in detail and depicted in Figure 5.5 for clear understanding.

A. *Artifical Intelligence (AI)*
    Even though AI is incorporated in existing 5G network, it is regarded as the prime feature of futuristic 6G communication technology. AI technologies are fragmented into a physical layer which consists of network structure, architecture, and computing layer consisting of software-defined network, network function virtualization, and edge/cloud computing [20].

B. *Quantum Communication*
    Quantum communication is an aspiring technology in 6G technology. Quantum communication provides security with crucial developments. It offers explanations and elevates communication which is unachievable through conventional communication techniques. However, it isn't the only panacea for each security peril.

C. *Blockchain*
    Blockchain is another prominent technology in 6G communication. Network decentralization, spectrum sharing, and distributed ledger technology are several uses of Blockchain technology. Blockchain technology could overcome spectrum monopoly and low-spectrum utilization. Privacy issues of Blockchain are related to authentication, communication, and access control [21].

D.   *Visible Light Communication (VLC)*

VLC uses visible light as a carrier for data transmission and can coexist with traditional radio-frequency (RF) communication systems. Hence, in order to address the constantly expanding demand for wireless communication, VLC is an aspiring technology. It has been deployed in indoor positioning systems and VANET. One of the key advantages of VLC in 6G communication is its high data rate. This high data rate can enable new applications, such as real-time streaming of ultra-high-definition videos and high-speed file transfer.

5.3.3   Case Studies Showing Advantages of 6G Communication

In this subsection, a few significant services and use cases which show the advantages of incorporating 6G communication technology are highlighted below:

- E-Healthcare
  Technological developments such as holographic communication, AI, and Virtual/Augmented Reality shall be available with the 6G network that will provide immense benefits to the healthcare system. By allowing telemedicine to remove the restrictions of time and distance, the futuristic 6G will optimize the e-healthcare's functions and workflow. However, ensuring the delivery of e-Health services will necessitate meeting demanding QoS standards such as authentic communication and robustness in mobility.

- Smart Environment
  The concept of a smart home/city aims to optimize various daily life processes and increase the quality of life. The 6G network is expected to combine ICT and a large number of IoT devices to further optimize processes like home security, waste management, transportation systems, traffic monitoring, and utilities-related operations. The 5G network is a step in this direction. Still, its strict requirements like high connection density, ultra-high reliability communications, tight security, and massive

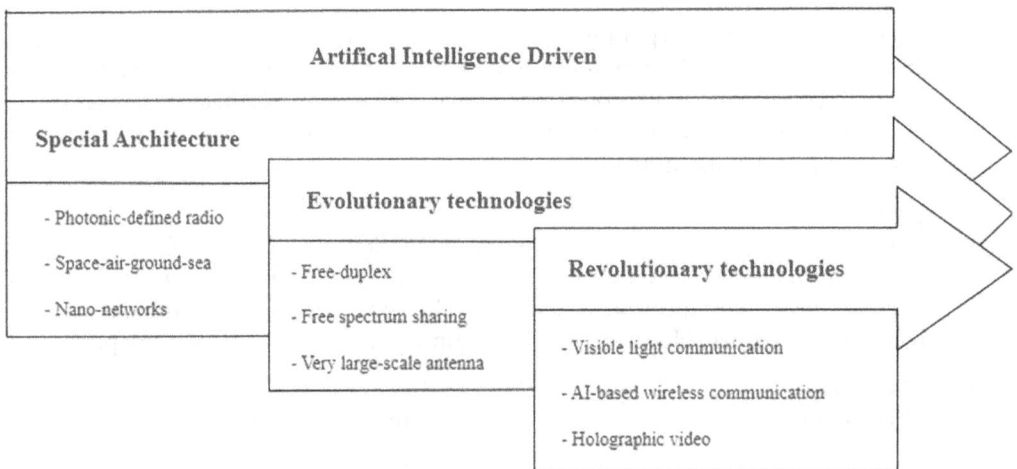

Figure 5.5 Key technologies of 6G communication.

data rates may limit the potential of these applications. Overall, a smart environment is a critical goal of modern networks, and the 6G network is expected to provide even greater optimization and efficiency in daily life.

- Automation in Industry 5.0
  The development of Industry 5.0 is a recent trend and with the arrival of the 6G network, it is expected to reach new heights of automation and optimization. The aim of Industry 5.0 is to reduce the dependency on human intervention in traditional manufacturing and industrial processes, replacing it with automatic control systems, communication networks, and CPSs. To achieve this goal, the 6G network must meet strict key performance indicators (KPIs) such as high reliability levels, low latency, and multiple connected links.

5.3.4 Amalgamation of 6G Communication with Industry 5.0

The amalgamation of 6G communication technology with Industry 5.0 is a hot topic. 6G in Industry 5.0 can offer more degree of freedom, a wireless link with more flexibility, and enables a higher capability of data transfer rate.

- *Prospect*
  The prospect for Industry 5.0 with 6G technology is stated as follows: Industry 5.0 is a recent shift in traditional production that makes use of smart technologies. Technically, the focus is on utilizing widespread machine-to-machine (M2M) communication with the support of Internet of Things (IoT) infrastructure and devices to achieve automation. Industry 5.0 will allow machine components to use AI algorithms to self-diagnose and predict possible future issues, preventing any failures. In line with this, Industry 5.0 will also allow robots to make autonomous decisions that significantly affect people's lives \cite{22}. Additionally, the core Industry 5.0 enablers include predictive maintenance of IoT devices such as smart sensors and 3D printers. The Cyber-Physical System (CPS) which allows for the connection is the medium through which Industry 5.0 first proposed the objective of the digital mutation of manufacturing. Industry 5.0 is thought to be greatly facilitated by effective automation systems for asynchronous communication. In this context, 6G is suggesting the usage of THz domains to yield the necessary QoS for intelligent integrated systems.

- *Opportunities*
  The opportunities or advantages of incorporating 6G into Industry 5.0 are listed below [23]:
  - i. It provides super high-definition (SHD) as well as extremely high-definition (EHD) video transmission that needs ultra-high throughput.
  - ii. It bears extremely low latency which can be up to 10 microseconds (ms).
  - iii. It ensures to drastically decrease the safety-related accidents in mining industries with the advanced sensing power and remote monitoring feature.
  - iv. It drastically increases the bounds of human activity such as space travel and exploration of the deep sea by enabling effective space and underwater transmissions.

v. It facilitates advanced services such as hyper high-speed railway (HSR).

vi. It improves 5G applications such as autonomous vehicles.

vii. It provides enhanced mobile broadband communication, low latency as well as high mobility and long-distance communication.

- *Security Issues*

Security as well as privacy are critical issues in 6G communication, and it is imperative to ensure both security with privacy to enable the technologies discussed in earlier sections. Access control, malicious activity detection, authentication, encryption, etc. are a few of the security and privacy concerns with its applications in healthcare (sensing devices), industrial automation, IoT applications, etc. As a result, it is crucial to address the recent privacy and security concerns; or else, 6G won't be able to achieve its performance boom. These challenges are considered in relation to many technologies, such as—blockchain, AI, and THz. Electromagnetic signatures are suggested as a solution to the authentication issues in THz communication [24].

- *Future Directions*

For the successful deployment of 6G technology, a variety of technical issues need to be addressed. A few of the significant security issues are mentioned below:

- **Atmospheric Absorption and High-Level Propagation of THz**

Although the *THz* frequency bands provide high throughput, it needs to address its data transmission over long-distance devices [25]. This causes high propagation loss because of the atmospheric absorption property. Hence, absorptive effects are witnessed. To address this issue, a novel transceiver architecture is proposed for the *THz* frequency bands with effectual antennas.

- **Complexity in Resource Management for 3D Networking**

The ability to build attractive 3D network diagrams for better knowledge is known as 3D networking. It is one of the most potent succeeding networking features that will be incorporated into 6G network communication. 3D networking has grown vertically, enabling the development of novel management of resources, mobility support, routing protocol, and channel-access techniques. The fusion of 3D networking into 6G communication will offer better network management and operations.

- **Heterogeneous Hardware Constraints**

A variety of heterogeneous devices, frequency bands, architectures, topologies, and operating systems are expected to merge into the 6G network. MIMO technology, which has a more complicated architecture than 5G communication, will be further incorporated into the advanced 6G [26]. Additionally, 6G will have to adapt to sophisticated communication algorithms and protocols. However, hardware operations may get more difficult with the incorporation of AI and machine learning methods into 6G. Therefore, it'll be extremely difficult to combine all of the contrasting components into a singular platform [27].

- **Spectrum and Interference Management**

Spectrum-sharing strategies must be effective and creative because of the scarcity of spectrum at higher frequencies and their interference issues. In order to maximize the Quality of Service (QoS) and to achieve optimal resource usage, effectual spectrum management strategies are essential. Researchers must address issues related to spectrum management for contrasting networks and devices inside the identical frequency range in order to efficiently deploy 6G.

○ **High-Capacity Backhaul Network Connectivity**
An effective backhaul connection is a key component of the network connectivity solution. For remote locations, finding the right backhaul design could be exceedingly difficult. High-level density access networks are drastically different from existing networks. These access networks which are common in 6G will have the ability to deliver data at extremely high speed. The backhaul network will be essential in the future for enabling extensive connectivity between core and access networks. Such possibilities have a lot of potentialities that will support the evolving satellite constellations including SpaceX's Starlink and OneWeb's Telesat's systems, but they are remarkably expensive in nature [26]. The advanced backhaul network is necessary since optical fiber and FSO networks are the more suitable alternatives and have the ability to enhance their performance in a 6G network.

○ **Beam Management in THz Communications**
High data rates may be supported by beamforming via sophisticated MIMO systems. The difficulty in managing beams for high-frequency bands like the THz band is due to the challenges posed by its propagation characteristics. Hence, to deal with enhanced MIMO's propagation characteristics of 6G, a successful beam management system is essential to propose.

5.4   Conclusion

In this chapter, the prospects, opportunities, and future aspects upraised by the amalgamation of 6G communication with Industry 5.0 are presented. 6G communication technology has the potential to revolutionize Industry 5.0, enabling more advanced automation, data analytics, and communication. However, it will still take time and significant investment before 6G is widely adopted and implemented in the industry. According to the study, based on current trends in 6G technology, in the future, 6G could have a great impact on Industry 5.0 with enhanced speed and precision of industrial robots, leading to more advanced automation. The combination of Industry 5.0 and 6G could bring about a new era of productivity, efficiency, and innovation. However, there are several limitations that are needed to be overcome before this can become a reality. A few of the prime limitations of 5G are: Security, integration with existing technologies, investment and adoption, etc., which shall be taken care of in 6G. Further, the proposed study put forward a few potential findings of evolving concepts which include ultra-fast data transmission, massive connectivity, etc. Inversely, there's a downside to every upside, therefore, the proposed study sums up with a peculiar view that the realization of amalgamation of 6G communication with Industry 5.0 may lead to increased energy consumption, higher carbon emissions, etc., which are unfavorable for humankind, and minimizing these can be considered as a future scope of the study.

Acknowledgment

The research work is supported by the Ministry of Education, the Government of India.

References

[1] Verma, A., Bhattacharya, P., Madhani, N., Trivedi, C., Bhushan, B., Tanwar, S. and Sharma, R., Blockchain for Industry 5.0: Vision, Opportunities, Key Enablers, and Future Directions, *IEEE Access*, **10**, 69160, 2022.

[2] Sisinni, E., Saifullah, A., Han, S., Jennehag, U. and Gidlund, M., Industrial Internet of Things: Challenges, Opportunities, and Directions. *IEEE Transactions on Industrial Informatics*, **14**, 4724, 2018.

[3] Yu, X. and Guo, H., A Survey on IIoT Security. *IEEE VTS Asia Pacific Wireless Communications Symposium (APWCS)*, **1**, 2019.

[4] Nahavandi, S., Industry 5.0—A Human-Centric Solution. *Sustainability*, **11**, 4371, 2019.

[5] Akbar, M. S., Hussain, Z., Sheng, Q. Z. and Mukhopadhyay, S., 6G Survey on Challenges, Requirements, Applications, Key Enabling Technologies, Use Cases, AI Integration Issues and Security Aspects. *arXiv preprint arXiv:2206.00868*, 2022.

[6] Upadhyaya, P., Dutt, S. and Upadhyaya, S. "6G Communication: Next Generation Technology for IoT Applications," *2021 First International Conference on Advances in Computing and Future Communication Technologies (ICACFCT)*, Meerut, India, 2021, pp. 23-26, doi: 10.1109/ICACFCT53978.2021.9837375.

[7] Rada, M., *INDUSTRY 5.0 Definition*, 2018. Retrieved from https://michaelrada.medium.com/industry-5-0-de?nition.

[8] Longo, F., Padovano, A. and Umbrello, S., Value-Oriented and Ethical Technology Engineering in Industry 5.0: A Human-Centric Perspective for the Design of the Factory of the Future. *Applied Sciences*, **10**, 4182, 2020.

[9] Friedman, B. and Hendry, D. G., *Value Sensitive Design: Shaping Technology with Moral Imagination*. MIT Press, 2019.

[10] Koch, P. J., van Amstel, M. K., Debska, P., Thormann, M. A., Tetzlaff, A. J., Bøgh, S. and Chrysostomou, D., A Skill-Based Robot Co-Worker for Industrial Maintenance Tasks. *Procedia Manufacturing*, **11**, 83, 2017.

[11] Chi, H. R., Wu, C. K., Huang, N. F., Tsang, K. F. and Radwan, A., "A Survey of Network Automation for Industrial Internet-of-Things Toward Industry 5.0," in *IEEE Transactions on Industrial Informatics*, 19(2):2065-2077, Feb. 2023, doi: 10.1109/TII.2022.3215231.

[12] Schneider, S., The Industrial Internet of Things (IIOT) Applications and Taxonomy. *Internet of Things and Data Analytics Handbook*, **41**, 2017.

[13] Mohsan, S. A. H., Mazinani, A., Malik, W., Younas, I., Othman, N. Q. H., Amjad, H. and Mahmood, A., 6G: Envisioning the Key Technologies, Applications and Challenges. *International Journal of Advanced Computer Science and Applications*, **11**, 9, 2020.

[14] Maiti, M., Vuković, D., Mukherjee, A., Paikarao, P.D. and Yadav, J.K. Advanced Data Integration in Banking, Financial, and Insurance Software in the Age of COVID-19. *Software: Practice and Experience*, **52**, 887, 2021.

[15] Yang, P., Xiao, Y., Xiao, M. and Li, S., 6G Wireless Communications: Vision and Potential Techniques. *IEEE Network*, **33**, 70, 2020.

[16] Sheth, K., Patel, K., Shah, H., Tanwar, S., Gupta, R. and Kumar, N., A Taxonomy of AI Techniques for 6G Communication Networks. *Computer Communications*, **161**, 279, 2020.

[17] De Alwis, C., Kalla, A., Pham, Q. V., Kumar, P., Dev, K., Hwang, W. J. and Liyanage, M., Survey on 6G Frontiers: Trends, Applications, Requirements, Technologies and Future Research. *IEEE Open Journal of the Communications Society*, **2**, 836, 2021.

[18] Mahmoud, H. H. H., Amer, A. A. and Ismail, T., 6G: A Comprehensive Survey on Technologies, Applications, Challenges, and Research Problems. *Transactions on Emerging Telecommunications Technologies*, **32**, e4233, 2021.

[19] Majumder, S., Ray, S., Sadhukhan, D., Khan, M. K. and Dasgupta, M, ESOTP: ECC-Based Secure Object Tracking Protocol for IoT Communication. *International Journal of Communication Systems*, **35**, e5026, 2021.

[20] Tataria, H., Shafi, M., Molisch, A. F., Dohler, M., Sjöland, H. and Tufvesson, F., 6G Wireless Systems: Vision, Requirements, Challenges, Insights, and Opportunities. *Proceedings of the IEEE*, **109**, 1166, 2021.

[21] Dong, W., Xu, Z. H., Li, X. X. and Xiao, S. P., Low-Cost Subarrayed Sensor Array Design Strategy for IoT and Future 6G Applications. *IEEE Internet of Things Journal*, **7**, 4816, 2020.

[22] Chatterjee, U. and Ray, S., Security Issues on IoT Communication and Evolving Solutions. *Soft Computing in Interdisciplinary Sciences*, **183**, 2022.

[23] Mukherjee, A., Goswami, P., Khan, M. A., Manman, L., Yang, L. and Pillai, P., Energy-Efficient Resource Allocation Strategy in Massive IoT for Industrial 6G Applications. *IEEE Internet of Things Journal*, **8**, 5194, 2020.

[24] Ji, B., Wang, Y., Song, K., Li, C., Wen, H., Menon, V. G. and Mumtaz, S., A Survey of Computational Intelligence for 6G: Key Technologies, Applications and Trends. *IEEE Transactions on Industrial Informatics*, **17**, 7145, 2021.

[25] Adhikari, S., Ray, S. (2019). A Lightweight and Secure IoT Communication Framework in Content-Centric Network Using Elliptic Curve Cryptography. In: Khare, A., Tiwary, U., Sethi, I., Singh, N. (eds) *Recent Trends in Communication, Computing, and Electronics*. Lecture Notes in Electrical Engineering, vol. 524. Springer, Singapore. https://doi.org/10.1007/978-981-13-2685-1_21

[26] Letaief, K. B., Shi, Y., Lu, J. and Lu, J., Edge Artificial Intelligence for 6G: Vision, Enabling Technologies, and Applications. *IEEE Journal on Selected Areas in Communications*, 40, 5, 2021.

[27] Ray, S. and Biswas, G. P., Design of Mobile Public Key Infrastructure (M-PKI) Using Elliptic Curve Cryptography. *International Journal on Cryptography and Information Security (IJCIS)*, 3, 25, 2013.

Chapter 6

Cyber-Physical System in AI-Enabled Smart Healthcare System

Jyoti Srivastava and Sidheswar Routray

Chapter Contents

6.1 Introduction

The last few years of the pandemic demanded a very strong as well as Smart Healthcare System being able to monitor each and every patient efficiently in order to handle the outbreak of rapidly increasing diseases [1]. Smart healthcare system enables patients as well as doctors to monitor and access healthcare data from a remote location without physically visiting the hospital. This automation in healthcare can be provided using Artificial-Intelligence-based IoMT framework [2]. These wide developments and automation in healthcare use CPSs to enhance the functionality of various physical and cyber medical components. Cyber-Physical System acts as a developing technology that is exponentially gaining demand every day in the research domain. Cyber-Physical System in SHS acts as an interface between the cyber and physical world in order to enhance medical treatments as shown in Figure 6.1. CPS is required to interconnect medical equipment (physical world) for patient monitoring and medical data (cyber world) analysis which is stored in the cloud [3].

CPS is an engineered system that is made up of continuously integrated computational algorithms and physical components. Besides healthcare, CPS is having wide applications in different domains like aerospace, civil infrastructure, energy, entertainment, manufacturing, and transportation [4] as shown in Figure 6.2. Collaborating the concept of IoT (Internet of Things) and IoS (Internet of Service) together with Cyber-Physical System brings the idea of "Industrial Internet" that is evolved in the United States and the Germany innovation

DOI: 10.1201/9781003376712-6

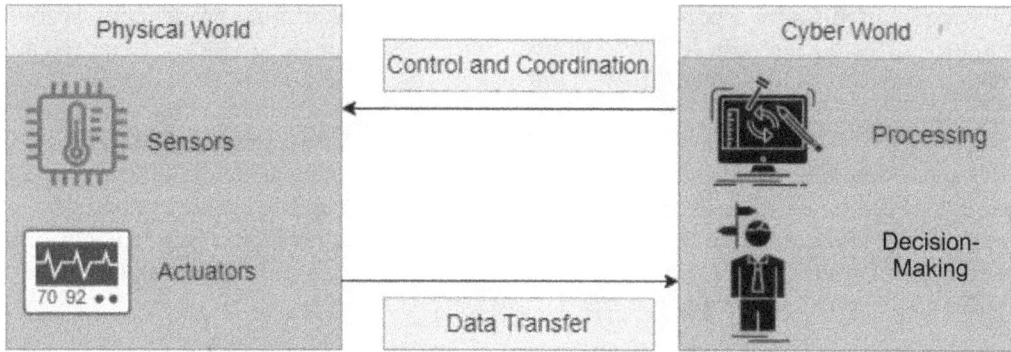

Figure 6.1 Cyber-physical system overview.

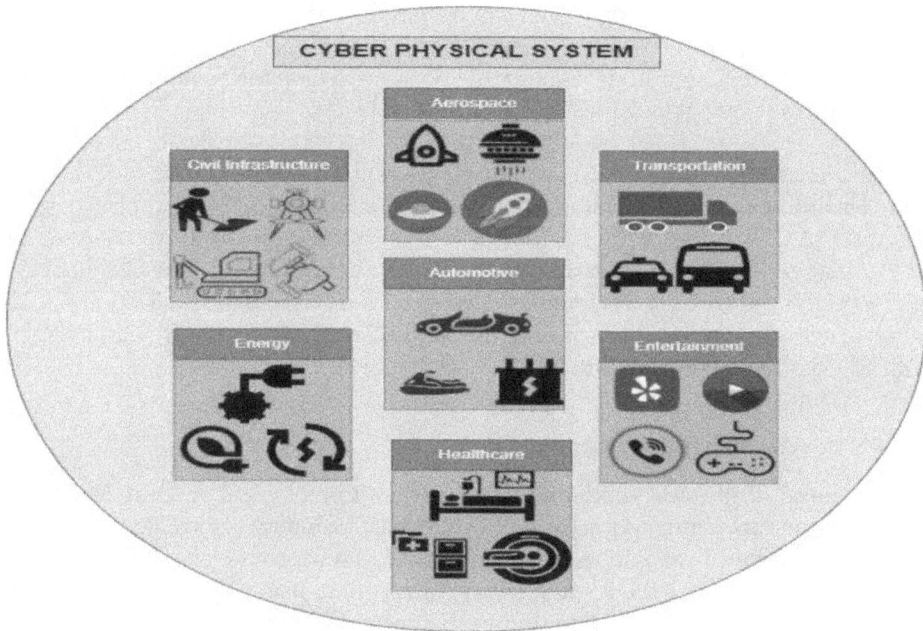

Figure 6.2 Cyber-physical system in different domains.

"Industry 4.0" [5]. CPS also plays an important role in transportation to enhance control-lability, competency, and consistency of physical equipment involved in the maintainability of transportation systems, e.g., vehicle accident prevention techniques and zero-net power creation systems [6].

Industry Foundation Classes (IFCs) is the traditional schema for monitoring and control-ling the structural health in civil infrastructure, and CPS here brings a further extension of IFC with enhanced automation facilities [7]. Aerospace cyber-physical systems are made-up of various intelligent equipment to collect, process, and communicate data with better reliability, security, and efficiency in order to accomplish advanced tasks and innovations [8]. In order to store and process huge amounts of data generating everyday, cloud-based

CPS can be used for incorporating Virtual Machines (VMs) for more energy conservation and thereby better QoS [9]. Automotive Engineering also uses CPS to secure advanced automobile inventions like cyber cars with intelligent automotive control units like Electronic Control Units (ECUs) [10].

Sighting the wide applications of CPS in different domains, we are motivated to incorporate CPS in AI-enabled Smart Healthcare Systems to enhance the functionality of different physical and cyber medical components that will help the society to be updated about their health parameters and also consult doctors remotely whenever required. It will also help to control the spreading of harmful viruses due to regular hospital visits for each and every small problem.

The major contributions of this research work are as follows:

i. To evaluate multiple CPS-engaged Smart Healthcare Systems architecture.
ii. To portray various applications of CPSs in SHS and characterize various CPS applications in Smart Healthcare Systems.
iii. To discover various technologies that can be used in CPS to enhance the performance of SHS.
iv. To compare various algorithms used for medical status monitoring utilizing CPS for SHS.
v. To plan various research challenges in CPS for SHS.

The rest of the chapter is organized as follows: Section 6.2 represents an overview of CPS in Smart Healthcare Systems including different characteristics of CPS in SHS, mostly used architectures of CPS in Smart Healthcare Systems, monitoring medical status using CPS in Smart Healthcare, and discusses various technologies that can be used for CPS in SHS. Section 6.3 explains various Big data analysis techniques in CPS for SHS. Section 6.4 shows digital security techniques in CPS for SHS. Section 6.5 explains various research challenges that can be encountered during the design of CPS in SHS.

6.2   CPS in AI-Enabled Smart Healthcare

CPS is composed of physical space and cyberspace where physical space includes various sensors and actuators contemporarily, and cyberspace includes various algorithms to analyze and predict the medical data received from medical sensors as elaborated in Figure 6.3. The bottommost layer, i.e., the sensing layer, collects data using various sensors from the human body and then stores this data to the cloud using suitable network transmission medium. Data from the cloud is then processed and analyzed in the cyberspace of CPS using various smart tools and techniques. Now the analyzed data can be used for various predictions and user-end experiences as shown in Figure 6.3.

Continuous work on CPS introduces various emerging technologies for pervasive sensing. Due to these inventions in the last few months, CPS becomes an extremely popular concept, where the original definitions have been extended and new concepts have been added. To understand the recent features and composition of CPS in this section, we will elaborate on different features of CPS in AI-enabled Smart Healthcare.

6.2.1   Characteristics of CPS in AI-Enabled SHS

Besides the cyber world, CPS is also composed of different physical components ranging from minor smart equipment like sensors or actuators to large equipment that can be controlled by smart devices (like smartphones). Depending on the working situations of

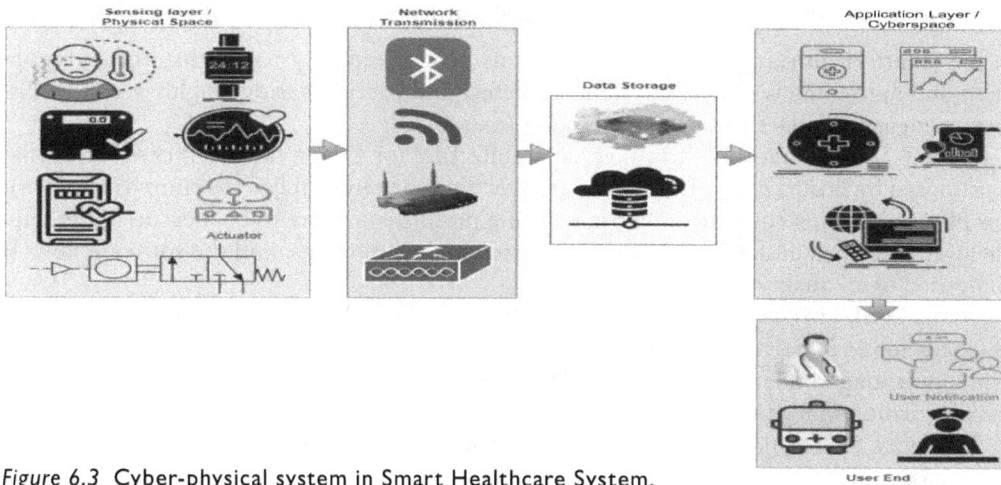

Figure 6.3 Cyber-physical system in Smart Healthcare System.

different equipment involved within CPS, the overall health of CPS can be defined, which is divided into four states, namely, critical, healthy, unhealthy, and nonworking [11]. The healthy state shows that the task is completed on time, the unhealthy state shows that the parameters are within the threshold value, the critical state represents failures within the system beyond a critical limit, and nonworking state shows faults and failures. Input and output to these physical devices can be of different formats, and they can be structured or unstructured; hence interfaces are required to convert these data into a fixed format that will be widely accepted [12]. CPS in AI-enabled Smart Healthcare Systems can be categorized among various levels depending on the requirement and utilization of these medical data representing different characteristics of CPS in Smart Healthcare.

- **Unit-level CPS:** This is the basic level of CPS in Smart Healthcare and used to monitor and control the health parameters of various patients admitted in different sections of the hospital. Continuous monitoring of the patient's physiological condition (like heart rate, blood pressure, body temperature) is required at the unit level, and then these healthcare data are transmitted to different smart systems for analysis and managing different sensors and actuators linked with the patient. Here, the healthcare staff also plays an important role to support and inform the experts about immediate patient treatment when required [1].
- **Integration-level CPS:** In this stage, various hospitals collaborate with smart homes for delivering healthcare services to various patients remotely. It can later integrate with smart ambulances for continuous monitoring and making necessary emergency arrangements at hospitals for transferring risky patients to hospitals whenever required [1].
- **System-level CPS:** At this stage, different autonomous systems support CPS in Smart healthcare thereby forming Healthcare Cyber-Physical System in Smart City. Smart hospitals, smart homes, smart ambulances, etc., will form a healthcare ecosystem providing rich, professional, and personalized healthcare services to patients [1].
- **Acceptance-level CPS:** At this stage, different researchers, engineers, academicians, scientists, and health experts collaborate to make an effective SHS based on different policies and standards [1].
- **Evolutionary-level CPS:** This stage is the ideal future CPS in Smart Healthcare having advanced properties like self-adaptability and self-manageability [1].

6.2.2 Architectures of CPS in Smart Healthcare

In the Smart Healthcare System, CPS works as an embedded system of healthcare equipment and autonomous network systems to detect and prevent various highly sensitive diseases such as COVID-19 as explained in Figure 6.4.

Figure 6.5 shows another three-layered architecture for CPS in the healthcare sector that is proposed by Bordel et al. [14]. In this architecture, the physical layer is composed of various physical sensing equipment whose working performances are to be measured, the middle layer is formed of different controlling hardware components, and the software layer is composed of all analysis techniques.

Fatima et al. [15] explained a general architecture that elaborates various architectural practices for IoT and CPS as shown in Figure 6.6; the interoperability among these two fields indicating similar processes like data preprocessing, element analysis, etc.; and the transformation that can be performed from IoT to CPS and vice versa.

Zhan et al. [16] again proposed a three-layered architecture where the bottommost layer is the data collection layer which is composed of different nodes and adapters that can be used to provide a combined interface for collecting diverse data from different hospitals, individual users, or from the Internet. Adapters here can be used to preprocess the collected raw data having different structures and formats before transmitting it to the data management layer. The next layer is the data management layer which is mainly composed of two modules: The DFS module (Distributed File Storage module) and the DPC module (Distributed Parallel Computing module) supported through various Big data techniques for enhancing data storage and processing capabilities of the system. Now, the topmost

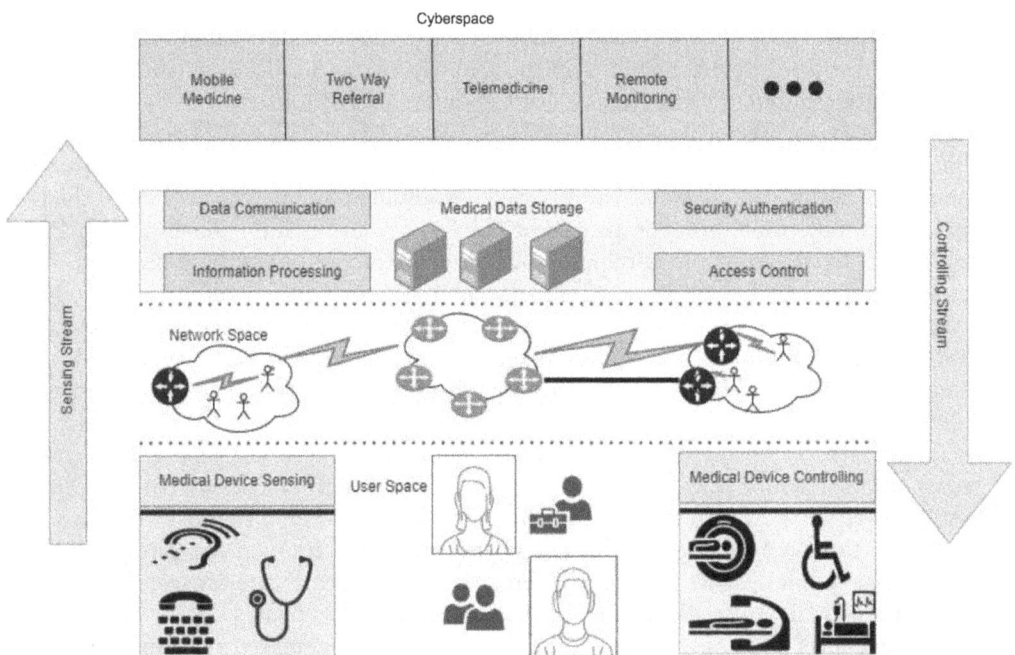

Figure 6.4 Medical cyber-physical system architecture [13].

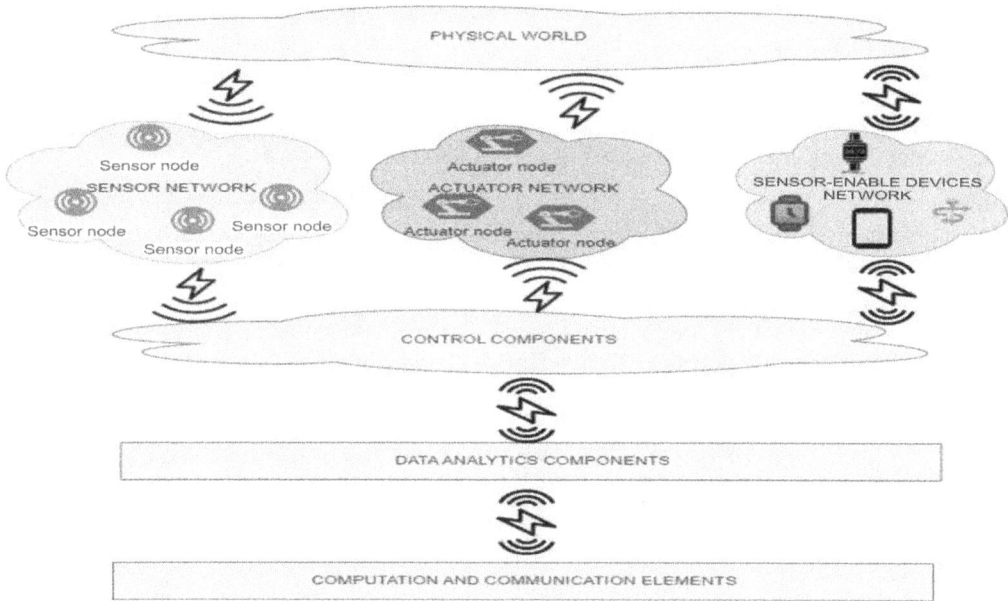

Figure 6.5 Architecture of CPS in pervasive sensing [14].

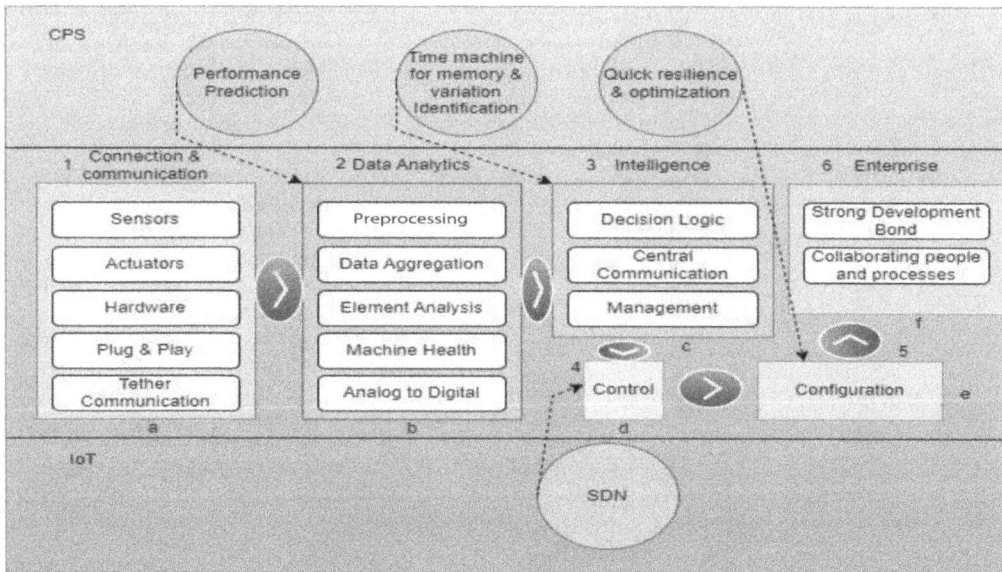

Figure 6.6 Architectural dataflow indicating same and different processes between CPS and IoT [15].

layer is the Application Service layer which enables various user-centric APIs to show the basic data analysis results. It will help to provide general user-adaptable and specialized healthcare services.

6.2.3 Monitoring Medical Status Using CPS in Smart Healthcare

Time is the most important factor to check the reliability and performance of Cyber-Physical System in Smart healthcare. Minor delays of milliseconds in the treatment of a patient can result in a risky or critical situation; hence, it will thereby reduce the efficiency and reliability of the system. CPS is used for continuous monitoring and control capabilities and thereby introduces a new term known as Health Monitoring Status (HMS). HMS can be used for a double purpose: To monitor the condition of each physical component of CPS and to perform a proactive approach regarding the feasibility of CPS components [1]. Monitoring the medical status of patients is very important mainly after a surgery or even for the routine activity of risky patients. Amin et al. [17] discusses various CPS solutions designed for e-diagnostics and monitoring high-risk pregnancy, elderly people, and edema patients as shown in Table 6.1.

Table 6.1 CPS Solutions for Health Monitoring

Sr. No	Study	Model Features	Functionality
1	Suzuki et al. [20]	- Wireless multipurpose sensors in surroundings - Patients position detection using fuzzy inference - Various pill box sensors - Web interface	- Estimated timings can be measured for medicine intake - Medicine incorrect intake dosage detection - Patient monitoring
2	Wu et al. [18]	- Standard agitation semantics - Decision graph for analysis - Drug package scanning through RFID tags	- Automation in notification frequency calculation - Reminder notification for medication - Detection of various patient activities - Keeping track of medicine consumption
3	Varshney [19]	- Wearable sensors - Patient behavior monitoring	- Monitoring of patient's health - Reminder for medication-related stuff
4	Ishak et al. [21]	- Web-based database management - Smart medicine cabinet linked with microcontroller	- Automatic detection and monitoring of medicine intake
5	Chen et al. [22]	- Wearable wrist band sensors	- Patient's motion detection
6	Kalantarian et al. [23, 24]	- Piezoelectric sensors attached in Smart necklace - Wearable sensors in wrist bands - Ultra low-power consuming Bluetooth protocol for android application - Bayesian networks	- Automatic pills' intake detection - Different swallow's classification and their display - Patient movement detection

6.2.4 Cyber-Physical Systems Technologies in Smart Healthcare System

Building a cyberinfrastructure having automation requires CPS to integrate with various advanced technologies in order to enhance performance. These technologies include Digital Twin, IoT, Blockchain, AI and ML, cloud computing etc., as explained in Table 6.2.

Table 6.2 CPS Technologies in SHS

Sr. No.	Technology	Study	Application Areas	Application in SHS
1	IoT	Verma [1] Wang et al. [4] Maleh et al. [8] Bordel et al. [14] Fatima et al. [15] Amin et al. [17]	- Aerospace technology - Control theory - Interoperability and transformation with CPS - Smart homes	- Machine to machine communication - Intrusion detection - Pervasive sensing - Data integrity - Confidentiality
2	Digital Twin	Khan et al. [25] Liu et al. [26] Verma [1]	- Data management - Industrial manufacturing - Precision analysis - Cloud data monitoring	- Medical data analysis and prediction - Elderly care - Real-time supervision - Resource graph creation
3	Robotics	Terashima et al. [27] Khan et al. [28] D'Auria et al. [29] Munir et al. [10] Verma [1]	- Automotive - Engineering - Smart homes - Manufacturing task in Industry 4.0 - Surveillance robots for security and safety	- Health monitoring - Medication adherence - Patient care - Surgery assistance
4	AI and ML	Maleh et al. [8] Wu et al. [18] Srivastava et al. [2] Wang et al. [33] Verma [1]	- Aerospace technology - Context-Aware Smart System - Mobility awareness in smart transportation - Medical engineering	- Task automation - Prediction Analysis - Decision-making - Intrusion detection - Medication adherence - Diagnosis
5	Blockchain	Nguyen et al. [30] Gupta et al. [31] Dedeoglu et al. [32] Verma [1]	- Security in smart homes - Robustness in CPS - Permission and access control management	- Security - Privacy - Authentication - Encryption
6	Cloud Computing	Qi et al. [9] Zhang et al. [16] Gupta et al. [31] Verma [1]	- QoS - Energy conservation - Virtual machine scheduling - Industry 4.0 standards for cloud as service, control as a service, machinery as a service, etc.	- Resource control and allocation - Easy access and sharing of electronic health record. - Distributed ledger for communication
7	Big Data Analytics	Zhang et al. [16] Wang et al. [33] Verma [1]	- Data Storage - Distributed systems - Smart transportation - Medical engineering	- Service-oriented architecture - Heterogenous data management - Reduce medical cost - Personalized medical services

In Smart Healthcare Systems, Digital Twin is basically a simulation model that provides virtual resources in cyberspace to plan, control, and coordinate according to the digital information received from the resources [25]. IoT can be helpful for machine-to-machine (M2M) communication using wireless communication, Bluetooth, radio communication etc., that is forwarded to cloud or remote servers for further controlling and analysis via AI and ML prediction algorithms [1]. In CPS for Smart healthcare different medical devices, M2M communication within diverse sensors continuously collects huge data that can be in different formats so it needs to be processed, stored, and analyzed for receiving medical information, and this is the situation where Big Data Analytics comes into its role [33].

CPS for Smart Healthcare can use cloud computing to digitally store the Electronic Health Records (EHR) at the cloud server in an encrypted format to enable easy sharing and access among various entities like patients, doctors, hospital staff, and insurance companies [9]. Besides storing encrypted data, Industry 4.0 standards introduce various advance service models for cloud manufacturing as explained in Table 6.2 [31]. Robots are basically autonomous machines that can perform assigned tasks with proper accuracy [27]. In CPS for Smart Healthcare Systems, medical robots can be used for assisting in patient care and surgery. Industrial robots can be used for manufacturing tasks and surveillance [28]. Blockchain technology helps in providing a decentralized and distributed database platform for secured and authenticated access to EHR maintained by cloud servers [31].

6.3   Big Data Analytics in CPS for Smart Healthcare System

Recent advancements in healthcare technologies envision automation in the cyber as well as physical world of Smart Healthcare Systems, but these advancements result in an exponential increase of healthcare data which is raising problems against the storage, handling, and processing of these huge datasets. SHS deals with different kinds of devices that are continuously generating data in different formats and accordingly require different storage schemes and formats and thereby introducing Big data problems. Big data analytics will be beneficial in reducing medical costs and also ensures personalized experiences as shown in Table 6.3. Zhang et al. [16] introduced a patient-centric healthcare CPS for advanced applications and services which is named Health-CPS that is managed by cloud and Big data techniques for better convenience. Sarosh et al. [35] introduces a cryptosystem using the logistic equation, Hyperchaotic equation, and Deoxyribonucleic Acid (DNA) encoding for maintaining the overall security of the system and uses a lossless Computational Secret Image Sharing (CSIS) technique to convert encrypted image into shares for distributed storage of Big data on multiple servers. Syed et al. [36] proposed a smart healthcare framework that predicts around 12 physical activities with 97.1% accuracy that will help to monitor human motion, especially for elderly people using IoMT and an intelligent machine learning algorithm. Figure 6.8 depicts the classifier performance in terms of data precision and recall evaluation when the person is performing various physical activities. Bansal et al. [37] combined the concepts of nanomaterial, IoT, and Big data for Electrocardiogram (ECG) monitoring to enhance system properties like conductivity, toxicity, electrical properties, reduced cost, and integration complexities. Here, Big data analytics is used for ECG data analysis, decision-making, and extracting useful information via algorithms and intelligent storage. Manogaran et al. [38] proposes a new architecture that is capable of processing and storing large scalable sensor data (Big data) for SHS and will be capable of predicting heart diseases. Figure 6.7 shows the performance

Table 6.3 Big Data Analytics in CPS for SHS

Sr. No	Study	Model Techniques and Features	Functionality
1	Zhang et al. [16]	- CPS - Distributed storage - Parallel computing - Cloud computing	- Patient-centric Healthcare applications and services - Better conveniency
2	Sarosh et al. [35]	- logistic equation - Hyperchaotic equation - Deoxyribonucleic acid - Computational Secret image sharing	- Security framework - Distributed storage - Cloud-based cryptosystem - Strong key sensitivity
3	Syed et al. [36]	- Parallel computing - Hadoop MapReduce techniques - Multinomial naïve bayes classifier	- Patient's activity monitoring specially for elderly people - High scalability - Overall accuracy 97.1 %
4	Bansal et al. [37]	- Nanoelectronics - IoT - Big data	- Patient ECG monitoring - Remote access by doctors - Helpful in dealing with emergency situations
5	Manogaran et al. [38]	- Meta-Fog redirection (Apache Pig, Apache HBase) - Grouping and choosing architecture (Data categorization function)	- Key management and security services - Predict heart diseases - Secure fog to cloud integration

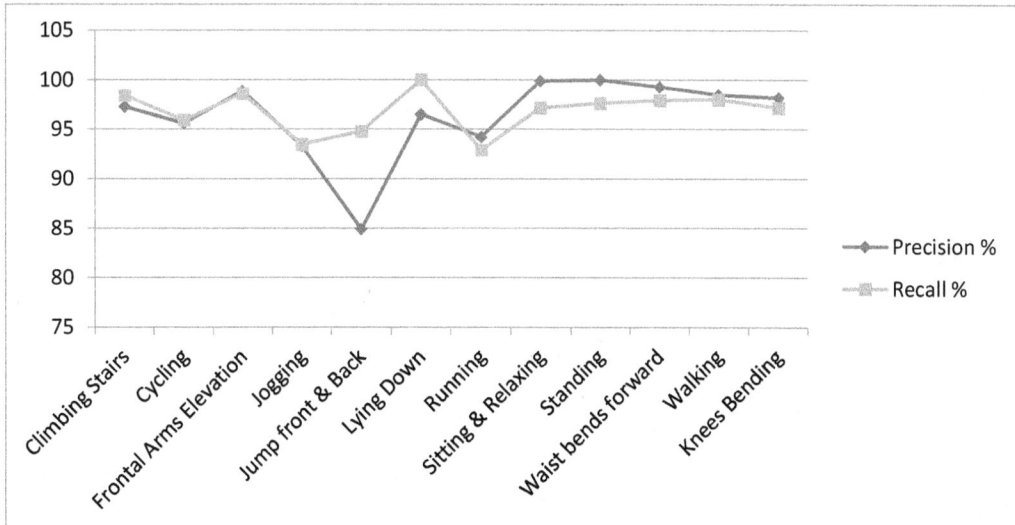

Figure 6.7 Classifier performance evaluation for various physical activities using precision and recall evaluation [36].

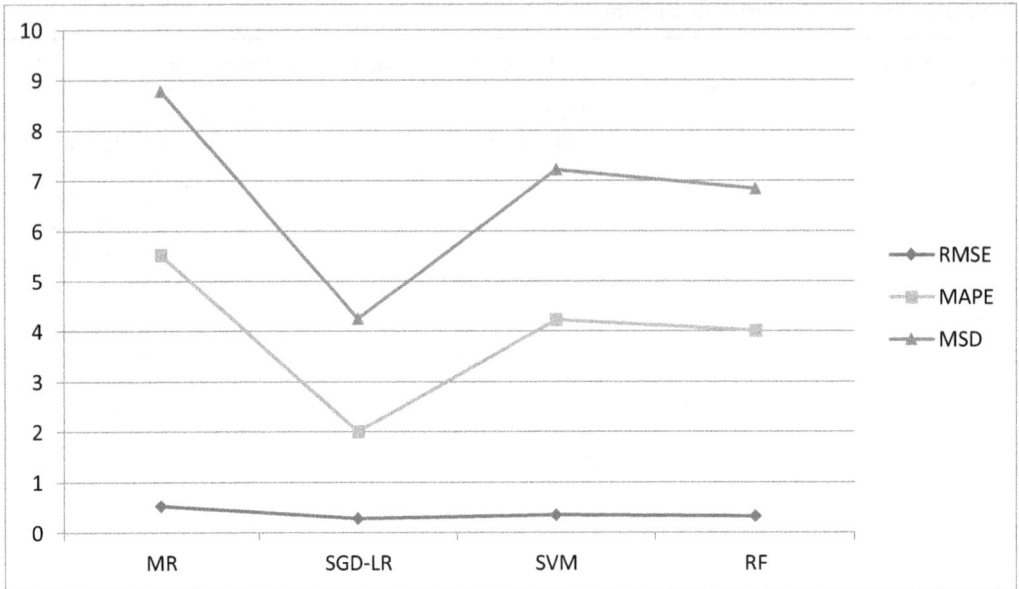

Figure 6.8 Performance comparison of stochastic gradient descent algorithm with Logistic Regression (SGD-LR), Multiple Regression (MR), Support Vector Machine (SVM), and Random Forest (RF) [38].

comparison of the proposed algorithm with various other Big data techniques, namely Multiple Regression (MR), Support Vector Machine (SVM), and Random Forest (RF). The proposed architecture is basically composed of two main subsystems, first Meta-Fog redirection that uses Big data techniques like Apache Pig and Apache HBase for collection and storage of the collected sensor data, and second Grouping and Choosing architecture for providing category-wise management and security services [44, 45].

6.4   Digital Security in CPS for Smart Healthcare System

CPS in Smart healthcare systems is emerging as an advanced patient-centric model that enables real-time healthcare data collection using various types of machinery equipped with smart sensors, encrypted data aggregation at mobile devices, and then transmitting this encrypted data to the cloud for storage and accessibility. Here, digital security of this healthcare data becomes a very challenging problem, and to solve such problems, few security techniques have been discussed in Table 6.4. To address the authentication of legitimate patient-wearable devices with secured data transmission, Adil et al. [39] proposed a crossbreed lightweight supervised machine learning (SML) decentralized authentication technique pursued by CPBE&D (cryptographic parameter-based encryption and decryption) scheme. Most of the current secured healthcare systems with the highest security are unable to maintain their efficiency under resource-constrained systems, and to deal with that situation, Kumar et al. [40] proposed a balanced, secured, and lightweight smart healthcare CPS that preserves the system privacy and maintains patient data integrity on the cloud with public data verifiability utilizing least communication and computation cost. To enhance the responsiveness of SHS, Ramasamy et al. [41] proposed an AI-enabled IOT-CPS algorithm which is having two sub-algorithms, namely, the disease classification

Table 6.4 Digital Security Techniques in CPS for SHS

Sr. No	Model Techniques and Features	Model Techniques and Features	Functionality
1	Adil et al. [39]	- Supervised Machine Learning (SML) - Cryptographic Parameter-Based Encryption and Decryption (CPBE&D)	- Validation of legal patient wearable devices - Decentralized Authentication - Secure data transmission
2	Kumar et al. [40]	- *Secure and lightweight smart healthcare CPS*	- Secured against chosen ciphertext attack - Secured against chosen message attack - Least communication and computation to achieve efficiency - Preserves system privacy - Healthcare data integrity on cloud
3	Ramasamy et al. [41]	- Disease classification algorithm - Disease prediction algorithm - AI–enabled IoT–CPS algorithm	- High accuracy - Precise result - Automated fall detection system - Low computation time
4	Challa et al. [42]	- Burrows–Abadi–Needham logic (BAN logic) - Automated Validation of Internet Security Protocols and Applications (AVISPA)	- Authentication between cloud server and user - Authentication between cloud and smart meter
5	Yang et al. [43]	- Lightweight Sharable and Traceable (LiST) algorithm - Pairing based cryptography	- End-to-end patient's data encryption - Efficient keyword search system - Supports tracing of traitors - Allows on-demand user revocation

algorithm to achieve high accuracy and precision in identifying the disease within minimum time and disease prediction algorithm for fall prediction of elderly patients. Challa et al. [42] proposed Burrows–Abadi–Needham logic (BAN logic) for providing authentication between a cloud server and a user followed by Automated Validation of Internet Security Protocols and Applications (AVISPA) tool to ensure authentication between the cloud server and smart meter. Figure 6.9 compares the total computation overhead of proposed algorithm with various other schemes like Remote User Authentication (RUA), Mutual Authentication Scheme (MAS), Ad-hoc Authentication Scheme (AAS), and Lightweight Authentication Protocol for IoT (LAP-IoT). For providing efficient and scalable sharing of encrypted data, Yang et al. [43] proposed a LiST (lightweight Sharable and Traceable) highly secured mobile health system that ensures end-to-end encryption of patient data for a mobile device to. the end user. This system also provides on-demand user abrogation, supports traitors' tracing and fine-grained control in data access, and enables efficient keyword search.

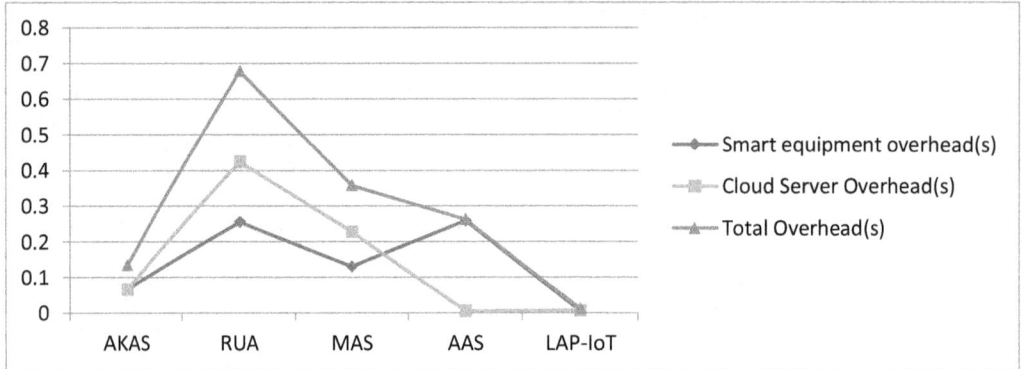

Figure 6.9 Performance comparison in terms of total computation overhead for Authenticate KeyAgreement Scheme (AKAS), Remote User Authentication (RUA), Mutual Authentication Scheme (MAS), Ad-hoc Authentication Scheme (AAS), and Lightweight Authentication Protocol for IoT (LAP-IoT) [42].

6.5   Application of SHS in Real-Time Scenario

CPS in Smart Healthcare Systems has various applications among which Joshi et al. [46] proposed a system used for monitoring diabetes levels among different COVID patients. As physical visits to such patients are prohibited, this technology helps the doctor and the patients to examine the periodic glucose level among COVID-19 patients under medication.Tiwari et al. [47] proposed a system that uses a plethysmography technique for pulse rate detection and digitally displays the results that make it a real-time health monitoring device. Here, the flow-based cardiac output can be calculated using finger blood flow examination with mobile light and cameras, which is proven comparatively safe for patients. Uddin [48] proposed a system that allows users to check their heart rate by just looking into the phone without thorough checking via hand. This system is the next step toward SHS, but it requires continuous heart monitoring as all the controlling of the cardiovascular disease sensing system is handled through smartphones. Malla et al. [49] proposed a microwave imaging-based technique to investigate and safely detect breast tumor among various patients. Soundararajan et al. [50] used BSN (body sensor network) to analyze symptoms of Parkinson's disease, and the early result displayed will help the doctor to take immediate action wherever required. It will be beneficial for the patients as well, as they can take preliminary actions if the disease is at an initial stage. Balasundaram et al. [51] proposed IoT-based SHS for efficient health monitoring and symptom diagnosis of patients in the emergency care unit. Sathish Kumar et al. [52] proposed an automated system that will automatically detect bone fractures in the human body with the help of extensively deep evaluation and experimental analysis on various available X-ray images.

6.6   CPS in Smart Healthcare Challenges

CPS in Smart Healthcare Systems brings various automation facilities with the help of various CPS technologies like Artificial Intelligence, Blockchain, Big data, and cloud computing. Despite such advance technologies, there are numerous challenges that developers and designers of cyber-physical systems still face during the implementation of a precise and accurate CPS in the healthcare sector, and a few of them are discussed below.

- **Instant Requirement and Time Management:** Data in cyber-physical system travels through various levels to reach up to the end user, and all these transmissions generally suffer from various real-world problems like packet loss, transmission delay, and network jitter, which lead to the meantime unavailability of CPS [34]. As we know, Smart Healthcare System deals with very sensitive patients and their instant requirements, thereby delays in taking action by caretakers due to CPS failure can adversely affect the patient's health. Incorporating such kind of real-world data with a patient medical history including complete field knowledge is the most challenging task of CPS, but it is the most required task for a smart health monitoring system.
- **Continuous Components' Coordination:** CPS architecture incorporates various components within different layers: Physical layer includes various sensors and actuators, the network layer includes various gateways and hubs, and the cyber layer includes processing and analysis techniques etc., and all these layers and components need continuous coordination and communication to accomplish the overall system task [12]. This is the most challenging task for CPS as different components can have different computational capabilities in terms of energy, efficiency, accuracy, precision etc., and this constraint leads to inconsistent information for treatment, signal distortion due to the close proximity, and various other such problems.
- **Data Diversity:** Data in Cyber-Physical System can be generated either from the physical resources or the cyber resources which are having completely diverse environments and data formats that need to be maintained and processed in a common platform to give the required information to the end user. Besides that, there is an exponential increase in the generation of these kinds of data, which again creates maintenance and management problems [33]. Though cloud computing and Big data techniques solved many such problems, security constraints of CPS again raise a big question about these techniques. Hence, diverse data management for CPS in the Smart healthcare system must be implemented in such a way that the patient information privacy scheme should not get violated.
- **Dynamic Aspects of CPS:** Dynamic aspects of a cyber-physical system aim for a system that must have a simple, collaborative, and friendly user interface so that different users can easily relate with the system and can easily communicate within internal healthcare units as per the real-time change in patients' medical status. The system should also provide e-learning techniques for the Smart healthcare system that include actions that are to be taken in different medical situations. Now the problem arises when such facilities degrade system effectiveness in terms of various parameters like energy, efficiency, accuracy, and security [13]. Therefore, a balanced UI must be developed that maintains both system performance and user comfortability.

6.7 Conclusion

CPS includes the physical as well as the cyber components of a system where physical components include various equipment with inbuilt sensors for data collection, and cyber components include various data-handling schemes that enhance the overall performance of the system with various automation techniques. CPS works within different domains like manufacturing, civil engineering, and aerospace. CPS plays an important role in the Smart Healthcare System in terms of enhanced security, efficiency, and accuracy from the system. CPS utilizes various high-performance technologies like cloud computing, Big data, and robotics for the betterment of the system performance. Big data analytics enables the processing and handling of the rapidly increasing healthcare dataset generated from different components of CPS having heterogeneous formats and environments, ensuring a personalized experience with reduced medical costs. This work will help an individual to discover different CPS characteristics and applications in various domains, technologies working with CPS, Big data analytics with CPS, digital security in CPS, and the research challenges in real-time implementation of CPS in Smart Healthcare Systems.

References

[1] Verma, Rupali. "Smart city healthcare cyber physical system: Characteristics, technologies and challenges." *Wireless Personal Communications* 122, no. 2 (2022): 1413–1433.

[2] Srivastava, Jyoti, Sidheswar Routray, Sultan Ahmad, and Mohammad Maqbool Waris. "Internet of Medical Things (IoMT)-based smart healthcare system: Trends and progress." *Computational Intelligence and Neuroscience* 2022 (2022).

[3] Balakrishnan, Amutha, Ramana Kadiyala, Gaurav Dhiman, Gokul Ashok, Sandeep Kautish, Kusum Yadav, and J. Maruthi Nagendra Prasad. "A personalized eccentric cyber-physical system Architecture for smart healthcare." *Security and Communication Networks* 2021 (2021): 1–36.

[4] Wang, Lihui, Xi Vincent Wang, Lihui Wang, and Xi Vincent Wang. "Latest advancement in CPS and IoT applications." *Cloud-Based Cyber-Physical Systems in Manufacturing* (2018): 33–61.

[5] Dafflon, Baudouin, Nejib Moalla, and Yacine Ouzrout. "The challenges, approaches, and used techniques of CPS for manufacturing in Industry 4.0: A literature review." *The International Journal of Advanced Manufacturing Technology* 113 (2021): 2395–2412.

[6] Xiong, Gang, Fenghua Zhu, Xiwei Liu, Xisong Dong, Wuling Huang, Songhang Chen, and Kai Zhao. "Cyber-physical-social system in intelligent transportation." *IEEE/CAA Journal of AutomaticaSinica* 2, no. 3 (2015): 320–333.

[7] Smarsly, Kay, Michael Theiler, and Kosmas Dragos. "IFC-based modeling of cyberphysical systems in civil engineering." In *Proceedings of the 24th International Workshop on Intelligent Computing in Engineering (EG-ICE)*. Nottingham, UK, vol. 7, no. 10, 2017.

[8] Maleh, Yassine. "Machine learning techniques for IoT intrusions detection in aerospace cyber-physical systems." *Machine Learning and Data Mining in Aerospace Technology* (2020): 205–232.

[9] Qi, Lianyong, Yi Chen, Yuan Yuan, Shucun Fu, Xuyun Zhang, and Xiaolong Xu. "A QoSaware virtual machine scheduling method for energy conservation in cloud-based cyberphysical systems." *World Wide Web* 23 (2020): 1275–1297.

[10] Munir, Arslan, and Farinaz Koushanfar. "Design and analysis of secure and dependable automotive CPS: A steer-by-wire case study." *IEEE Transactions on Dependable and Secure Computing* 17, no. 4 (2018): 813–827.

[11] Shangguan, Lantian, and Swaminathan Gopalswamy. "Health monitoring for cyber physical systems." *IEEE Systems Journal* 14, no. 1 (2019): 1457–1467.

[12] Napoleone, Alessia, Marco Macchi, and Alessandro Pozzetti. "A review on the characteristics of cyber-physical systems for the future smart factories." *Journal of Manufacturing Systems* 54 (2020): 305–335.

[13] Chen, Fulong, Yuqing Tang, Canlin Wang, Jing Huang, Cheng Huang, Dong Xie, Taochun Wang, and Chuanxin Zhao. "Medical cyber—physical systems: A solution to smart health and the state of the art." *IEEE Transactions on Computational Social Systems* 9, no. 5 (2021): 1359–1386.

[14] Bordel, Borja, Ramón Alcarria, Tomás Robles, and Diego Martín. "Cyber—physical systems: Extending pervasive sensing from control theory to the Internet of Things." *Pervasive and Mobile Computing* 40 (2017): 156–184.

[15] Fatima, Iqra, Saif U. R. Malik, Adeel Anjum, and Naveed Ahmad. "Cyber physical systems and IoT: Architectural practices, interoperability, and transformation." *IT Professional* 22, no. 3 (2020): 46–54.

[16] Zhang, Yin, Meikang Qiu, Chun-Wei Tsai, Mohammad Mehedi Hassan, and Atif Alamri. "Health-CPS: Healthcare cyber-physical system assisted by cloud and big data." *IEEE Systems Journal* 11, no. 1 (2015): 88–95.

[17] Amin, Sara, Tooba Salahuddin, and Abdelaziz Bouras. "Cyber physical systems and smart homes in healthcare: Current state and challenges." In *2020 IEEE International Conference on Informatics, IoT, and Enabling Technologies (ICIoT)*, pp. 302–309. IEEE, 2020.

[18] Wu, Qiong, Zhiwei Zeng, Jun Lin, and Yiqiang Chen. "AI empowered context-aware smart system for medication adherence." *International Journal of Crowd Science*, vol. 1, no. 2, pp. 102-109, June 2017.

[19] Varshney, Upkar. "Smart medication management system and multiple interventions for medication adherence." *Decision Support Systems* 55, no. 2 (2013): 538–551.

[20] Suzuki, Takuo, and Yasushi Nakauchi. "Intelligent medicine case for dosing monitoring: Design and implementation." *SICE Journal of Control, Measurement, and System Integration* 4, no. 2 (2011): 163–171.

[21] S. Ishak, H. Z. Abidin, and M. Muhamad, "Improving medical adherenceusing smart medicine cabinet monitoring system." *The Indonesian Journal of Electrical Engineering and Computer Science* 9, no. 1 (2018).

[22] Chen, Chen, Nasser Kehtarnavaz, and Roozbeh Jafari. "A medication adherence monitoring system for pill bottles based on a wearable inertial sensor." In *2014 36th Annual International Conference of the IEEE Engineering in Medicine and Biology Society*, pp. 4983–4986. IEEE, 2014.

[23] Kalantarian, Haik, Nabil Alshurafa, Ebrahim Nemati, Tuan Le, and Majid Sarrafzadeh. "A smartwatch-based medication adherence system." In *2015 IEEE 12th International Conference on Wearable and Implantable Body Sensor Networks (BSN)*, pp. 1–6. IEEE, 2015.

[24] Kalantarian, Haik, Babak Motamed, Nabil Alshurafa, and Majid Sarrafzadeh. "A wearable sensor system for medication adherence prediction." *Artificial Intelligence in Medicine* 69 (2016): 43–52.

[25] Khan, Sagheer, Tughrul Arslan, and Tharmalingam Ratnarajah. "Digital twin perspective of fourth industrial and healthcare revolution." *IEEE Access* 10 (2022): 25732–25754.

[26] Liu, Ying, Lin Zhang, Yuan Yang, Longfei Zhou, Lei Ren, Fei Wang, Rong Liu, Zhibo Pang, and M. Jamal Deen. "A novel cloud-based framework for the elderly healthcare services using digital twin." *IEEE Access* 7 (2019): 49088–49101.

[27] Terashima, Kazuhiko, Kazuhiro Funato, and Takuyuki Komoda. "Healthcare robots and smart hospital based on human-robot interaction." In *Human-Robot Interaction Perspectives and Applications*. IntechOpen, 2022.

[28] Khan, Zeashan Hameed, Afifa Siddique, and Chang Won Lee. "Robotics utilization for healthcare digitization in global COVID-19 management." *International Journal of Environmental Research and Public Health* 17, no. 11 (2020): 3819.

[29] D'Auria, Daniela, and Fabio Persia. "A collaborative robotic cyber physical system for surgery applications." In *2017 IEEE International Conference on Information Reuse and Integration (IRI)*, pp. 79–83. IEEE, 2017.

[30] Nguyen, Gia Nhu, Nin Ho Le Viet, Mohamed Elhoseny, K. Shankar, B. B. Gupta, and Ahmed A. Abd El-Latif. "Secure blockchain enabled Cyber—physical systems in healthcare using deep belief network with ResNet model." *Journal of Parallel and Distributed Computing* 153 (2021): 150–160.

[31] Gupta, Brij B., Kuan-Ching Li, Victor C. M. Leung, Kostas E. Psannis, and Shingo Yamaguchi. "Blockchain-assisted secure fine-grained searchable encryption for a cloudbased healthcare cyber-physical system." *IEEE/CAA Journal of Automatica Sinica* 8, no. 12 (2021): 1877–1890.

[32] Dedeoglu, Volkan, Ali Dorri, Raja Jurdak, Regio A. Michelin, Roben C. Lunardi, Salil S. Kanhere, and Avelino F. Zorzo. "A journey in applying blockchain for cyberphysical systems." In *2020 International Conference on COMmunication Systems & NETworkS (COMSNETS)*, pp. 383–390. IEEE, 2020.

[33] Wang, Lidong, and Cheryl Ann Alexander. "Big data analytics in medical engineering and healthcare: Methods, advances and challenges." *Journal of Medical Engineering & Technology* 44, no. 6 (2020): 267–283.

[34] Shah, Tejal, Ali Yavari, Karan Mitra, Saguna Saguna, Prem Prakash Jayaraman, Fethi Rabhi, and Rajiv Ranjan. "Remote health care cyber-physical system: Quality of service (QoS) challenges and opportunities." *IET Cyber-Physical Systems: Theory & Applications* 1, no. 1 (2016): 40–48.

[35] Sarosh, Parsa, Shabir A. Parah, G. Mohiuddin Bhat, and Khan Muhammad. "A security management framework for big data in smart healthcare." *Big Data Research* 25 (2021): 100225.

[36] Syed, Liyakathunisa, Saima Jabeen, S. Manimala, and Abdullah Alsaeedi. "Smart healthcare framework for ambient assisted living using IoMT and big data analytics techniques." *Future Generation Computer Systems* 101 (2019): 136–151.

[37] Bansal, Malti, and Bani Gandhi. "IoT & big data in smart healthcare (ECG monitoring)." In *2019 International Conference on Machine Learning, Big Data, Cloud and Parallel Computing (COMITCon)*, pp. 390–396. IEEE, 2019.

[38] Manogaran, Gunasekaran, Ramachandran Varatharajan, Daphne Lopez, Priyan Malarvizhi Kumar, Revathi Sundarasekar, and Chandu Thota. "A new architect7ure of Internet of Things and big data ecosystem for secured smart healthcare monitoring and alerting system." *Future Generation Computer Systems* 82 (2018): 375–387.

[39] Adil, Muhammad, Muhammad Khurram Khan, Muhammad Mohsin Jadoon, Muhammad Attique, Houbing Song, and Ahmed Farouk, "An AI-Enabled Hybrid Lightweight Authentication Scheme for Intelligent IoMT Based Cyber-Physical Systems," in *IEEE Transactions on Network Science and Engineering,* vol. 10, no. 5, pp. 2719-2730, 1 Sept.-Oct. 2023.

[40] Kumar, Mahender, and Satish Chand. "A provable secure and lightweight smart healthcare cyber-physical system with public verifiability." *IEEE Systems Journal* 16, no. 4 (2021): 5501–5508.

[41] Ramasamy, Lakshmana Kumar, Firoz Khan, Mohammad Shah, Balusupati Veera Venkata Siva Prasad, Celestine Iwendi, and Cresantus Biamba. "Secure smart wearable computing through artificial intelligence-enabled internet of things and cyber-physical systems for health monitoring." *Sensors* 22, no. 3 (2022): 1076.

[42] Challa, Sravani, Ashok Kumar Das, Prosanta Gope, Neeraj Kumar, Fan Wu, and Athanasios V. Vasilakos. "Design and analysis of authenticated key agreement scheme in cloud-assisted cyber—physical systems." *Future Generation Computer Systems* 108 (2020): 1267–1286.

[43] Yang, Yang, Ximeng Liu, Robert H. Deng, and Yingjiu Li. "Lightweight sharable and traceable secure mobile health system." *IEEE Transactions on Dependable and Secure Computing* 17, no. 1 (2017): 78–91.

[44] Srivastava, Jyoti, and Sidheswar Routray. "AI enabled internet of medical things framework for smart healthcare." In *Innovations in Intelligent Computing and Communication: First International Conference, ICIICC 2022, Bhubaneswar, Odisha, India, December 16–17, 2022, Proceedings*, pp. 30–46. Springer International Publishing, 2023.

[45] Rajasoundaran, S., A. V. Prabu, Sidheswar Routray, S. V. N. Santhosh Kumar, Prince Priya Malla, Suman Maloji, Amrit Mukherjee, and Uttam Ghosh. "Machine learning based deep job exploration and secure transactions in virtual private cloud systems." *Computers & Security* 109 (2021): 102379.

[46] Joshi, Amit M., Urvashi P. Shukla, and Saraju P. Mohanty. "Smart healthcare for diabetes during COVID-19." *IEEE Consumer Electronics Magazine* 10, no. 1 (2020): 66–71.

[47] Tiwari, Anurag, Viney Dhiman, Mohamed A. M. Iesa, Haider Alsarhan, Abolfazl Mehbodniya, and Mohammad Shabaz. "Patient behavioral analysis with smart healthcare and IoT." *Behavioural Neurology* 2021 (2021).

[48] Uddin, Md Zia. "A wearable sensor-based activity prediction system to facilitate edge computing in smart healthcare system." *Journal of Parallel and Distributed Computing* 123 (2019): 46–53.

[49] Malla, Prince Priya, Sudhakar Sahu, and Sidheswar Routray. "Investigation of breast tumor detection using microwave imaging technique." In *2020 International Conference on Computer Communication and Informatics (ICCCI)*, pp. 1–4. IEEE, 2020.

[50] Soundararajan, Rajasoundaran, A. V. Prabu, Sidheswar Routray, Prince Priya Malla, Arun Kumar Ray, Gopinath Palai, Osama S. Faragallah et al. "Deeply trained real-time body sensor networks for analyzing the symptoms of Parkinson's disease." *IEEE Access* 10 (2022): 63403–63421.

[51] Balasundaram, A., Sidheswar Routray, A. V. Prabu, Prabhakar Krishnan, Prince Priya Malla, and Moinak Maiti, Internet of Things (IoT)-Based Smart Healthcare System for Efficient Diagnostics of Health Parameters of Patients in Emergency Care," in *IEEE Internet of Things Journal,* vol. 10, no. 21, pp. 18563-18570, 1 Nov.1, 2023.

[52] Sathish Kumar, L., A. V. Prabu, V. Pandimurugan, S. Rajasoundaran, Prince Priya Malla, and Sidheswar Routray. "A comparative experimental analysis and deep evaluation practices on human bone fracture detection using x-ray images." *Concurrency and Computation: Practice and Experience* 34, no. 26 (2022): e7307.

Chapter 7

Service-Oriented Distributed Architecture for Sustainable Secure Smart City

Sourav Banerjee, Sudip Barik, Arijit Sil, and Jerry Chun-Wei Lin

Chapter Contents

7.1 Introduction

Smart cities and urban computing are exciting research areas full of new challenges and opportunities. A large urban area can be called a smart city if various services and operations such as governance, healthcare, education, fuel, and energy distribution are supported by an ICT infrastructure that ensures greater efficiency and simpler operations. Prominent examples of such services include intelligent traffic monitoring, efficient and sustainable distribution of various energy needs (electricity, gas, fuel, etc.), ICT-enabled public transportation, sensor-based pollution monitoring and control systems, online services for individual citizens (e.g. buying a new car or property, applying for a passport or driver's license, reporting a crime, and seeking health services), etc. All of these services need to be integrated into an architectural model to facilitate the daily lives of individuals. The design should also ensure that everyone benefits from the system regardless of their ICT capabilities. With the launch of the iPhone and Android handsets, substantial changes were made to both the business models and the technological infrastructures on online platforms, serving as precursors to this progression. Additionally, ICT industry-specific equipment and solutions now need to communicate with telecommunications networks. Meanwhile, telecom companies want to provide more cutting-edge business solutions than they provide at the moment. Thus, it seems sensible that as new business models are put forth, existing system architectures will change to match those models. The business models proposed for smart cities are no exception: By definition, they imply an evolution of the value chain as well as the related system architecture that underpins both of these businesses. Table 7.1 demonstrates the impact of Smart City Business Models [41]. The vast majority of Smart City Business Models fall into one of several types, as shown Below [41].

DOI: 10.1201/ 9781003376712-7

Table 7.1 Impact of Smart City Business Models

Class	Example	Technological Significance
Environmental enhancement	Smart metres, smart grid, air quality monitoring	New network-connected gadgets
Economic growth	Green growth initiatives, incubators, and smart education	Data aggregation, open data
Cost effectiveness	Removing data silos across government agencies	Open data, cloud computing
Safety	Detecting firemen and adjusting transport to avoid a collision	New devices, new data sources, data aggregation, and open data are all being introduced
The standard of living	Data-driven feedback loops in urban planning	Data collection and information management
Citizens who are linked	Apps for transport for a "connected commute"	Privacy, data aggregation, and open data
Innovative business models	Using smartphone data from throughout a city to generate new advertising and revenue streams for local businesses	Data aggregation, privacy, open data, and data provenance

Categories of Smart City Business Models:

- Public–Private Partnerships (PPPs): This model involves collaboration between local governments and private companies to develop and implement smart city solutions. The public sector provides the regulatory framework, while the private sector invests in the technology and infrastructure needed to make the city "smart."
- Software as a Service (SaaS): Here, smart city solutions are provided to customers on a subscription basis. The SaaS provider is responsible for developing and maintaining the software, while the customer pays a fee to use it.
- Infrastructure as a Service (IaaS): In this model, the smart city infrastructure is provided as a service, similar to cloud computing. The IaaS provider is responsible for building and maintaining the infrastructure, while the city pays a fee to use it.
- Build–Operate–Transfer: In this model, a private company builds and operates the smart city infrastructure for a set period, typically 20–30 years. After this time, ownership is transferred to the government.
- Joint Ventures: A joint venture involves a partnership between the government and private companies, with both parties sharing risks and rewards. This model is often used for large-scale projects that require significant investment.
- Service Contracts: In this model, the government contracts with private companies to provide specific smart city services, such as waste management, transportation, or energy management.
- User Pays: The user pays model involves charging users for the use of smart city services, such as toll roads, parking, or public transportation. This model can help to generate revenue and encourage more sustainable use of resources.

- Outsourcing: In an outsourcing model, the government outsources the operation and maintenance of smart city infrastructure to private companies, which are responsible for ensuring its efficient functioning.
- Data Monetization: This model involves collecting data from sensors and other sources throughout the city and then selling that data to other businesses or organizations. This can include information about traffic patterns, air quality, energy consumption, and other topics.
- Value-Added Services: This model involves offering additional services on top of the smart city infrastructure, such as maintenance and repair, consulting, or other services that can help customers get the most out of their smart city solutions.
- Smart City Platforms: This model involves building a platform that integrates different smart city technologies and services, allowing customers to access multiple solutions from a single provider.

The main motivation for building a smart city is to use the innovations of the digital and information technology revolution to support the daily operations of a metropolis. This also facilitates the adoption of an information society that continuously gathers information from various sources (sensors, citizens, government agencies, social media, etc.) to better understand and serve various stakeholders, paving the way for sustainable development in the long run. Most importantly, smart city progress leads to a citizen-centric approach to public administration that can eliminate corruption and free citizens from the stranglehold of bureaucracy [1,6,13,14,17].

Over the years, various aspects of building smart cities have been extensively studied by researchers, and many have developed various definitions, frameworks, and solutions for smart cities [12,18,23,25,42]. The goal is to identify and solve problems that arise in an urban space with the increasing complexity of urban operations and the growing population by taking advantage of recent advances in information and communication technology (Figure 7.1). Waste management, smart distribution of scarce and limited resources, environmental degradation, healthcare system, traffic congestion, and reconstruction and reuse of obsolete infrastructure are some of the fundamental problems faced by cities around the world [10,15,30].

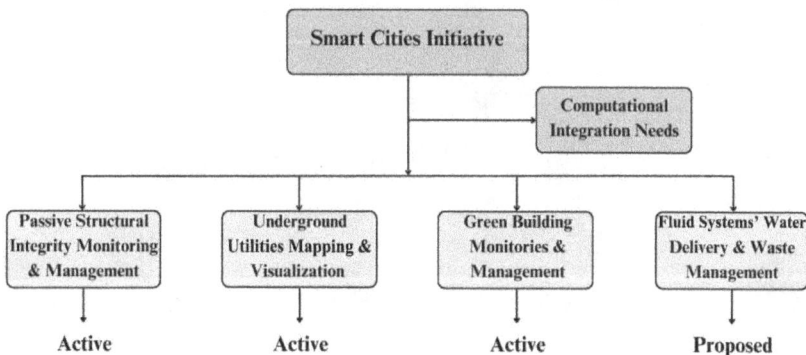

Figure 7.1 Smart cities initiatives.

Therefore, smart city today can be visualised as a massive information system consisting of multiple subsystems that effectively cooperate in collecting, storing, retrieving, and visualising data and enable citizens to access relevant information on demand using modern network and communication mediums as depicted in Figure 7.2. There are some challenges along the way to achieving these goals. Establishing adequate physical infrastructure and secure and intelligent data processing and finally creating appropriate interfaces for citizens to adopt smart city services are some of the obstacles that need to be overcome. For example implementing a smart transportation and traffic management system holds tremendous potential for improving traffic conditions and road safety in a city, but the system may require new types of networks to enable communication between vehicles. The implementation of intelligent traffic management requires extensive infrastructure throughout the city, and special care must be taken to ensure the security and privacy of the data collected and used. Banerjee et al. [11] survey the possibilities of how automated vehicles can be embedded into the traffic management system to bring about the next range of revolution in the transport system. In [29] to determine an effective way, the images captured as sequence from a camera are analysed using edge detection algorithm, object counting tool, and queue length determination. Following that, the number of vehicles at the intersection is assessed, and the flow of traffic is effectively controlled. Radio Frequency Identification (RFID), a brand-new technology that can be combined with the current signalling system to provide intelligent real-time traffic control, is introduced in [38]. In [53], the suggested method finds the vehicle candidates from the foreground image during vehicle detection, solving issues like headlight effects. Vehicles are tracked using the tracking technique over a series of frames. Finally, the tracking capabilities are improved by using a mechanism that makes up for error cases in busy environments.

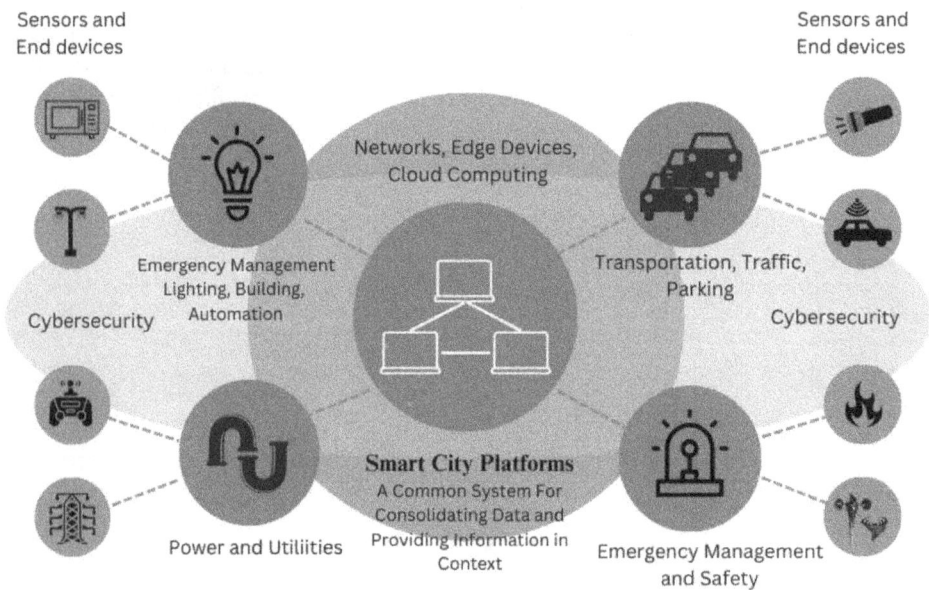

Figure 7.2 Smart city as information system.

Similarly, Ghosh et al. in [44] analyse security, privacy, and efficient information management in a healthcare system backed by IOT as the sector is vulnerable to cybersecurity threats. Finally, event-driven architecture (EDA) is mentioned in [10], although it has yet to be implemented. Table 7.2 demonstrates that architecture is independent of the smart city organisation [public organisation, state-owned enterprise (SOE), project coalition, or private company], as multi-tier architecture is observed in a variety of situations that follow alternative organisational forms [7]. Layer selection in multi-tier architectures, on the other hand, is unaffected by ICT smart city selection (e.g. Trikala is a digital city and Kyoto is an online city).

In order to effectively solve the difficulties of an information management system specifically created for smart cities, this chapter suggests a new paradigm. The following are the prerequisites for creating a smart city:

- Establishing a strong network and sensor systems to enable continuous monitoring and thorough real-world data collection.
- Effective methods for processing, retrieving, and storing data.
- Making it possible for data to be used by a variety of applications.

This will eventually convert the theory of a smart city to a reality. The system must also meet additional criteria for scalability and security, which are all important cosiderations.

Table 7.2 Preferred Architectures in Various Examined Cases

Case	Findings	
	Architecture	*Organisation*
European Smart Cities	Urban Intelligence Measurement System	Project (various European cities)
Two cities in the Netherlands	SOA	Two cities in the Netherlands
Fifty Two cities	n-tier architecture (four layers): Network, content, intelligence, services	Public organizations [i.e. Gdansk (Poland), Masdar (UAE)] Public–private partnership (PPP) [i.e. Amsterdam (the Netherlands)] Private Companies [Malaga (Spain), New Songdo (Korea)]
Helsinki, Kyoto	n-Tier architecture (three layers): Information, interface, interaction	State-Owned-Enterprise (SOE) run by the municipality
Dubai	n-Tier architecture (three layers): Infrastructure, data, application	Public organisation (government)
Trikala, Greece	n-Tier architecture (six layers): Data, infrastructure, interconnection, business, service and user	State-Owned-Enterprise (SOE) run by the municipality

Given the size of a modern metropolis and the rate of urbanization, these difficulties are exacerbated in real life.

In light of these difficulties, our study suggests a high-end architecture based on the effective distribution and storage of urban data to ease the operation of smart cities. The model is based on hierarchical and service-oriented storage of data that facilitates data visualization and decision-making. While the hierarchical model introduces data redundancy, it also effectively reduces network latency. Service-oriented distribution of data also enables improved data security and privacy. Third-party applications can access appropriate service-related datasets to provide services with enriched functionality and useful information to end users. The main contributions of this chapter are as follows:

1. Highlight key challenges in smart city implementation.
2. Propose a service-oriented architecture for smart city management in order to address these key challenges.
3. Provide an open yet secure data model to encourage third-party application development.

7.2   Background

In recent years, the concept of a smart city has grown in popularity as a means of addressing the issues of urbanisation and the need for sustainable development. A smart city is one that uses new technologies to optimise its functioning, increase its people' quality of life, and lessen its environmental effect. A smart city accomplishes this by utilising a variety of systems and devices that collect and analyse data to provide insights into urban activities [50].

So, as the number of devices and systems in a smart city grows, so does the complexity of managing them. Moreover, ensuring the security of these systems and protecting the privacy of citizens' data have become a major concern. Therefore, designing effective and sustainable architecture for a smart city is critical to its success.

Service-Oriented Architecture (SOA) [19,35,54] is a well-established architectural style that has been widely used in the design of large-scale distributed systems. SOA promotes the creation of modular and reusable software components that can be combined to create complex systems. By adopting an SOA-based approach, a smart city can create a flexible and adaptable architecture style that can evolve over time as new technologies emerge.

In addition to being service-oriented, smart city architecture should also be distributed. A distributed architecture allows for the decentralisation of data processing and reduces the risk of system failure or downtime. Furthermore, a distributed architecture enables the use of edge computing and the deployment of sensors and devices closer to the source of data, thus reducing latency and improving the accuracy of data analysis.

Sustainability: Sustainability [24,48,55] is a crucial component of any smart city initiative. The proposed architecture incorporates sustainability principles by leveraging distributed computing and service-oriented architecture principles. By distributing services across multiple nodes, the architecture reduces energy consumption and increases fault tolerance. Furthermore, the architecture encourages the use of renewable energy sources like solar and wind power.

Security: Security [46,47,57] is also a crucial component of any smart city initiative. The proposed architecture incorporates security principles by using distributed computing and service-oriented architecture principles. By distributing services across multiple nodes, the architecture reduces the impact of security breaches. In addition, the architecture uses secure communication protocols such as SSL/TLS to ensure the confidentiality and integrity of data.

7.3 Related Work

Since there are very different definitions of a smart city in the literature, researchers have developed various smart city architectures that focus on criteria such as technology, data flow, and interaction with stakeholders.

Information and communication technology (ICT) will spread from the desktop PC out into the real world and become pervasive, according to Wright in [45]. This vision is similar to that described by Weiser [52] and others [9,21,56]. The authors of [28] argue that as computer and network technology improve, digital cities will also undergo change. No digital city can continue to exist in its existing form. In order to grasp these digital cities' present situation and potential, various studies examine them. Komninos [33] proposed architecture from a technological point of view in Figure 7.3. They divided the architecture into three different layers as shown in Figure 7.3. The lowest layer, the information storage layer, is responsible for capturing and storing digital content. The application layer above it provides all the services needed to interpret and organise digital data. The top layer is the user interface, which makes data accessible to end users via various web applications using different digital tools. [34] provides a terminological foundation for treating a digital city as a self-organizing meaning-producing system intended to aid social or spatial navigation. Kryssanov et al. examine the concept of a digital city from a semiotic standpoint.

From the standpoint of user–system interaction, Al-Hader et al. [2] developed a five-level pyramid framework for smart cities (Figure 7.4). The lowest level is the smart infrastructure layer, which consists of sensor networks, various natural resources, communication networks, etc. The layer above is the intelligent data layer, which contains various types of spatial relational databases and data stores. The third layer is for providing various

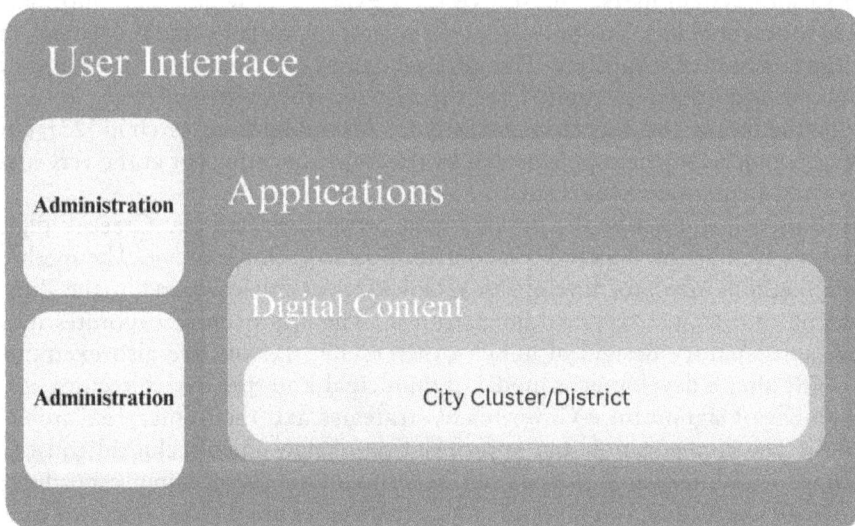

Figure 7.3 Three-layer smart city structure from technology perspective.

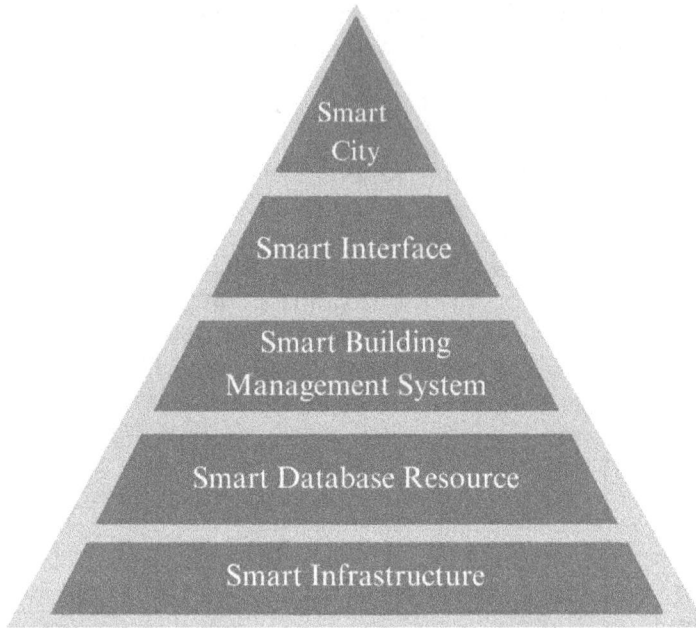

Figure 7.4 Five-level pyramid structure from user–system interaction perspective.

management systems for smart cities. Above it is the standard user interface layer, which contains the user interaction tools. The smart city layer is on top, and it unifies and streamlines interaction with the previous four layers.

Anthopoulos and Fitsilis have presented [4] a five-layer smart city architecture that seeks to integrate the physical and conceptual parts of such a system. The stakeholder layer identifies the various system users. The service layer explains how data and information are distributed to stakeholders. The business layer focuses on the rules and procedures that will control how a smart city operates. The infrastructure layer consists of the basic network and communication devices required for the system, while the data layer is responsible for defining the means and ways to manage data. According to research in [22], cities and face-to-face contacts may be supplemented by telecommunications or at the very least, they aren't a powerful alternative for them.

The authors of [5] describe the development process for Trikala, Greece's digital city (e-Trikala), as well as the digital city's contribution to e-Government. The methodology shows the methods used to develop an e-Government environment in the digital city that offers more than just service administration. The approach incorporates difficulties related to participatory design, although other related factors are also examined. The authors of [8] offer a development model of municipal Enterprise Architecture using data from a number of significant e-Government strategies and their enterprise architectures (Figure 7.5). The configuration and appropriation of new socio-technical constituencies are examined and afterwards interpreted in terms of social learning, according to the authors of [50]. According to Filipponi et al., smart cities are divided into two categories: Knowledge Processors (KPs) and Semantic Information Brokers (SIBs) [20]. SIB is where information is saved. Once the KP is linked to the SIB, operations are initiated via the

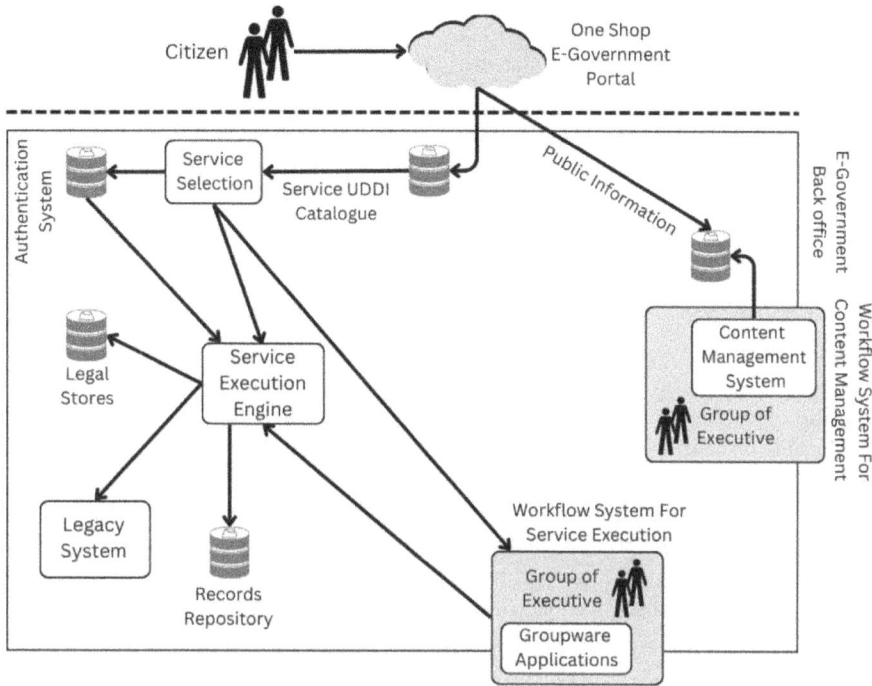

Figure 7.5 A logical view of common central e-government system.

Smart Space Access Protocol (SSAP). Through SSAP, KP manages all of the different user sessions. Harrison et al. [27], on the other hand, proposed a framework using infrastructure and stakeholders to collect the necessary information about the evolving behaviour of a smart city. In addition, Lugaric et al. [37] proposed a smart city architecture that consists of three parts: the physical network, the communication devices, and the data flow. To improve the reliability of smart city IoT devices, Ghosh et al. [39] build a dynamic correlation between nodes in a single cluster based on their statistical behaviour when performing smart communication in Cognitive Radio Sensor Networks. Chourabi et al. [17] proposed a framework that attempts to understand a smart city by identifying external and internal categories of factors that act as influencers. External factors include governments, people, various other communities, natural resources, physical infrastructure, and the economy. Internal factors include technologies, rules, and related policies. In [38], Banerjee et al. propose an energy-efficient cloud computing mechanism through VM selection that is specifically designed for an environment where large-scale virtualized data centres need to communicate via the cloud and facilitate service-oriented computing. From the above discussion, it is clear that recent advances in Big data management and information technology will play a critical role in the development of future smart cities. Taking these factors into account, this chapter proposes a distributed, service-oriented smart city architecture that (i) allows a free yet secure flow of usable data and information between different modules, (ii) is highly flexible to adapt to rapidly evolving technologies, and (iii) follows an open data model that enables an interface for data sharing with third-party developers and agencies.

7.4 Key Challenges

A smart city can be thought of as a vast information system made up of multiple information subsystems that are primarily based on information and communication technologies. Recent advancements in cloud computing, the Internet of Things, the Semantic Web, and Big data analytics offer an infrastructure base and successful smart city development solutions. Innovative ICT technology deployment can vastly boost the potential of smart city services, which can benefit inhabitants. However, when all of these technologies are combined in complex smart city architecture, some inherent challenges arise that must be addressed, which are mentioned here.

ICT Infrastructure: Developing and deploying the necessary ICT infrastructure in all designated locations in a city remain a major obstacle. The lack of a fast and stable Internet connection throughout the city is one of the biggest challenges. The infrastructures already in place are not scalable, which is another challenge. Before a smart city initiative can be implemented, a reliable, scalable, and high-quality infrastructure must be installed [40].

Security and Privacy: As we move towards a smart city, citizens carrying sophisticated handheld devices must provide personal data to smart city services [43]. For example to know traffic conditions, a person must share his or her current location with a traffic management system. Meeting the necessary security requirements for processing a huge amount of private and sensitive data is a fundamental challenge in any smart city architecture. Sensitive data is always vulnerable to attacks by hackers, viruses, worms, and Trojans, and if compromised, it can lead to a total disruption of the system, resulting in large losses. The system must be carefully protected at every step of the collection, storage, and retrieval of private, confidential data, not only to make data available but also to build stakeholder confidence in the use of the services.

Big Data Management: A huge amount of data will always be generated by all smart city systems [3]. The data can be generated in both structured and unstructured formats. In order to handle this huge amount of data, efficient data management is an essential requirement. Continuously collecting, storing, and retrieving data generated by a huge number of smart city sensors are indeed formidable challenges. For example, the sensor data from all vehicles operating in a city at GPS can be used to predict traffic flow, but it requires an efficient handling of the enormous amount of data [31,32,51].

Financial Cost: Since a smart city requires enormous IT infrastructure, the financial cost of setting things up is enormous [26]. To run this system efficiently, millions of sensors, network devices, and computers will be needed. There will also be a high cost to employing IT professionals and maintenance staff. Maintaining the flow of operations and regularly updating the systems will be necessary, for which separate budgetary resources will need to be allocated. Maintaining the required level of efficiency and ensuring security/privacy will incur additional overhead costs.

Interoperability: Smart city architecture provides a heterogeneous environment where different categories of applications, devices, and platforms must communicate seamlessly. The network devices and connectivity required for smart traffic management are different from the type of networking used for short-range wireless communications. Integrating the existing applications and network infrastructures into smart city architecture is a major challenge [16].

Efficiency, Availability, and Scalability: Smart city systems are mission-critical and require a high availability rate. As a system grows in size and complexity, its availability becomes critical. A smart city, which is currently a massive and complicated system, will expand by leaps and bounds as it is deployed. With vast volumes of data flowing in all the time, assuring data availability, scalability, and efficiency remains a vital concern. A smart city provides benefits such as performance optimisation, interoperability, effective planning, and fast response to service demands. In times of crisis, for example, all civilian agencies should be able to collaborate to respond to an emergency.

A smart city appears to be the appropriate solution to the majority of the difficulties caused by rapid urbanisation and population increase. However, research has identified challenges due to inequity, the digital divide, and residents' diverse cultural habits. The proposed architecture addresses the technological aspects of smart city deployment.

7.5   Proposed Architecture

Figure 7.6 shows the different layers in smart city architecture. From the centre to the periphery, these layers are the physical infrastructure layer (PI), the central data management and knowledge processing layer (CDM&KP), the service layer, and the application layer.

The physical infrastructure in the form of communication and network devices is the core layer of the proposed architecture. All other layers depend on it, and this layer physically connects our architecture to the real world. This layer consists of high-speed wired and wireless communication devices, sensors, actuators, routers, gateways, and all other necessary hardware devices. Initially, this layer receives all data collected in the PI layer. The data can be in either a structured or unstructured format. This layer is then responsible for effectively classifying, processing, and analysing the raw data, followed by storing the relevant data in a central repository from which the data can be easily retrieved. Effective data management is thus one of the main tasks performed here. Another important task of this layer is to respond to service requests coming from a higher layer. Sometimes, it also forwards a service request in the form of a command to the PI layer to possibly recalibrate or reorganise some of the physical devices. This layer is sort of the core of the entire architecture and controls

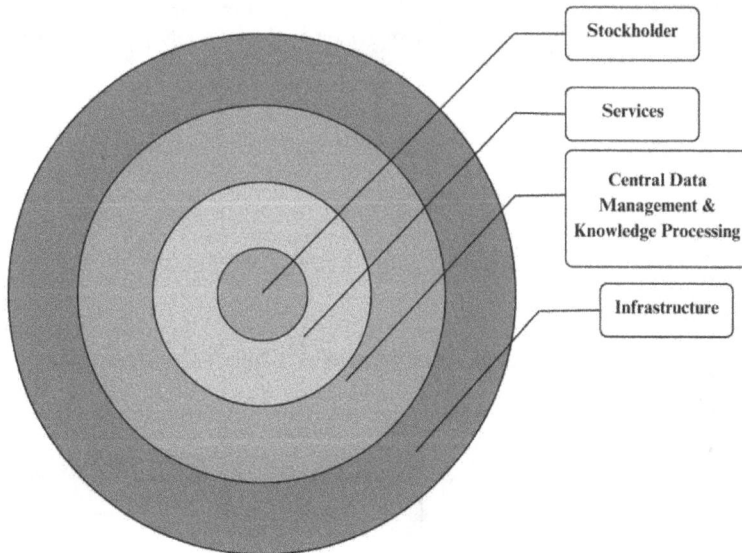

Figure 7.6 Smart city systems.

the flow of data and service requests that come in as needed. The service layer facilitates the development of strategic links between different departments of public and private organisations responsible for the operation of various services related to urban life. Typically, all of these departments operate in isolation with little or no coordination between them. This is problematic because it hinders effective response and service in different circumstances. The departments responsible for a particular service essentially use all their resources (ICT or otherwise) to gather information relevant to their work. This leads to many problems and delays in processing service requests. It is very tedious for citizens to get a response to a service request that involves different departments. Even in disaster management, where rapid response is an absolute priority, a lack of coordination between different units can severely hamper the effectiveness of the response.

The architecture focuses on data sharing to enable mutual coordination between different sectors. It also offers open data services to third parties who wish to use the information required to provide a service. This approach even allows cross-platform applications to use real-time collected data from different sectors for decision-making, which can lead to shorter response times and effectively increase the overall efficiency of city services. By continuously sharing information from different domains, service providers can gain insights into existing problems or even predict impending problems to take countermeasures before the problem escalates or occurs.

Similarly, the availability of real-time data across different domains enables efficient resource management, allowing effective coordination and sharing of information, and the response to an event or service request is effective and almost instantaneous.

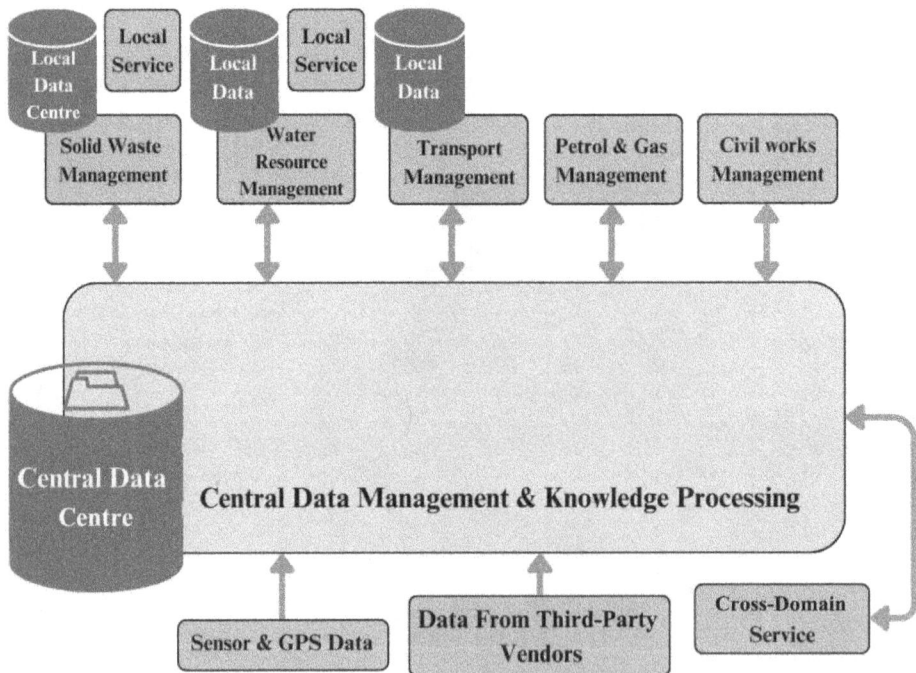

Figure 7.7 Holistic view of proposed architecture.

Figure 7.7 shows a holistic view of our proposed architecture. The architecture is designed as a system of subsystems, with each subsystem fully integrated and connected to other subsystems through the CDMS&KP layer. The CDMS&KP enables data sharing between different services operating at the service level. A subsystem can share information and insights through CDMS&KP while benefiting from cross-domain knowledge and services accessible through the architecture. CDMS&KP acts as an integration point for information flow in different directions. In addition, CDMS&KP can also use the data and information flowing through it to make decisions in real-time.

7.5.1 Preliminaries

The traditional city is divided into different zones of administration. Each zone administration is responsible for providing services in that area through a local office that manages the records for that area. Therefore, the proposed architecture is also based on such a scenario. Each subsystem is responsible for a specific service and collects data from the zone sites, where the data is managed in the cloud of the local data centre. For electricity management, the system maintains a number of cloud data centres throughout the city, each of which stores its local data. Similar facilities can be implemented for water management, waste management, gas supply management, etc. Given the complexity and challenges at the technical and social levels, a smart city can only be developed in a phase-wise manner. By facilitating zone-level services, this architecture enables the gradual and ever-expanding deployment of smart city services. Each zone is considered an autonomous system consisting of a local data centre, network facilities, and any other related infrastructure capable of collecting, storing, retrieving, and processing data. Zonal systems can also interoperate via web services to provide cross-zone services.

7.5.2 Data Management

Figure 7.8 depicts a city's electricity management system. Zone A stores data collected by sensors and other devices in a local relational database. Zone A data centres provide services to zone A clients. Because of its established ACID property and query optimisation capability, a relational database is selected. A fast and reliable network connects all zonal data centres to the central data centre. Data from all zonal data centres is consolidated in a centralised data management facility. The volume of data created across all zones will be enormous. To deal with the enormity and complexity of the acquired data, sophisticated Big data management solutions must be in place.

Service-Oriented Architecture: The smart city architecture described in this study is essentially a large-scale distributed system that adheres to the decentralisation principle. To enable interoperability, modularity, software reuse, and application integration, the system must leverage open standards like XML, WSDL, SOAP, and UDDI across multiple platforms, heterogeneous networks, and a range of service requirements. Water and waste management services provided by another application can thus be used by the disaster management application via smooth interoperability.

Open Data Model: Since any data centre will generate a huge amount of data, it is only prudent to make that data available to anyone who is interested. Therefore, the architecture follows an open data model that allows researchers, app developers, and private agencies

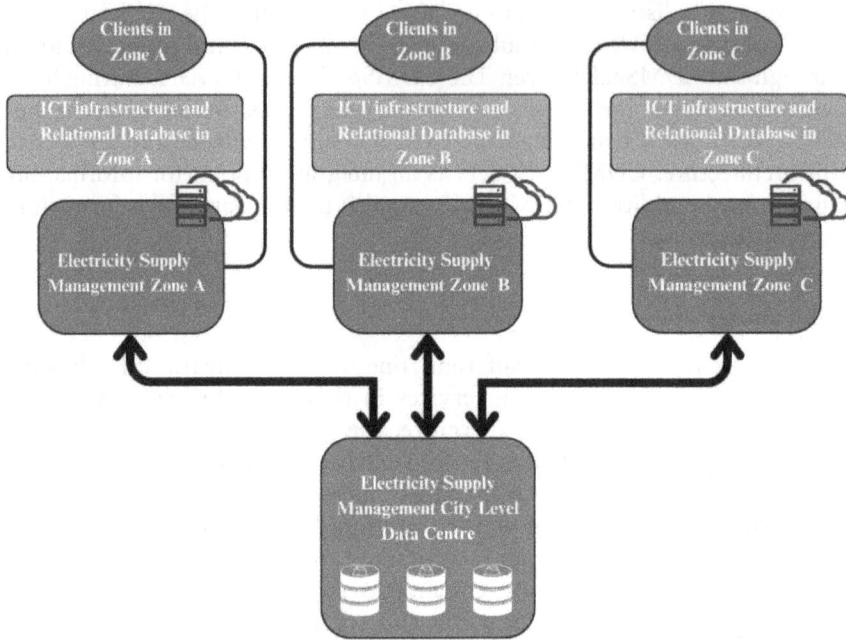

Figure 7.8 Electricity management system for a city.

to access the central repository through a revenue-based system in a selective manner after ensuring privacy and data security. This allows researchers and app developers not only to gain new insights into city operations but also to develop new apps to better serve citizens.

7.6   Proposed Model and Challenges

In this section, we list all the challenges in a smart city identified here and the corresponding solutions offered by the proposed model. The infrastructure of a smart city is characterised by the heterogeneity of networks, applications, and platforms. It also requires the construction of extensive infrastructure in the form of sensors, wireless and wired networks, communication devices, etc. Since the architecture is based on zonal partitioning of infrastructure and public services, the entire infrastructure does not need to be operational at all times. Newer and advanced services can be easily integrated into this infrastructure. The open data model allows data to be managed and shared among different categories of customers. This is especially beneficial as the transition to new technologies and smarter cities will increase the amount of data to be processed, managed, and analysed many times over, as this mission-critical system requires high data availability and can tolerate shorter downtimes. The architecture is flexible enough to grow horizontally to accommodate newer services while growing vertically to provide an improved quality of service in each zone. In the proposed model, each public service operates its zonal centre. Although they operate independently, they sometimes exchange information through message passing. We propose the implementation of end-to-end encryption standards to ensure the integrity and security of messages. Since the system operates on a zone-by-zone basis, an attack on one of the units/data centres cannot possibly bring down the entire building. The model also encourages obtaining necessary permissions from the clients before releasing data to various web services. The model also facilitates the efficient distribution of resources in real time by enabling instant collaboration between different departments. This approach improves the processing and sharing of real-time data for planning the efficient distribution of resources.

Following is a discussion of how the proposed model addresses key challenges:

ICT Infrastructure: A zone-by-zone implementation means that not all infrastructure needs to be brought online at the same time. The model supports a phased rollout of smart city services. Nevertheless, the model still requires adequate infrastructure on the way to a smart city.

Security and Privacy: The model provides data as part of various web services it offers. To ensure the security of the data, it uses the standard of end-to-end encryption at the message level. The model also collects private and sensitive data from its users. It shares private information only after obtaining the necessary consent from the user. Data is shared only with authenticated agencies in accordance with the rules and regulations set by the relevant authority.

Big Data Management: Zone-wise distribution of data helps with effective collection, storage, and analysis. Each public service department stores data collected in a given area in a zonal data centre. As a result, decisions and services that affect a particular zone do not have to be routed through central data storage. If a service needs to be provided at the city level, the zonal data centres can always share the information with the central data repository, where it can be further processed to find a solution.

Cost: The initial cost of installing all the ICT infrastructure and setting up zonal data centres along with a centralised data repository may seem very expensive, but in the long run, the efficiency and ease of use built into a smart city architecture will tilt the balance in its favour by increasing the speed and efficiency of public services and reducing costs for users.

Interoperability: The model provides a service-oriented architecture in which data and information are available through web services. This approach makes data available to different clients. Web services seamlessly integrate different applications, regardless of the underlying platform.

Efficiency, availability, and scalability: The smart city architecture proposed here eliminates the need for many manual operations in city administration. The division of the city into self-sufficient operational zones results in the separation of daily activities. Automation also provides greater efficiency and shorter response times.

Because the system is divided into separate operational zones, the unavailability of services in a particular zone for any reason does not affect the entire system. Since most data is also replicated in a central repository, a failed data service can be quickly restored using the central data centre.

The system supports the easy introduction and integration of new services. It also enables the replacement of old services with new ones. Even trusted third parties can use data through web services to find solutions to the city's problems. This horizontal growth cycle is one of the unique features of the architecture. In addition, each zonal entity can improve its services and expand its resources as needed, which also enables vertical growth.

7.7   Conclusion and Future Direction

This chapter presents a distributed service-oriented architecture for smart cities. The architecture mainly identifies and addresses the technical aspects of a smart city. The model uses a hierarchical approach to data management and knowledge processing, which facilitates information flow and availability. The distributed nature of the system enables the decentralisation of data and services, improving efficiency, availability, and reliability at the zone level. The service-oriented nature of the proposed architecture adapts well to the heterogeneity of a modern metropolitan environment. The open data model is important to encourage third-party developers and researchers to access the necessary information of the system and develop innovative services that can improve city life. In the future, the challenges associated with cities in specific geographic areas (e.g. a Hill Station) can be identified, and the capabilities of the current architecture to deal with such unique conditions can be explored.

Bibliography

[1] Mahmoud Al-Hader and Ahmad Rodzi. The smart city infrastructure development & monitoring. *Theoretical and Empirical Researches in Urban Management*, 4(2 (11):87–94, 2009.

[2] Mahmoud Al-Hader, Ahmad Rodzi, Abdul Rashid Sharif, and Noordin Ahmad. Smart city components architicture. In *2009 International Conference on Computational Intelligence, Modelling and Simulation*, pages 93–97. IEEE, 2009.

[3] E. Al Nuaimi, H. Al Neyadi, N. Mohamed and J. Al-Jaroodi, Applications of big data to smart cities. *Journal of Internet Services and Applications*, 6, pp. 1-15, 2015.

[4] L. Anthopoulos, and P. Fitsilis, "From Digital to Ubiquitous Cities: Defining a Common Architecture for Urban Development," *2010 Sixth International Conference on Intelligent Environments*, Kuala Lumpur, Malaysia, pp. 301-306, 2010.

[5] Leo G. Anthopoulos and Ioannis A. Tsoukalas. The implementation model of a digital city. The case study of the digital city of trikala, greece: E-trikala. *Journal of e-Government*, 2(2):91–109, 2006.

[6] Leonidas Anthopoulos and Panos Fitsilis. From online to ubiquitous cities: The technical transformation of virtual communities. In *Next Generation Society. Technological and Legal Issues: Third International Conference, e-Democracy 2009, Athens, Greece, September 23–25, 2009, Revised Selected Papers 3*, pages 360–372. Springer, 2010.

[7] Leonidas Anthopoulos and Panos Fitsilis. Exploring architectural and organizational features in smart cities. In *16th International Conference on Advanced Communication Technology*, pages 190–195. IEEE, 2014.

[8] Leonidas G. Anthopoulos. Collaborative enterprise architecture for municipal environments. In *Advances in Government Enterprise Architecture*, pages 392–408. IGI Global, 2009.

[9] B. Azvine, N. Azarmi, and K. C. Tsui. Soft computing: A tool for building intelligent systems. *BT Technology Journal*, 14(4):37–45, 1996.

[10] Tuba Bakici, Esteve Almirall, and Jonathan Wareham. A smart city initiative: The case of barcelona. *Journal of the Knowledge Economy*, 4(6), 2012.

[11] Sourav Banerjee, Chinmay Chakraborty, and Sumit Chatterjee , A Survey on IoT Based Traffic Control and Prediction Mechanism. In: Balas, V., Solanki, V., Kumar, R., Khari, M. (eds) *Internet of Things and Big Data Analytics for Smart Generation*. Intelligent Systems Reference Library, vol. 154. Springer, Cham. https://doi.org/10.1007/978-3-030-04203-5_4

[12] Iain Begg. Cities and competitiveness. *Urban Studies*, 36(5–6):795–809, 1999.

[13] Jennifer Belissent. The core of a smart city must be smart governance. *Forrester Research Inc.* Cambridge Google Scholar, 2011.

[14] Christine Bellamy and G. D. Garson. The politics of public information systems. *Public Administration and Public Policy*, 77:85–98, 2000.

[15] Zhang, Abraham, V. G. Venkatesh, Jason X. Wang, Venkatesh Mani, Ming Wan, and Ting Qu. "Drivers of industry 4.0-enabled smart waste management in supply chain operations: a circular economy perspective in china." *Production Planning & Control* 34, no. 10, pp. 870-886, 2023.

[16] Arianna Brutti, Piero De Sabbata, Angelo Frascella, Nicola Gessa, Raffaele Ianniello, Cristiano Novelli, Stefano Pizzuti, and Giovanni Ponti. Smart city platform specification: A modular approach to achieve interoperability in smart cities. In *The Internet of Things for Smart Urban Ecosystems*, Springer, Cham. pages 25–50, 2019.

[17] H. Chourabi, T. Nam, S. Walker, J. R. Gil-Garcia, S. Mellouli, and K. Nahon, T. A. Pardo, and H. J. Scholl "Understanding smart cities: An integrative framework." In *2012 45th Hawaii international conference on system sciences,* IEEE, pp. 2289-2297, 2012.

[18] Stefanie Dulhr. Potentials for Polycentric Development in Europe: The Espon 1.1. 1 Project Report. *Planning, Practice & Research*, 20(2):235–239, 2005.

[19] Hisham Elhoseny, Mohamed Elhoseny, Samir Abdelrazek, Hazem Bakry, and Alaa Riad. Utilizing service oriented architecture (SOA) in smart cities. *International Journal of Advanced Computer Technology (IJACT)*, 8(3):77–84, 2016.

[20] Luca Filipponi, Andrea Vitaletti, Giada Landi, Vincenzo Memeo, Giorgio Laura, and Paolo Pucci. Smart city: An event driven architecture for monitoring public spaces with heterogeneous sensors. In *2010 Fourth International Conference on Sensor Technologies and Applications*, pages 281–286. IEEE, 2010.

[21] Margaret Fleck, Marcos Frid, Tim Kindberg, Eamonn O'Brien-Strain, Rakhi Rajani, and Mirjana Spasojevic. From informing to remembering: Ubiquitous systems in interactive museums. *IEEE Pervasive Computing*, 1(2):13–21, 2002.

[22] Jess Gaspar and Edward L. Glaeser. Information Technology and the Future of Cities. *Journal of Urban Economics*, 43(1):136–156, 1998.

[23] Rudolf Giffi nger, Christian Fertner, Hans Kramar, Robert Kalasek, and Natasıa Pichler Milanovic. Meijers, evert-smart cities. *Ranking of European Medium-Sized Cities, Vienna University of Technology,* Final Report, 2007, Online: https://www.researchgate.net/publication/261367640_Smart_cities_-_Ranking_of_European_medium-sized_cities

[24] Pierpaolo Girardi and Andrea Temporelli. Smartainability: A methodology for assessing the sustainability of the smart city. *Energy Procedia*, 111:810–816, 2017. https://www.researchgate.net/publication/241977644_The_vision_of_a_smart_city#fullTextFileContent

[25] R. Hall, B. Bowerman, Joseph Braverman, J. Taylor, Helen Todosow, and U. Wimmersperg. The vision of a smart city. *2nd International Life Extension Technology Workshop Paris, France September 28, 2000*, 2000.

[26] Steve Hamilton and Ximon Zhu. Funding and financing smart cities. *The Journal of Government Financial Management*, 66(1):26–33, 2017.

[27] Colin Harrison, Barbara Eckman, Rick Hamilton, Perry Hartswick, Jayant Kalagnanam, Jurij Paraszczak, and Peter Williams. Foundations for smarter cities. *IBM Journal of Research and Development*, 54(4):1–16, 2010.

[28] Ishida T, Isbister K, editors. Digital cities: technologies, experiences, and future perspectives. Springer Science & Business Media; 2000 Mar 2.

[29] Supriya Kamoji, Aswathi Nambiar, Karishma Khot, and Ravi Bajpai. Dynamic vehicle traffic management system. *IJRET International Journal of Research in Engineering and Technology*, vol. 4. pp. 352-356. 10.15623/ijret.2015.0404063. 2015.

[30] Yin, C., Xiong, Z., Chen, H., Wang, J., Cooper, D. and David, B., "A literature survey on smart cities," *Science China*. Information Sciences, 58(10):1-18, 2015.

[31] Zaheer Khan, Ashiq Anjum, and Saad Liaquat Kiani. Cloud based big data analytics for smart future cities. In *2013 IEEE/ACM 6th International Conference on Utility and Cloud Computing*, pages 381–386. IEEE, 2013.

[32] Zaheer Khan, Ashiq Anjum, and Saad Liaquat Kiani. Cloud based big data analytics for smart future cities. In *2013 IEEE/ACM 6th International Conference on Utility and Cloud Computing*, pages 381–386. IEEE, 2013.

[33] Nicos Komninos. The architecture of intelligent cities: Integrating human, collective and artificial intelligence to enhance knowledge and innovation. In *2006 2nd IET International Conference on Intelligent Environments-IE 06*, volume 1, pages 13–20. IET, 2006.

[34] Victor V. Kryssanov, Masayuki Okabe, Koh Kakusho, and Michihiko Minoh. Communication of social agents and the digital city—a semiotic perspective. In *Digital Cities II: Computational and Sociological Approaches: Second Kyoto Workshop on Digital Cities Kyoto, Japan, October 18–20, 2001 Revised Papers 2*, pages 56–70. Springer, 2002.

[35] Dilshodbek Kuryazov, Bekmurod Khujamuratov, and Khursand Sherkhanov, "Sustainable Service-Oriented Architecture for Smart City Development," *2019 International Conference on Information Science and Communications Technologies (ICISCT)*, Tashkent, Uzbekistan, pp. 1-5, 2019. doi: 10.1109/ICISCT47635.2019.9011877.

[36] Ninad Lanke and Sheetal Koul. Smart traffic management system. *International Journal of Computer Applications*, 75(7), 2013.

[37] Luka Lugaric, Slavko Krajcar, and Zdenko Simic. Smart city—platform for emergent phenomena power system testbed simulator. In *2010 IEEE PES Innovative Smart Grid Technologies Conference Europe (ISGT Europe)*, pages 1–7. IEEE, 2010.

[38] Riman Mandal, Manash Kumar Mondal, Sourav Banerjee, and Utpal Biswas. An approach toward design and development of an energy-aware vm selection policy with improved sla violation in the domain of green cloud computing. *The Journal of Supercomputing*, 76:7374–7393, 2020.

[39] Li Manman, Pratik Goswami, Proshikshya Mukherjee, Amrit Mukherjee, Lixia Yang, Uttam Ghosh, Varun G. Menon, Yinan Qi, and Lewis Nkenyereye. Distributed artificial intelligence empowered sustainable cognitive radio sensor networks: A smart city on-demand perspective. *Sustainable Cities and Society*, 75:103265, 2021.

[40] Catherine A. Middleton, Andrew Clement, and Graham Longford. *ICT Infrastructure as Public Infrastructure: Exploring the Benefits of Public Wireless Networks*. TPRC, 2006.

[41] Catherine E. A. Mulligan and Magnus Olsson. Architectural implications of smart city business models: An evolutionary perspective. *IEEE Communications Magazine*, 51(6):80–85, 2013.

[42] Michael Parkinson, Greg Clark, Mary Hutchins, James Simmie, and Hans Verdonk. *Competitive European Cities: Where Do the Core Cities Stand?* Office of the Deputy Prime Minister London, 2004.

[43] Daniela Popescul and Laura-Diana Genete. Data security in smart cities: Challenges and solutions. *Informatica Economicffa*, 20(1), 2016.

[44] Pradip Kumar Sharma, Uttam Ghosh, Lin Cai, and Jianping He. Guest editorial: Security, privacy, and trust analysis and service management for intelligent internet of things healthcare. *IEEE Transactions on Industrial Informatics*, 18(3):1968–1970, 2021.

[45] A. Steventon and S. Wright. Intelligent spaces—the vision, the opportunities, and the barriers. In *Intelligent Spaces: The Application of Pervasive ICT*, pages 1–17. Springer-Verlag, 2006.

[46] Kehua Su, Jie Li, and Hongbo Fu. Smart city and the applications. In *2011 International Conference on Electronics, Communications and Control (ICECC)*, pages 1028–1031. IEEE, 2011.

[47] Chai K. Toh. Security for smart cities. *IET Smart Cities*, 2(2):95–104, 2020.

[48] Angeliki Maria Toli and Niamh Murtagh. The concept of sustainability in smart city definitions. *Frontiers in Built Environment*, 6:77, 2020.

[49] Zhao Tong, Feng Ye, Ming Yan, Hong Liu, and Sunitha Basodi. A survey on algorithms for intelligent computing and smart city applications. *Big Data Mining and Analytics*, 4(3):155–172, 2021.

[50] M. J. Van Lieshout. Configuring the digital city of amsterdam: Social learning in experimentation. *New Media & Society*, 3(2):131–156, 2001.

[51] Ignasi Vilajosana, Jordi Llosa, Borja Martinez, Marc Domingo-Prieto, Albert Angles, and Xavier Vilajosana. Bootstrapping smart cities through a selfsustainable model based on big data flows. *IEEE Communications Magazine*, 51(6):128–134, 2013.

[52] Mark Weiser. The computing for the twenty-first century. *Scientific American*, 94–104, 1991.

[53] Bing-Fei Wu, Chih-Chung Kao, Jhy-Hong Juang, and Yi-Shiun Huang. A new approach to video-based traffic surveillance using fuzzy hybrid information inference mechanism. *IEEE Transactions on Intelligent Transportation Systems*, 14(1):485–491, 2012.

[54] Zhang Xiong, Yanwei Zheng, and Chao Li. Data vitalization's perspective towards smart city: A reference model for data service oriented architecture. In *2014 14th IEEE/ACM International Symposium on Cluster, Cloud and Grid Computing*, pages 865–874. IEEE, 2014.

[55] Tan Yigitcanlar and Md Kamruzzaman. Does smart city policy lead to sustainability of cities? *Land Use Policy*, 73:49–58, 2018.

[56] Lotfi A. Zadeh. The roles of fuzzy logic and soft computing in the conception, design and deployment of intelligent systems. *BT Technology Journal*, 14(4):32–36, 1996.

[57] Kuan Zhang, Jianbing Ni, Kan Yang, Xiaohui Liang, Ju Ren, and Xuemin Sherman Shen. Security and privacy in smart city applications: Challenges and solutions. *IEEE Communications Magazine*, 55(1):122–129, 2017.

Chapter 8

A Comprehensive Security Risk Analysis of Cliff Edge on Cyber-Physical Systems

Koustav Kumar Mondal, Asmita Biswas, and Deepsubhra Guha Roy

8.1   Introduction

To create the physical world function appropriately and more straightforwardly, the Cyber-Physical System (CPS) attempts to track physical systems' activity and enable actions to alter their behavior. A cyber-physical arrangement (CPS) normally consists of a pair of fundamental parts, a physical mechanism, and cyber conformity. The motive state is usually tracked or managed via a specific cyber device, a networked arrangement of many small devices with (often wireless) capability for sensing, computation, and communication. A spontaneous occurrence (e.g., an inoperative volcano) and a human-made dynamic structure (e.g., an operational room), rather a more confused mixture of those two may be the dynamic mechanism involved. As the physical and virtual systems' link grows, the physical systems become more and more vulnerable to computer system security vulnerabilities. For instance, several hackers have hacked into the U.S. Federal Aviation Agency's air traffic control expedition networks in many incidents in contemporary ages, according to an Inspector

DOI: 10.1201/9781003376712-8

General's research proposed to the FAA in 2009 (Mills, 2009). Any hacker can now even hack specific medical devices with wireless communications embedded in the human body. A CIA report (O'Connell, 2008) shows that hackers have breached energy grids in many neighborhoods just outside of the United States and triggered a power failure that involved several towns in at least one instance. Within 2010, the antagonists exhibited a software intermediary summoned CarShark (Koscher et al., 2010) that could remotely disable a car engine, switch off individual brakes so that the car does not stop, and create tools by tracking messages among the electronic control units (ECUs) and injecting fake datagrams to send out strikes and provide false readings. Hackers have formulated a virus 2023 that can effectively target the Siemens plant-control arrangement.

The bulk of cybersecurity event datasets are confidential, usually combined with the additional caveat that they are either kept hidden or just exchanged within acutely technological circles as bugs are identified. Consequently, there is an incentive provided before and after the vulnerable discovery stage for presumably silent zero-day penetration by offenders who may be highly competent. Therefore, it can be argued that the emphasis is not directly on the existence and substance of an exploit but on how it is capable of weaving between computer security systems to accomplish the ultimate purpose of exploitation. In reality, redemption vulnerabilities are located in even deeper cyber-physical regularities, such as electrical energy systems, conveyance infrastructure, and healthcare regularities. Researchers are beginning to think about CPS stability. If we have more intelligent and highly knowledgeable cyber-physical networks, these systems' potential flaws should be carefully considered. In reality, security is a relatively recent environment for CPSs and more research needs to be done within this field. Similar to any other emerging area, the greatest initiative appears to concentrate on mapping clarifications from subsisting contexts, previously mentioned as sensor networks that bestow networked activities and low-ability features among CPS (Roy et al., 2018a). Typically, though, these resolutions have yet to be established to address CPSc. As an illustration, the gas department's cyber-physical machine would communicate with the one that controls the wounded individual's well-being to accomplish the rescue operation, imagining an example of gas leakage in a bright house. These applications are, under normal circumstances, separate. Nevertheless, all these applications need to collaborate and exchange resources to achieve the same purpose if there is an incident. For interoperation between heterogeneous systems, conventional safe communication solutions are not planned. Cyber-physical networks guarantee that the device is always protected when communicating with another system is a fundamental challenge. Some modern CPS security challenges need to be discussed as well.

In this chapter, we first discuss the background of CPS, which consists of CPS layers and components and CPS model type; second, the flaws, attacking models, and forms of adversaries are identified. After that, we also discuss some recommendations and failures in CPS.

8.2 Background of CPS

We introduce the CPS structure—its key panels, modules, including the main models of the CPS in this section.

8.2.1 CPS Layers and Components

The CPS device design consists of various courses and modules that depend on diverse communication etiquette and technology to interact across multiple courses.

8.2.1.1 CPS Layers

Each design of the CPS consists of three fundamental courses—a specific course of vision, the course of transmission, furthermore the course of operation, as manifested and defined in Figure 8.1. A specific study (Ashibani and Mahmoud, 2017) is a foundation for reviewing the different CPS course protection problems.

- **Perception Layer:** It is often referred to since unless every layer of identification or the course of sense. This involves cameras, actuators, aggregators, tags for Radio Frequency Identification (RFID), Global Positioning System (GPS), and many other devices. To track, monitor, and analyze the physical environment, these instruments capture real-time data. Examples of such collected data include vibrational and visual signals, as well as data related to electrical usage, heat, position, chemistry, and biology, depending on the sensor type. Once collected, this data is then processed through treatment panels, allowing sensors to analyze real-time data within large regional interface specialties. Besides, actuator securing relies on official authorizations to assure that all input plus command groups remain error-free and safeguarded. In general, increasing the degree of protection includes applying a framework of end-to-end encryption on each course. Heavyweight reckonings and high-retention specifications will then be set in place. There is a need for reliable and lightweight protection protocols to be built in this context, taking into account the capacities of the equipment and safety specifications.
- **Transmission Layer:** This is the second layer of CPS, which is additionally assigned as the course of transport or the network layer. Layer as mentioned earlier, reciprocations furthermore prepare knowledge among the levels of interpretation and implementation. Managing Local Area Networks' (LANs') further networking standards, including Bluetooth, 4G, 5G, InfraRed (IR), ZigBee, Wi-Fi, LTE, different innovations onward, data transfer, and interaction are accomplished across the Internet. Notwithstanding the reason mentioned above, separate protocols, such as Internet Protocol version 6 (IPv6)(Wu and Lu, 2010), imply proficiency to manage the growth during the particular number of machines connecting to the Internet. The course, as

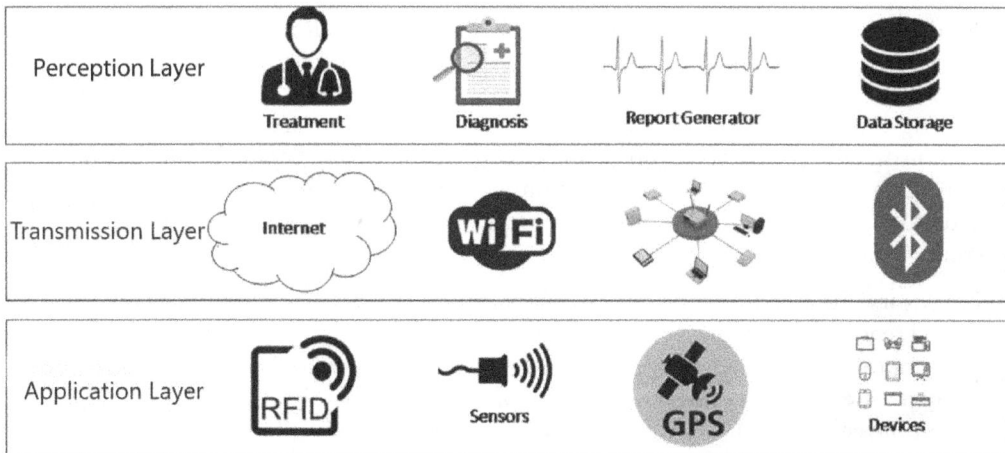

Figure 8.1 Cyber-physical system layers.

mentioned above, guarantees information routing and communication via fog foundation services, networking modules, swapping and Internet gateways, firewalls, and Intrusion Detection/Prevention Mechanisms (IDS/IPS) (Sommestad et al., 2010). To bypass interruptions and vicious muggings, including ransomware, spiteful cryptogram injection, Denial of Service/Distributed Denial of Service (DoS/DDoS), eavesdropping, and furthermore, unlawful admittance initiatives, it is important to ensure their delivery until outsourcing data content (Weiss, 2010). This poses a provocation, principally concerning resource-constrained machines, owing to the overhead commanded in phases of the processor and energy sources demanded. Overhead includes both computational demands on the processor and the energy resources required.

- **Application Layer:** This signifies a single layer that is threefold, moreover, and social. This layer devises some transmission layer information and furthermore carries out the physical device's instructions, including those of the sensors and actuators. This is accomplished by the application of complicated decision-making algorithms concentrated on aggregated information (Saqib et al., 2015). Besides, before evaluating the correctly invoked automatic behavior, this layer receives and processes knowledge from the perception layer (Khan et al., 2012). Algorithms for cloud storage, middleware, and information tunneling are utilized to handle this layer's data. Protecting and retaining privacy involve protecting the leakage of sensitive knowledge. Anonymization, information masking (camouflage) (Roy et al., 2019a), and isolation protection, including hidden distribution, are among the most well-known protective methods. Besides, this layer often includes a substantial multifactor authentication classification to avoid unauthorized admittance and privilege intensification. The scale of the data produced has become a big concern because of the rise in Internet-connected devices (Kumar and Patel, 2014). Thus, securing large data calls for appropriate encryption strategies to handle vast data volumes in a timely and reliable manner.

8.2.1.2 CPS Components

For sensing erudition (Gries et al., 2017) concerning monitoring signals, CPS components are used (Figure 8.2). CPS components are divided in this respect into a couple of influential classes: Sensing Components (SC), which handle and sense erudition, and Controlling Components (CC), which track and regulate beacons.

- **Sensing Components:** The sensors remain essentially positioned in the perception layer and furthermore serve the purpose concerning sensors to collect and transmit information upon aggregators. This information is then assigned for further analysis to some actuators to obtain a reliable strategic thinking. These main protective-service-detecting components are listed below.

  ○ **Sensors:** To determine the quality of the data gathered, meaningful data is collected and reported following a correlation method called "calibration." Sensing knowledge is important since the options can focus on the interpretation of this information.

  ○ **Aggregators:** The statistics obtained from sensors is mainly analyzed at the transmission layer (i.e., routers, sockets, and access points) until the subsequent decision is issued (s). In reality, the aggregation of data is analyzed, and the results are gathered for a given goal, where this information is collected and presented after a statistical

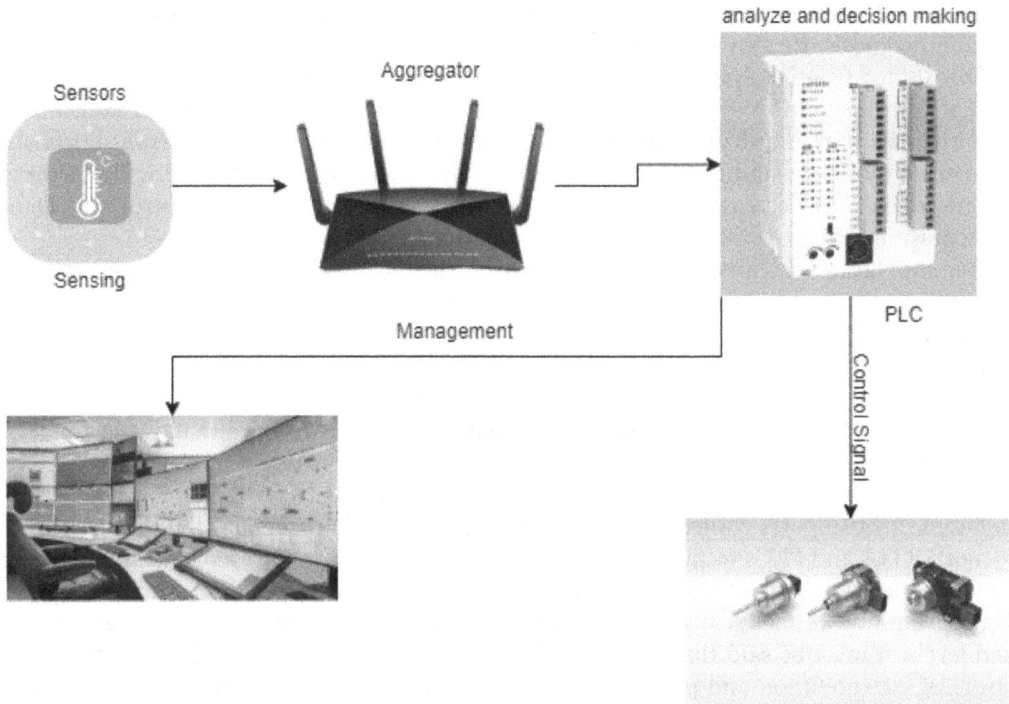

Figure 8.2 Infrastructure of cyber-physical system.

review. Online Analytical Processing (OLAP) is a main cluster formation method used as an online knowledge production reporting method.

○ **Actuators:** To render the details available to the external world based on the judgments taken by the aggregators, the implementation layer is located. Because actuators are highly dependent on certain network devices, any operation executed by the CPS is focused on a previous data aggregation sequence (Roy et al., 2019b). Following the definitions, actuators often process electrical signals as an input, generating physical behavior as an output.

- **Controlling Components:** To attain greater precision and security degrees from disruptive attacks or collisions, primarily signal blasting, noise, and intrusion, they imply managing beacons and operating an absolute position in beacon administration, tracking, and superintendence. As a consequence, it has become indispensable to focus on Programmable Logic Controllers (PLCs) and Distributed Control System (DCSs) onward including their elements [i.e., Programmable Automation Controller (PAC) (Mazur et al., 2012), Operational Technology/Information Technology (OT/IT) (Morelli et al., 2019), Control Loop/ Server (Vogel and Zack, 2006), and Human Machine Interface (HMI)/Graphical User Interface (GUI) (Ardanza et al., 2019)]. First, we mention the numerous forms of control systems in use inside CPS structures.

- **Programmable Logic Controllers (PLCs):** Primarily conceived to substitute wired DPDT and SPDT relays, they are called the research-automated computers that monitor the production processes, including the output of robotic systems and/or the handling of fault diagnosis, thereby achieving greater flexibility and resilience.

- **Distributed Control Systems (DCSs):** Computer-controlled management systems are virtual instrumentation systems that enable the deployment of decentralized controllers within the system utilizing formal authority from a central operator. There is improvement in DCSs' reliability due to the item and management phase, while its implementation cost has decreased. DCS may be equivalent to Supervisory Control and Data Acquisition (SCADA) schemes in certain situations.
- **Remote Terminal Units (RTUs):** These are electronic devices which are powered by a microprocessor, such as that of the Master Terminal Unit (MTU) (Stouffer et al., 2011) or "Remote Telemetry Uni."(Roy et al., 2019c). They do not endorse any feedback loop and control algorithm, unlike the PLC(s). Therefore, they are more suited for wireless interactions in larger regional telemetry regions. The key role of the RTUs is to use a supervisory messaging framework to communicate SCADA to the specific object(s) that govern(s) these artifacts via the transmission of sensor information by the system.

In fact, a tiny computer-controlled "artificial brain" [Central Processing Unit (CPU)] is used by both RTUs and PLCs to process different components from smart sensors and pumping devices; thus, the usage of IEDs (Intelligent Electronic Devices) to relay streams of data or, in the event of an attack, causes an alarm. Concerning the interaction between instruments and levels, it must be said that sensing instruments are implemented predominantly at the layers of interpretation and propagation. In contrast, the regulating modules are installed at the level of operation.

8.2.2   CPS Model Types

It is possible to classify CPS models into three major types:

- **Timed-Actor CPS:** In comparison to nonfactors focused on efficiency and timing, this model helps with the effective aspects centered on action and correctness. A (Geilen et al., 2011), a principle was implemented with a practical and traditional convergence that limits some collection of activities and increases efficiency by reducing sophistication. The primary emphasis is on the optimization focused on the concept of "earlier-the-better" as it allows the opportunity to recognize probabilistic abstract concepts of semi systems. As is experienced, such time-deterministic frameworks imply undersized susceptibility over problems with phase bang, making it easier to extract analytical bounds.
- **Event-Based CPS:** In these models, before another sensor decision is being made, an occurrence necessity does sense moreover observed with the right CPS components. However, based on the nondeterministic device delay induced by the numerous CPS behaviors, namely sensing, actuating, communicating, and computation, the individual part-timing constraints differ. Hu et al. (2016a) suggested that season limits may occur controlled by utilizing an event-based method that uses CPS activities to ensure the device's coordination, calculation, and control processes. This makes it possible for the CPS to become more relevant and more effective for spatiotemporal knowledge.
- **Lattice-Based Event Model:** In Tan et al. (2010), together with local and global event attributes, CPS events are portrayed as per the type of event. If these events are mixed, they

may be used to describe any event's spatiotemporal property while additionally defining all those elements that a particular conclusion has identified.

- **Hybrid-Based CPS Model:** Hybrid CPS schemes remain complex structures consisting of two different types of interactive processes—a persistent state (physical functional networks) and a discrete state (discrete computer systems). The reaction of isolated transient events defined by neural networks and the complex action represented by the equation(s) of differential/difference depend on growth and evolution. Unlike different CPS variants, hybrid CPS is integrated through an interface, rendering it susceptible to procrastination. Also, no hierarchical simulation is supported by hybrid CPS frameworks, and they are not appropriate for the modeling of linear models. The simulation problems of hybrid systems caused by CPS have been addressed (Roy et al., 2016). In reality, Kumar et al. (2012) discussed and solved CPS device network latency problems using real-time hybrid authentication. In contrast, Tidwell et al. introduced a customizable actual hybrid structural test for it. Finally, Jianhui (2011) proposed an event-driven control of CPS focused on hybrid automatons.

8.3 Cyber-Physical System Attacks and Vulnerabilities

Protection services also weren't built into CPS networks by default, compared to other communication networks, leaving room for criminals to manipulate numerous bugs and risks to initiate security breaches. The aforementioned is additionally attributable to just individual independent existence of CPS designs, as all use multiple technologies and protocols to work in separate IoT domains and interact.

8.3.1 Cyber-Physical System Attacks

The numerous types of threats that threaten the various facets of CPS networks, particularly cyber and physical ones, are presented in subsequent subsections.

8.3.1.1 Physical Attacks

In previous years, direct attacks have been more successful, in particular toward industrialized CPSs. Most of these assaults have already been responded to. Nonetheless, this essay presents a wider variety of forms of material intervention:

- **Infected Items:** Aforementioned group covers infected CDs, USBs, computers, also drives, and before-mentioned while individual Stuxnet worm, which, when embedded within a cyber-physical computer, piles up on clandestine malware comprising malicious programs.
- **Abuse of Privilege:** If rogue or dissatisfied staff obtain access to the CPS domain's server rooms and implementation zones, they are committing abuse of privilege attack. This can help them implant a rogue USB for exploitation or catch sensitive details by running harmful security software or as a keystroke.
- **Wire Cuts/Taps/Dialing:** Since the contact lines of certain cyber-physical headquarters (HQs), including telephony and Wi-Fi, are already physically observable, attackers may break the cables or wiretap them to decrypt the individual transmitted data.
- **False Identity:** As perpetrators try to portray themselves as legitimate workers, this assault happens with ample expertise to deceive others. To achieve easy access and greater

contact with other staff, they mostly serve as cleaners. The Maroochy Water Leak in Australia in 2000 (Slay and Miller, 2007) is an outstanding illustration of this.

- **Stalkers:** These would be typically legitimate workers who behave curiously (with sinister intent) by peering over the heads of CPS supervisors and designers to gain their trust to coerce or offer them positions in additional CPS organizations.
- **Surveillance of CCTV Camera Images:** This involves intercepting videos from closed-circuit news cameras that are used to protect access and important aspects within CPS zones. For carrying out a physical attack to be undetected, the aforementioned can conspire by collapsing camera beacons, condensing transmission cables, removing the video, obtaining admittance over some cloud handle and surveillance division, and so on.
- **Key-Card Hijacking:** This entails copying valid cards taken from workers or generating authentic lookalike counterparts to achieve full/partial admittance and breach specific field of the CPS.
- **Physical Breach:** this assault involves the procurement of unauthorized physical entry to the infrastructure, specifically by a physical breach, such as the 2011 Springfield Pumping Station case (Fillatre et al., 2017); a loophole, such as the 2013 Georgia Water Treatment Plant case (Credeur, 2013); or a security breach exploited, such as the 2012 Canadian Telvent Company case. This enables an intruder to disrupt and shut down CPS equipment and network-connected production networks, resulting in a lack of accessibility and efficiency.
- **Malicious Third-Party Information Provider:** A particular primary aim of the aforementioned assault transpires to threaten the organization's CPS by breaching specific legal software "Industrial Control Systems," a before-mentioned essentially individual 2008 shutdown concerning a particular Georgia Nuclear Power Plant. This involved swapping legal files in their libraries with software intended to include remote access capability to monitor or exploit a device.
- **Misuse of Privilege:** Acts usually accompanied by insiders, practically "whistle-blowers," to carry out or aid in the implementation of a (cyber)-attack from inside. These high privileges offer them the power to carry out these threats by disclosing useful information regarding the flaws and shortcomings in CPS programs. This exploitation of authority will come in several shapes and sizes.

  o **Physical Tampering:** This includes obtaining unauthorized or spoofing-approved entry through confined fields to disrupt CPS arrangements and equipment, alter their operating convention, insert malicious information, or appropriate sensitive reports.
  o **Unauthorized Actions:** These are focused on conducting suspicious functions, such as opening/closing pumping stations, increasing/decreasing power voltage, opening closed ports, interfering with an external agency, redirecting network traffic, or leaking information.

- **Social Engineering:** It may demand several manipulative manners as before-mentioned essentially invert architecture (impersonating the tech-savvy), baiting (selling disruptive USBs or software), tailgating (following approved staff), or arranging quid pro quo (impersonating technical support teams) and, furthermore, remains focused upon the science of exploiting individuals (psychologically rather than emotionally) to disclose sensitive erudition through influencing their emotions.

Hackers have lately shifted their focus to CPS networks for espionage, infiltration, warfare, terrorism, and service stealing (Mahato et al., 2021), especially cyber-warfare (Ray, 2020), cybercrimes (Choraś et al., 2016), and (cyber)-terrorism (Haimes, 2002) in, countries like Lebanon (Barakat, 2019). (cyber)-sabotage (Alenius and Warren, 2012) (for example, cyber intrusions toward Estonia in 2007 (Kaeo, 2007) moreover Georgia in 2008 (Donovan Jr, 2009)), about (cyber)-espionage (Yeboah-Ofori et al., 2019). The lack of (cyber)-security exposed a major problem among potentially catastrophic implications, especially in the United States.

8.3.1.2 Cyberattacks

There has been a spike in the specific number of cyber intrusions against CPS and IoCPT in recent years, with rather significant implications. CPS is particularly vulnerable to malicious SQL injection attacks (Francillon and Castelluccia, 2008) and script attacks (Roemer et al., 2012), simultaneously including false declaration intrusion offenses (Alemzadeh et al., 2016), zero-control info strikes (Hu et al., 2016b), and furthermore ultimately Control-Flow Attestation (C-FLAT) interventions, as per existing research carried out by Abera et al. (2016). As seen in Table 8.1, such assaults might lead to a complete blackout striking CPS industrial equipment and systems.

- **Eavesdropping:** Eavesdropping happens when nonsecure CPS web transactions are detected to access confidential details (passwords, usernames, instead of unspecified separate CPS knowledge). Eavesdropping may include any of the following methods: receptive by monitoring the transmission of CPS network messages, and, furthermore, dynamic by examining, searching, or tampering with some communication, pretending to obtain a valid source.
- **Cross-Site Scripting (XSS):** Happens meanwhile third-party network tools are utilized to insert malicious Coding Script inside a website's database, causing malicious scripts to

Table 8.1 Real-Time CPS Attack

Country	Target	Attack Name	Type	Date	Motives
The USA	Ohio Nuke Plant Network	Slammer Worm	Malware-DoS	January 25, 2003	Criminal
The USA	Georgia Nuclear Power Plant Shutdown	Installed Software Update	Undefined Software	March 7, 2008	Unclear
The USA	Springfield Pumping Station	Backdoor	Unauthorized Access	November 8, 2011	Criminal
Iran	Iranian nuclear facilities	Stuxnet	Worm	November, 2007	Political
Iran	Power plant and other industrie	Stuxnet-2	Worm	December 25, 2012	Political
Saudi Arabia	Saudi infrastructure in the energy industry	Shamoon-1	Malware	August 15–17, 2012	Religion-Political
Qatar	Qatar's RasGas	Shamoon	Malware	August 30, 2012	Political
Australia	Maroochy Water Breach	Remote Access	Unauthorized Access	March, 2000	Criminal
Canada	Telvent Company	Security Breach	Exploited Vulnerability	September 10, 2012	Criminal

run in the intended sufferer mesh browser (typically a targeted CPS programmer, constructor, or worker). XSS will hijack a victim's session and, in certain situations, register keystrokes and obtain remote access to their computer.

- **SQL Injection:** SQLi is a weakness that enables hackers to read and/or change sensitive data on CPS database-driven websites, as well as to conduct administrative tasks like database shutdown, particularly while CPS policies remain nevertheless utilizing SQL concerning information management.
- **Password Cracking:** This includes attempt to break the passwords of CPS users (primarily designers and administrators) utilizing brute force, dictionary (Narayanan and Shmatikov, 2005) (mitigated through key shift), rainbow table (Papantonakis et al., 2013), birthdays (mitigated via hashing), or online/offline password presuming interventions to obtain entry into individual identification database, or the incoming/outgoing transactions. As a consequence, it's necessary to stop more intensification.
- **Phishing:** This is accomplished in several ways, as before-mentioned in the forms of email phishing, vishing, spear-phishing or whaling, targeted at any complete CPS customer[such as programmers, experts, executives, Chief Executive Officers (CEOs), Chief Operating Officers (COOs), furthermore Chief Financial Officers (CFOs)], by imitating company peers or service providers.
- **Replay:** This involves impersonating ICSs, RTUs, and PLCs to intercept transmitted/received packets between them to trigger impediments that can impact CPSs' present-time regulations plus availability. These intercepted packets might, in certain circumstances, be changed, which would significantly disrupt regular processes.
- **DoS/DDoS:** DoS attacks exploit the infrastructure of the cyber-physical framework and moreover are initiated by a great range concerning computers that are locally compromised. DDoS attacks are normally carried out by Botnets, which use many infected computers to initiate a DDoS assault simultaneously from multiple geological positions. DoS interventions can gain various modes [i.e., blackhole (Al-Shurman et al., 2004), teardrop], whereas DDoS can reap significant subsequent patterns, both of which threaten CPS structures [i.e., ping-of-death, smurf (Kumar, 2007), including Black Energy sequence (BE-1, BE-2, and BE-3 (Khan et al., 2016)].
  - o **TCP SYN Flood:** This takes advantage of the TCP handshake mechanism by submitting requests to the server continuously without answering, forcing the server to reserve a room in expectation of an answer (Roy et al., 2018b). This results in an overload of the buffer, which creates the cyber-physical machine to fail.
- **Malicious Third-Party:** This involves malware that remotely attacks information-gathering networks and breaches them, specifically through utilizing botnets, Trojans, or worms to penetrate erudition via a CPS-encrypted tunnel of a constitutional device (i.e., PLC, ICS, or RTU) to a botnet Command-and-Control server through the use of a trustworthy third party within the mask. As a consequence, CPSs (Antonioli et al., 2018) and AMIs (Sgouras et al., 2017) are being attacked.
- **Watering-Hole Attack:** The intruder looks for some flaws in cyber-physical protection. Once a loophole has been found, a "watering leak" will be set up on the selected CPS website, where ransomware will be distributed by abusing the intended CPS framework, primarily via backdoors, rootkits, or zero-day exploits.
- **Malware:** Malware denotes ways to hack CPS machines to capture or drip data, damage appliances, or circumvent admittance command arrangements. This malware may

necessitate several different types of assault, but the most popular ones that threaten CPS are concisely described and conferred below.

- ○ **Botnets:** Aforementioned entails manipulating the flaws of CPS machines to transform them into bots or zombies, specifically to carry out DDoS assaults that are barely traceable [i.e., Ramnit (2015) (De Carli et al., 2017), Mirai (2016)(Kolias et al., 2017), Smominru botnet (2017) (De Carli et al., 2017), Mootbot (2020) (Seering et al., 2018), Wild-Pressure furthermore VictoryGate (2020)].
- ○ **Trojan:** Trojan implies a malicious mask that looks genuine but fools users into installing that. The Trojan infects the computer when it is downloaded and provides indirect admittance to withdraw data credentials including track user actions. Turla (2008), MiniPanzer/MegaPanzer (2009) (Scott and Chen, 2013), Ghost RAT (2009)(Boinapally et al., 2017), Shylock (2011)(Murdoch and Leaver, 2015), Coreflood (2011)(Hendraningrat et al., 2013), DarkCornet (2012) (Farinholt et al., 2017), MEMZ (2016), Tiny-Banker (2016) (Gostev et al., 2016), Banking.BR Android, and Botnet (2020) are examples of Remote Access Trojans that can transform a computer into a bot.
- ○ **Virus:** It may reproduce and propagate to other machines with the aid of humans or nonhumans. Viruses infect CPS computers and steal knowledge by linking themselves to other executable codes and programs.
- ○ **Worms:** These propagate by leveraging OS vulnerabilities to damage owner webs by bringing payloads to withdraw, alter, or erase info or overwhelm web-servers [aside from Stuxnet, Flame, and Duqu, e.g., Code Red/Code Red II (2001) (Cowie et al., 2001), Nimda (2001) (Machie et al., 2001), Triton (2017) (Di Pinto et al., 2018)].
- ○ **Rootkit:** Rootkit means to enter or manipulate a device remotely and covertly to achieve records, strip data by altering machine settings (i.e., Moonlight Maze (1999), and introduce Blackhole exploit kit (2012).
- ○ **Polymorphic Malware:** the identifiable entity shifts repeatedly and regularly to prevent getting identified and enhances indistinctness toward every pattern-matching identification method.
- ○ **Spyware:** Spyware is a harmful program remotely mounted on a computer for surveillance purposes having an outwardly exceptional understanding of the consumer or consent (e.g., surveillance, reconnaissance, or scanning). In reality, they [(ProjectSauron (2011) (Adams et al., 2020), Dark Caracal (2012), Red October (2013) (Chavez et al., 2015), Warrior Pride (2014)(Marquis-Boire et al., 2015), Fin-Fisher (2014), furthermoreCOVID-19 spyware) may be used for potential cyberattacks.
- ○ **Ransomware:** Ransomware signifies a spiteful malware that utilizes CPS vulnerabilities to retain and encrypt CPS data as a hostage, targeting oil refineries, power grids, industrial plants, and medicinal stations, also encrypting complete data reinforcements before a ransom is charged. Siskey (2016), SamSam (2016), Locky (2016), Jigsaw (2016), Hitler-Ransomware (2016), WannaCry (2017), Petya (2017), Bad-Rabbit (2017), Maze (2019), and Ekans (2020) ransomware are only a few cases.

- • **Side-Channel:** This is focused on data gathered of individual CPS devices, as beforementionedfor measuring information and energy usage, including electromagnetic losses that can be manipulated.

In fact, as early as the 1980s, Do et al. provided a far more thorough explanation of the attack (Fillatre et al., 2017). On the other hand, this chapter attempts to characterize these attacks' frequency as early as 2000, focusing on several reasons including, though not limited to, national, social, and criminal intentions.

8.3.2 Cyber-Physical System Vulnerabilities

A weakness is a software weakness that can be used for the intent of manufacturing reconnaissance (reconnaissance preferentially into effective interventions). Each vulnerability evaluation thus entails defining and assessing the available vulnerabilities of the CPS while at the same time identifying effective improvement plus precautionary steps to minimize, alleviate, and instead likewise remove the unspecified vulnerability. There exist three major categories of CPS vulnerabilities:

- **Network Vulnerabilities: They involve** shortcomings in preventive protection mechanisms, as well as breaching transparent wired/wireless contact and links, such as man-in-the-middle, eavesdropping, replay, sniffing, spoofing, communication-stack (network/transport/application layer) attacks (Zhu et al., 2011), backdoors (Nash, 2005), DoS/DDoS, and packet manipulation attacks (Amin et al., 2013).
- **Platform Vulnerabilities:** They involve vulnerabilities in device, applications, setup, and databases.
- **Management Vulnerabilities:** The absence of security protocols, processes, and practices are examples of management vulnerabilities.

Vulnerabilities exist regardless of several causes. There are, however, three primary triggers of vulnerability as explained here:

- **Assumption and Isolation:** In most CPS designs, they are focused on the trend of "defense through anonymity." Consequently, the emphasis here is on developing a dependable and stable framework but still considering the deployment of required security facilities without implying that networks are entirely disconnected from the outside world.
- **Increasing Connectivity:** As the network grows more connected, the attack surface becomes greater. Manufacturers have developed CPSs by introducing open networks and open wireless applications, as CPS devices have become more linked in recent years. Up until 2001, the bulk of ICS assaults were focused on internal attacks. This was before the Internet, which changed the concept of threats from the outer world.
- **Heterogeneity:** CPS solutions contain heterogeneous modules from third parties combined to construct CPS implementations. Consequently, CPS has developed into a multi-vendor scheme, with separate protection weaknesses for each product.
- **USB Usage:** Aforementioned stands a big trigger concerning vulnerabilities in CPS, before-mentioned since the particular matter concerning this Stuxnet strike against Iranian energy plants, as unique USB is within specific malware. The malware spread through many computers after it was plugged in due to manipulation and duplication.
- **Poor Practice:** This is mainly due to bad coding/weak abilities, which allow the code to perform endless loops or becoming too simple to be changed by an intruder.
- **Spying:** CPS networks are often susceptible to spying/surveillance assaults, which are carried out mostly by the use of spyware (malware) forms that obtain enigmatic admittance and furthermore stay undetected for extended periods, among the particular primary purposes of eavesdropping, stealing, and collecting sensitive data including knowledge.
- **Homogeneity:** Related forms of cyber-physical networks enduring specific similar flaws that can impact any of the equipment in their proximity until activated, a prime

illustration being the Stuxnet worm assault upon Iranian nuclear energy plants (Iasiello, 2013).

- **Suspicious Employees:** Through undermining moreover changing specific code writing, or through allowing distant entrance to hackers via specific availability regarding secured ports or obstructing an infected USB device, suspicious employees may deliberately or unintentionally damage or hurt CPS computers.

CPS weaknesses may also be of three kinds, namely electronic, human, furthermore meanwhile fused, all occur in any physical, cyber hazard.

8.3.2.1 Cyber Vulnerabilities

ICS implementations are vulnerable to security breaches considering that ICS depends massively upon public criterion etiquettes, including the Inter-Control Center Communications Protocol (ICCP) (Gungor et al., 2011) and also the Transmission Control Protocol/Internet Protocol (TCP/IP). In reality, ICCP experiences some crucial vulnerability to buffer surplusage (Zhu et al., 2011) and lacks fundamental protection measures. In reality, the Remote Procedure Call (RPC) protocol and ICSs are susceptible to different types of attack, including Stuxnet (1 and 2) (Karnouskos, 2011) and Duqu malware (1.0, 1.5, and 2.0) (Bencs´ath et al., 2012), Gauss malware (Bencs´ath et al., 2012), plus RED October malware (Chavez et al., 2015), as well as Shamoon malware (1, 2, and 3) (Dehlawi and Abokhodair, 2013).

Interception, sniffing, eavesdropping, wiretapping, wardialing also wardriving attacks (Francia III et al., 2012) and furthermore, meet-in-the-middle attacks are all feasible with open/non-secure wired/wireless communications like Ethernet. Wireless short-range communications are often sensitive, as insiders may intercept, evaluate, harm, erase, or even exploit them. Besides, if not safe, employees' linked tools to the ICS transpacific interface are susceptible to the botnet, indirect admittance Trojan, and furthermore rootkit interventions. Their devices are managed remotely by an attacker. Eavesdropping, replay assaults, and unintended connection attacks are all feasible for long-range wireless communications. However, SQL injection endures greatest significant web-related weakness. By inserting a malicious code that proceeds to operate indefinitely until executed without the user's awareness, attackers may enter any application database without authorization.

Since certain medicinal tools rely tediously upon radio transmissions, all are susceptible to a range of radio interventions such as jamming, alteration, and furthermore, replay due to a deficiency concerning encryption. Furthermore, GPS moreover particular machine microphones are already being used as a monitoring mechanism, enabling the target's position to be calculated or in-car conversations to be overheard by eavesdropping.

In contrast, ICS relies upon the Modbus plus DNP3 protocols to track sensors and actuators and command them. According to Humayed et al. (2017), the Modbus protocol lacks simple protection mechanisms before-mentioned such as encryption, authentication, including authorization. The aforementioned becomes rendered it vulnerable to eavesdropping, wiretapping, and port-scan, with the possibility of spoofing the controller by fake insertion of information. The DNP3 protocol is often vulnerable to the same bugs and threats, with the Cyclic Redundancy Review (CRC) integration essentially a probity measure being one of the key distinctions. Besides, Windows Server is susceptible to remote code execution, with further assaults being achieved in every running operating device by manipulating buffer overflow vulnerabilities (OS).

Furthermore, since smart grid power system architecture is built on the same protocols as ICS, Modbus, and DNP3, it is subject to the same bugs as ICS, Modbus, and DNP3. Consequently, the IEC 61850 etiquette was adopted in substation interfaces, which historically ignored protection features and were susceptible to eavesdropping therefore, culminating in attacks of interference or attacks of fake knowledge injection (Wang and Lu, 2013). Santamarta et al. reviewed specific convenient smart meter documentation and inhabited a "factory login" reckoning to execute simple arrangements. The aforementioned allows the customer total leverage covering an intelligent meter, resulting in energy outages, incorrect judgments, and smart meters targeting the same network. Furthermore, multiple computers are susceptible to battery-draining attacks (Rushanan et al., 2014).

Gollakota et al.(2011) and Halperin et al. (2008) manipulated specific Implantable Cardioverter Defibrillator (ICD) by injection assaults. The inventors have additionally demonstrated that intelligent cars are susceptible to multiple forms of assault. Another weakness was found by Radcliffe with Continuous Glucose Monitoring (CGM) sensors becoming defenseless to replay assaults. The CGM unit had been spoofed amidst inaccurate preferences being injected. The aforementioned is attributed to a specific point that when smart cars were built, security concerns were not considered. In fact, the protocol of the Controller Area Network (CAN) suffers from several vulnerabilities, which could rise while aggression against intelligent cars is exploited. A DoS assault would be more probable as a consequence of this. Because of the absence of security, a Tire-Pressure Monitoring Device (TPMS) is often vulnerable to eavesdropping as well as spoofing. Also, it is possible to explicitly manipulate Adaptive Cruise Control (ACC), which regulates replacement regarding the CAN system. A well-equipped intruder is capable of interrupting the activities of ACC sensors by introducing turbulence or spoofing. Consequently, you can monitor the vehicle by slowing it down, speeding it up, or even crashing it.

8.3.2.2 *Physical Vulnerabilities*

Tangible tampering may trigger data in cyber-physical segments to be corrupted. In reality, it is physical attacks that had a cyber effect. Due to the inadequate physical protection offered for these components, ICS components' physical visibility is regarded as a vulnerability. As a consequence, they're susceptible to physical tampering, change, adjustment, and even sabotage. CPS area systems (i.e., smart grids, power grids, supply chains, etc.) are subject to significant related ICS flaws when exposed without physical protection to a vast range of physical elements, rendering them liable for physical damage. As a consequence, Mo et al. emphasized identification and preventive solutions in their paper. Humayed et al. suggested (2017) that medicinal tools are susceptible to dynamic entry, accompanied by a significant risk regarding downloading malware inside them or constantly altering the system's settings, compromising the welfare of the user. Besides, dynamic admittance to any medicinal system is often a risk since the serial number of the device may be retrieved by an intruder to conduct targeted interventions.

Essentially mentioned earlier, CPS arrangements endure different flaws, rendering them vulnerable to varying standards of assault.

8.4 Characteristics of Adversaries

This section explains many key forms of future adversaries:

- Experienced hackers are advanced programmers having special expertise to recognize complex bugs in current applications furthermore produce operating exploit codes.

- Dissatisfied insiders having spiteful intention do not necessarily have a great understanding of cyber intrusion expertise, and their information of the game operation additionally empowers them to acquire unrestricted admittance to overthrow the conformity or to appropriate system data, which are known to be the key source of cybercrime and sabotage; workers, contractors, or business associates may be the types of insiders.
- The ransom will be the prime motive for a criminal organization conducting an assault on a cyber-physical infrastructure (O'Connell, 2008).
- Many terrorists want higher-impact objectives in a single nation, before-mentioned essentially aero orderliness or energy network systems, and all could acquire the particular capability to pull these vital cyber-physical installations down. Also, by employing highly trained coders, hiring control device developers, and bribing insiders, they can presumably attempt to accomplish the target (Wan et al., 2010).

The CPS defenders should follow the necessary policies or tactics to react to the threat based on adversaries' types. Furthermore, to develop hazard models, researchers may obtain a deeper comprehension of foe features and the potential to predict an adversary.

8.5 Failure of CPS

It is necessary to illustrate the key failures experienced by CP systems, considering the various risks, interventions, and vulnerabilities that a specific CPS field experiences. These errors may be small (causing just minor damage) or severe (causing serious damage) (severe damage). More knowledge is available in Avizienis et al. (2004), as they offered a properly established plus thorough description.

- **Content Loss:** This indicates that each content concerning specific information transmitted is unreliable, occurring in the failure of the functional structure. The lack of content may imply both numerical or non numerical mode (i.e., alphabets, graphics, sounds, or colors).
- **Timing Failure:** This indicates that specific timing (transmission/receipt) of the processing of information is postponed or disrupted (collected/transmitted too short-stemmed or exceedingly belated). The aforementioned will change the specific methods of decision-making and could create difficulties with data administration.
- **Sensors Failure:** This suggests that specific sensors imply no long-drawn working correctly, presenting a significant challenge to decision-making owing to incorrect input or forcing the CPS device to a standstill. In 2005, at the Taum Sauk Hydroelectric Power Station, a related event occurred.
- **Silent Failure:** This happens in a distributed device where there is no communication sent or received.
- **Babbling Failure:** This happens as the data is provided, allowing the device to crash and work in a babbling fashion.
- **Budget Failure:** This happens where the significant expense of installing a cyber-physical device reaches the resources allocated ere the system is checked. This is largely attributed to a lack of preparation.
- **Schedule failure:** This happens when due to more upgrades, further tests, or insufficiency for customer requirements, the schedule provided for preparing, testing, and reviewing a specified CPS is not met.

- **Application Failure:** This happens where the service interface propagates an error and impacts its decision-making capacity or/and usual output quality. This failure will result in a partial or total failure of the CPS device, which can be temporary or permanent.
- **Consistent/Inconsistent Failures:** A consistent failure arises when all CPS customers view a specified service identically. If all CPS customers interpret an erroneous service differently (i.e., Bohrbugs, Mandelbugs, Heisenbugs, and Byzantine defeats), this is referred to as an incomplete failure.

8.6   Recommendation for CPS

Various monitoring mechanisms should be enforced and strengthened to increase defense against various risks and assaults. This involves, among others:

- **Prioritization and Classification:** Prioritization and classification of sensitive CPS elements and properties before evaluating, handling, and reviewing threats to ensure adequate expenditure on the right protection initiatives (basic, normal, or advanced) concerning their costs versus the probability of an incident happening and its effects.
- **Careful Financial Preparation and Management:** To safeguard critical/non critical CPS properties and materials, they must be carried out for feasible expenditure and required costs/resources.
- **Lightweight Dynamic Key-Based Cryptographic Algorithms:** Certain methods may continue to be utilized to maintain a range of security resources, including message secrecy, uprightness, and authentication, both of which transpire needed throughout protected CPS interactions. The aforementioned can imply accomplished through utilizing the latest contemporaries regarding cryptographic algorithms outlined inside the following sections (Roy et al., 2018c). The value of these strategies is that they will strike a reasonable compromise between the degree of protection and efficiency. Since a dynamic key is used per post, the strength against attacks has been demonstrated (preferentially an assemblage of information, depending on utilization compulsions also specifications). Besides, to generate a collection of cryptographic primitives and to change cryptographic primitives, the aforementioned changing key is utilized. As a consequence of the various cryptographic primitives used, the unconventional ciphertext may be produced for specific identical plaintext. However, these algorithms' reliability is illustrated because they only need one round of iteration and employ basic operations while preventing diffusion. The latest contemporaries regarding certain cryptographic algorithms decrease significant latency, energy, moreover aloft processing needed, allowing CPS tools to manage their key functionalities better.
- **Defining Privileges:** Aforementioned can act known essentially individual various effective admittance management strategy that grants permits furthermore exemptions based about specific roles/tasks/attributes of the users meanwhile this appears toward locating CPS and when finishing the job or on leave of the employee, withdrawing certain access rights. The usage of the least privilege strategy comes under this as well. The privilege concept could then be focused on the Attribute-Based Access Control (ABAC) framework, where access permissions are specified by policies coupled with attributes. It's worth remembering that ABAC allows access control judgments dependent on attribute values' Boolean conditions. It offers a high degree of granularity required to render the access scheme for CPS control more reliable.

- **Powerful Multi-factor Authentication Entity:** Sadly, entity authentication systems focused upon each individual authentication determinant imply being nonadequately immune to authentication assaults, which imply exponentially enhancing severe effects. Each entity authentication mechanism is the primary route of protection under each conformity, and essentially any actual authentication assault will contribute to an attack on secrecy, credibility, plus availability. The principle of multi-factor authentication has recently been adopted by incorporating two or more further factors: 1) "you are" involving system fingerprint, user fingerprint, guidance geometry, iris scan, retina scan, etc., and 2) "you have" involving cryptographic keys to improve its robustness toward authentication interventions before-mentioned essentially those mentioned in (Melki et al., 2019). In addition to the usage of geographical positions, this mechanism should be an important prerequisite for CPS programs. These solutions benefit from reducing false negatives and complicating authentication attacks by compromising several variables rather than only one. This also restricts access to designated agencies and employees (devices/users) only.

- **Solid Password and Complex Hashing Process:** Passwords are called the authentication element for "you know." However, it is necessary to enforce multiple assaults, before-mentioned in terms of rainbow and hash table advances. To avoid them from happening, passwords' obligation continue to be re-hashed including a unique dynamic Nonce concerning a particular individual after a periodic interval. Furthermore, a stable cryptographic hash function, before-mentioned being SHA-3 or SHA-2, should be used (variant 512). Birthday attacks are avoided, furthermore, rainbow/hash table attacks are minimized.

- **Safe and Secured Audit:** This can be achieved with the aid of an audit manager framework that gathers and preserves logs in a distributed system. A potential approach that can be implemented in this sense has recently been proposed (Noura et al., 2020). This restricts any insider effort toward a cyber-physical structure and to track them back, and it protects the digital records of internal and external threats.

- **Enhanced Non-Cryptographic Solutions:** Though these, expect mutated IDS/IPS or AI-based IDS/IPS systems (using Machine Learning algorithms), coupled by sophisticated firewalls (i.e., Device as well as Next-Generation Firewalls) and customizable honeypots, to avoid any possible vulnerability-based protection breach. This can be accomplished by the usage of lightweight IDS/IPS moreover anomalybased individuals in particular. In particular, the anomaly detection algorithm can be chosen according to specific CPS' limitations, which may obtain analytical restricted devices instead of those that are based upon computer algorithms, as before-mentioned essentially random forests, for strong CPS tools. Approaching particular hand, signature-based techniques should be continued through a unique Gateway (GW), wherever total web transfer can be transpired and evaluated.

- **Safe and Checked Backups:** Theses remain important to ensure specific availability of CPS data and prevent information elimination either modification through securing robustness toward DoS/DDoS including Ransomware strikes, remarkably since before-mentioned interventions may direct to complete blackouts while within the specific fact regarding the US. These can be achieved with the aid of lightweight data security strategies like those mentioned in Noura et al. (2019).

- **Forensic Efforts:** It is necessary to recover the traces of any assault that happens. It is therefore important to incorporate modern solutions to anti-forensic strategies to retain some digital proof (Noura et al., 2020). This is achieved by retrieving records including tracking channels plus device actions, effectively restricting multiple reconnoitering endeavors. Nevertheless, those fresh advanced forensic instruments remain compatible with the software/hardware of numerous CPSs, particularly resource-constrained machines, and moreover, necessity likewise remains immune to anti-forensic trials.

- **Enhanced Approach to Incidents:** This requires the capacity to recognize, warn, and react to a specific event. To minimize threats, proposals for event response and investigation can be enforced. This offers defense against accidental technological and organizational failures (power loss, blackout) through contingency arrangements, and against malicious failures (cyber initiatives), by teams of CERT (Computer Emergency Response) and CSIRT (Computer Security Incident Response), including IRCF (Incident Response and Computer Forensics). Essentially to maintain an improved and productive cyber, physical, and device ecosystem with safe computing and communications, CPS experts plus designers undertake additional training and practice.

- **Real-Time Monitoring:** Utilizing advanced forensics or non-forensic software and techniques to operate real-time applications is vital to avoiding any cyber-physical device malfunction, whether unintentional or not. This helps CPSs' actions to be continuously tested and tracked and, thus, identify unspecified cyber threat endeavors under their immature platforms.

- **Security Check:** Staff monitoring necessity imply carried out ere including throughout specific work for each employee to remove moreover accommodate unspecified potential effort by a whistle-blower. Accordingly, it is strongly advisable to sign agreements such as the Non-Disclosure Agreement (NDA), the Confidentiality Agreement (CA), the Confidential Disclosure Agreement (CDA), the Proprietary Knowledge Agreement (PIA), or the Privacy Agreement (SA). These security checks are particularly relevant in vulnerable areas like nuclear power plants.

- **Periodic Employee Training:** This requires periodic ICS and PLC employee sensitivity training about the best practices of data protection depending on their degree and experience and the potential to spot any unusual action or operation. Employees must also be educated on a range of technology risks and poor behaviors, such as stopping downloading any app upgrades, avoiding social engineering and phishing attempts, and preserving transparency in the event of misconduct.

- **Daily Pen Monitoring and Vulnerability Assessment:** This must be conducted on a routine basis to implement device auditing, identify intimidations, including resolving them in real-time until all are detected furthermore misused via an intruder beneath zero-days misuse statuses.

- **Periodic Risk Evaluation:** The potential and effect regarding a proffered danger upon a critical/noncritical cyber-physical framework based upon a qualitative/quantitative danger assessment and a Cost–Benefit Analysis (CBA) must both be enforced to study and identify the danger based upon an adequate/non-adequate mode, including minimizing this soon while practicable.

- **Up-to-Date Devices:** Cyber-physical policies necessarily remain up-to-date in the areas of applications, firmware, including hardware by implementing checks and upgrades daily. Furthermore, such structures must be safe at various stages of deployment (layered protection), including the significant potential to alleviate and furthermore react to an intervention to minimize its effect including avoiding additional intensification and furthermore harm. Also, to avoid any payload injection, USB ports obligation signify corporally including inevitably disabled. The actions and behavior of PLC devices must be continuously checked concerning unspecified questionable behavior (Serhane et al., 2018).

- **AI Encryption Solutions:** In IDS/IPS aberration identification mechanisms or in "you are" or "you do" object authentication schemes, artificial intelligence is used. AI has immediately been seen essentially as a game-changing approach toward a range of cyber-physical strikes targeting CPS networks, computers, and, moreover, features of contact. Given the experience and resources needed to train an AI device, the accuracy of identification and avoidance is superior to human intervention. CPSs can be rendered to be more stable, resilient, and immune to cyber-physical attacks by recent developments in machine learning, furthermore unexpectedly in DL.

- **Defense In-Depth:** Most current technologies have a defense against a particular feature of attack or a necessity for security. Instead, a multipurpose security approach is needed for each CPS operating layer (perception, transmission, and application). For example, ISO 26262 and IEC 61508/Edition2, the two most accepted international requirements for practical protection in the automobile industry, should be valued and implemented. This guarantees a stable CPS deployment focused on technical protection, which requires the Protection Integrity Standard (SIL) basics, which in twist depend proceeding each Likelihood regarding Failure on Demand (PoFoD) moreover specific Risk Mitigation Factor (RMF) to maintain a far innumerable reliable moreover effective Danger and Risk Analysis (RA), primarily in those Electronic Control Units (ECU).

- **CPS Protection and Privacy Life Cycle:** Ultimately, our chapter provides a combined life cycle of Organizational and Practical Safety/Security (OPSS) to summarize this task, ensuring a good and secure CPS task. This system is based on the ISO 26262 and IEC 61508/Edition2 etiquette and their guide to maintaining CPS functional safety and protection. The system is composed of six major stages:

  o **Step 1:** Designing a strategy to build a CPS framework in compliance with the appropriate budget and associated costs by implementing a well-defined timetable. It often requires the help of various individuals (people in business, mechanics, managers, etc.) including non human capital (vehicles, machines, etc.).

  o **Step 2:** Step 2 necessitates a detailed risk and threat review, which involves adequate risk assessment and asset recognition, as well as the reciprocal interaction among a particular couple, to maintain correct decision-making on specific execution regarding specifically required protection countermeasures.

  o **Step 3:** Elaborates the necessary specifications for practical protection, security, and durability along with the main elements that are important to minimize danger and, in the event of their occurrence, to reduce their chance and effects.

  o **Step 4:** In terms of the newly implemented practical protection, protection, and reliability steps, the performance of CPS is assessed in an organizational manner in which performance monitoring and review would be carried out to ensure proper/mutual security output, protection enforcement, and reliability-performance contracts-offs.

  o **Step 5:** The cyber-physical device is checked and validated until the performance is measured to find any residual software/hardware error, protection gap, or performance problem and implement the expected alterations that were being commissioned. If the specific trial is ineffective, the specific method restarts repeatedly to assess when the concern necessitated position. If effective, before being officially deployed, the CPS would head for further commissioning.

    ○ **Step 6:** The implemented CPS device will undergo a trial phase after satisfactory tests to ascertain its operating state, thus tracking its actions and efficiency before being completely operational.

8.7　Present Work and Future Research Direction

CPSs are gradually being embraced (installed) and applied in new sectors by various vehicle businesses. This implies that future CPSs will confront a growth in functional and extra-functional needs. Condition monitoring and growing CPS automation capabilities are examples of functional needs. Safety, cyber security, and ecological responsibility are examples of extra-functional criteria. Many emerging CPS solutions are inherently inter-connected in nature (the capacity to access or communicate information to various security domains seamlessly). For example, autonomous autos that use robotics technology and telecom networks for smart devices both present significant prospects and problems. Technological advancements enable new sorts of coordination and interaction in CPS in the following areas:

- Physical, integrated, networked, and systems engineering are examples of technological domains; consider cloud and edge computing.
- Intelligent transportation systems (vehicular and infrastructures integration) are examples of stand-alone systems.
- Full cycle stages, in general, make data accessible throughout the life span and allow software updates. These are DevOps (Development–Operations integration) ideas related to continuous software development, integration, and deployment with input from operational systems.

CPS is rapidly emerging because of increased levels of automation and intelligence, especially data analytics. These qualities Increased levels of automation, Intelligence, and data analytics are determined through artificial intelligence approaches such as machine learning. According to a report from the National Science and Technology Council in the United States (Cowie et al., 2001), AI and data analytics potential are viewed as game changers. Context awareness is critical for new sorts of AI-based CPSs, particularly the capacity to identify which entities are now present in the relatively close and deduce their intentions. The advancement of AI technology across a variety of application sectors is expected to generate higher degrees of automation. Because of their potential to address social concerns and create cash, future CPSs will be assigned more complex jobs in open environments. To counteract conventional manufacturing uses, extremely sophisticated technologies are being applied and disseminated across the society—for example, self-driving automobiles on public highways with no human involvement and robot–human collaboration systems.

Smart CPS is in charge of dynamic changes in the environment (e.g., highly changeable traffic situations that alter fast in response to changing human behaviors and CPS infrastructures). It typically indicates that not all operational circumstances are known a priori (at the time of system creation), as shown in the example by Branquinho (2018). The broader implications include that open CPS faces a variety of current and new types of uncertainty, including largely unknown surroundings, weaknesses in safety and assaults (predict attack by establishing "attacker models"), and altering CPS itself (due to partial failures). Uncertainty can apply to features of a CPS as well as to all stages of the life cycle.

Cliff Edge on Cyber-Physical Systems 173

It is obvious that the potential qualities of future CPSs would necessitate new methods of thinking about system-level attributes and composability.

8.8 Conclusion

CPS technologies are essential parts of Enterprise v4.0, and by combining the real and cyber environments, they are now explaining how people communicate with the either external surroundings. The aim of introducing CPS programs, whether inside or outside of IoT (IoCPT), is to increase the standard of goods and systems' availability and functionality. On the other hand, CPSs are plagued by a bunch of protection and data issues that jeopardize their dependability, protection, and performance, as well as obstructing their widespread adoption.

This chapter looks at the past of the CPSs and the security risks and vulnerabilities in cyber-physical structures and the guidelines for current CPS security models. We hope that these problems and concerns can provide ample impetus for potential discussions and study interests in CPS protection aspects.

In this chapter, we discussed the protection threats including concerns in a cyber-physical system. We divided this book chapter into three parts; in the first part, we discussed the background of the CPSs. After that, we recognized the reasonable vulnerabilities, attack arguments, opponent features, and an assemblage of provocations that necessitate being inscribed. We furthermore discussed some recommendations and failures in CPSs.

References

Tigist Abera, Tigist, N. Asokan, Lucas Davi, Jan-Erik Ekberg, Thomas Nyman, Andrew Paverd, Ahmad-Reza Sadeghi, and Gene Tsudik. C-flat: Control-flow attestation for embedded systems software. In *Proceedings of the 2016 ACM SIGSAC Conference on Computer and Communications Security*, pages 743–754, 2016. https://www.witpress.com/Secure/ejournals/papers/TDI040103f.pdf

Adams, N.P.H., Chisnall, R.J., Pickering, C., and Schauer, S. How port security has to evolve to address the cyber-physical security threat: Lessons from the sauron project. *International Journal of Transport Development and Integration*, 4(1):29–41, 2020.

Homa Alemzadeh, Homa, Daniel Chen, Xiao Li, Thenkurussi Kesavadas, Zbigniew T. Kalbarczyk, and Ravishankar K. Iyer. Targeted attacks on teleoperated surgical robots: Dynamic model-based detection and mitigation. In *2016 46th Annual IEEE/IFIP International Conference on Dependable Systems and Networks (DSN)*, pages 395–406. IEEE, 2016.

Kari Alenius, Kari, and M. Warren. An exceptional war that ended in victory for estonia, or an ordinary e-disturbance? Estonian narratives of the cyber attacks in 2007. *The Institute Ecole Supérieure en Informatique Electronique et Automatique, Laval, France 5–6 July 2012 Edited by*, page 18, 2012. https://www.researchgate.net/publication/290231968_An_exceptional_war_that_ended_in_victory_for_estonia_or_an_ordinary_e-disturbance_Estonian_narratives_of_the_cyber-_Attacks_in_2007

Mohammad Al-Shurman, Mohammad, Seong-Moo Yoo, and Seungjin Park. Black hole attack in mobile ad hoc networks. In *Proceedings of the 42nd Annual Southeast Regional Conference*, pages 96–97, 2004. https://doi.org/10.1145/986537.986560

Saurabh Amin, Saurabh, Xavier Litrico, Shankar Sastry, and Alexandre M. Bayen. Cyber security of water scada systems—part i: Analysis and experimentation of stealthy deception attacks. *IEEE Transactions on Control Systems Technology*, 21(5):1963–1970, 2013.

Daniele Antonioli, Daniele, Giuseppe Bernieri, and Nils Ole Tippenhauer. Taking control: Design and implementation of botnets for cyber-physical attacks with cpsbot. *arXiv preprint arXiv:1802.00152*, 2018.

Aitor Ardanza, Aitor, Aitor Moreno, Álvaro Segura, Mikel de la Cruz, and Daniel Aguinaga. Sustainable and flexible industrial human machine interfaces to support adaptable applications in the industry 4.0 paradigm. *International Journal of Production Research*, 57(12):4045–4059, 2019.

Yosef Ashibani, Yosef and Qusay H. Mahmoud. Cyber physical systems security: Analysis, challenges and solutions. *Computers & Security*, 68:81–97, 2017.

Algirdas Avizienis, Algirdas, J.-C. Laprie, Brian Randell, and Carl Landwehr. Basic concepts and taxonomy of dependable and secure computing. *IEEE Transactions on Dependable and Secure Computing*, 1(1):11–33, 2004.

Kristofas Barakat, Kristofas. *Does Lebanon Possess the Capabilities to Defend Itself from Cyber-Theats? Learning from Estonia's Experience.(c2019)*. PhD thesis, Lebanese American University, 2019.

Boldizsár Bencsáth, Boldizsár, Gábor Pék, Levente Buttyán, and Mark Felegyhazi. The cousins of stuxnet: Duqu, flame, and gauss. *Future Internet*, 4(4):971–1003, 2012.

Vamshika Boinapally, Vamshika, George Hsieh, and Kevin S. Nauer. Building a gh0st malware experimentation environment. In *Proceedings of the International Conference on Security and Management (SAM)*, pages 89–95. The Steering Committee of The World Congress in Computer Science, Computer ..., 2017.

Marcelo Ayres Branquinho, Marcelo Ayres. Ransomware in industrial control systems. What comes after wannacry and petya global attacks? *WIT Transactions on the Built Environment*, 174:329–334, 2018.

Lorenzo De Carli, Lorenzo De, Ruben Torres, Gaspar Modelo-Howard, Alok Tongaonkar, and Somesh Jha. Botnet protocol inference in the presence of encrypted traffic. In *IEEE INFOCOM 2017-IEEE Conference on Computer Communications*, pages 1–9. IEEE, 2017.

Raymond Chavez, Raymond, William Kranich, and Alex Casella. Red october and its reincarnation. *Boston University—CS558 Network Security*, 2015. https://www.cs.bu.edu/~goldbe/teaching/HW55815/presos/redoct.pdf

Michal Choraś, Michal, Rafał Kozik, Adam Flizikowski, Witold Hołubowicz, and Rafał Renk. Cyber threats impacting critical infrastructures. In *Managing the Complexity of Critical Infrastructures*, pages 139–161. Springer, 2016.

James Cowie, James, A. Ogielski, B.J. Premore, and Yougu Yuan. Global routing instabilities triggered by code red ii and nimda worm attacks. *Technical Report*, Renesys Corporation, 2001.

Mary Jane Credeur, Mary Jane. FBI probes georgia water plant break-in on terror concern, 2013.

Zakariya Dehlawi, Zakariya and Norah Abokhodair. Saudi arabia's response to cyber conflict: A case study of the shamoon malware incident. In *2013 IEEE International Conference on Intelligence and Security Informatics*, pages 73–75. IEEE, 2013.

Alessandro Di Pinto, Alessandro, Younes Dragoni, and Andrea Carcano. Triton: The first ICS cyber attack on safety instrument systems. In *Proceedings of the Black Hat USA*, 2018:1–26, 2018.

George T Donovan Jr, George T. Russian operational art in the russo-georgian war of 2008. *Technical Report*, Army War Coll Carlisle Barracks PA, 2009.

Brown Farinholt, Brown, Mohammad Rezaeirad, Paul Pearce, Hitesh Dharmdasani, Haikuo Yin, Stevens Le Blond, Damon McCoy, and Kirill Levchenko. To catch a ratter: Monitoring the behavior of amateur darkcomet rat operators in the wild. In *2017 IEEE Symposium on Security and Privacy (SP)*, pages 770–787. IEEE, 2017.

Lionel Fillatre, Lionel, Igor Nikiforov, Peter Willett, et al. Security of scada systems against cyber–physical attacks. *IEEE Aerospace and Electronic Systems Magazine*, 32(5):28–45, 2017.

Guillermo Francia III, Guillermo, David Thornton, and Thomas Brookshire. "Cyberattacks on SCADA systems." In *Proc. 16th Colloquium Inf. Syst. Security Educ.*, pp. 9-14. 2012.

Aurélien Francillon, Aurélien and Claude Castelluccia. Code injection attacks on harvard-architecture devices. In *Proceedings of the 15th ACM Conference on Computer and Communications Security*, pages 15–26, 2008. https://doi.org/10.1145/1455770.1455775

Marc Geilen, Marc, Stavros Tripakis, and Maarten Wiggers. The earlier the better: A theory of timed actor interfaces. In *Proceedings of the 14th International Conference on Hybrid Systems: Computation and Control*, pages 23–32, 2011. https://doi.org/10.1145/1967701.1967707

Shyamnath Gollakota, Shyamnath, Haitham Hassanieh, Benjamin Ransford, Dina Katabi, and Kevin Fu. They can hear your heartbeats: non-invasive security for implantable medical devices. In *Proceedings of the ACM SIGCOMM 2011 Conference*, pages 2–13, 2011. https://doi.org/10.1145/2018436.2018438

Alexander Gostev, Alexander, Roman Unuchek, Maria Garnaeva, Denis Makrushin, and Anton Ivanov. It threat evolution in q1 2016. *Kapersky 2015 Report*, Kapersky L, 2016.

Stefan Gries, Stefan, Marc Hesenius, and Volker Gruhn. Cascading data corruption: About dependencies in cyber-physical systems: Poster. In *Proceedings of the 11th ACM International Conference on Distributed and Event-based Systems*, pages 345–346, 2017. https://doi.org/10.1145/3093742.3095092

Vehbi C Gungor, Vehbi C., Dilan Sahin, Taskin Kocak, Salih Ergut, Concettina Buccella, Carlo Cecati, and Gerhard P. Hancke. Smart grid technologies: Communication technologies and standards. *IEEE Transactions on Industrial Informatics*, 7(4):529–539, 2011.

Yacov Y Haimes, Yacov Y. Risk of terrorism to cyber-physical and organizational-societal infrastructures. *Public Works Management & Policy*, 6(4):231–240, 2002.

Daniel Halperin, Daniel, Thomas S. Heydt-Benjamin, Benjamin Ransford, Shane S. Clark, Benessa Defend, Will Morgan, Kevin Fu, Tadayoshi Kohno, and William H. Maisel. Pacemakers and implantable cardiac defibrillators: Software radio attacks and zero-power defenses. In *2008 IEEE Symposium on Security and Privacy (sp 2008)*, pages 129–142. IEEE, 2008.

Luky Hendraningrat, Luky, Shidong Li, and Ole Torsæter. A coreflood investigation of nanofluid enhanced oil recovery. *Journal of Petroleum Science and Engineering*, 111:128–138, 2013.

Fei Hu, Fei, Yu Lu, Athanasios V. Vasilakos, Qi Hao, Rui Ma, Yogendra Patil, Ting Zhang, Jiang Lu, Xin Li, and Neal N. Xiong. Robust cyber—physical systems: Concept, models, and implementation. *Future Generation Computer Systems*, 56:449–475, 2016a.

Hong Hu, Hong, Shweta Shinde, Sendroiu Adrian, Zheng Leong Chua, Prateek Saxena, and Zhenkai Liang. Data-oriented programming: On the expressiveness of non-control data attacks. In *2016 IEEE Symposium on Security and Privacy (SP)*, pages 969–986. IEEE, 2016b.

Abdulmalik Humayed, Abdulmalik, Jingqiang Lin, Fengjun Li, and Bo Luo. Cyber-physical systems security—a survey. *IEEE Internet of Things Journal*, 4(6):1802–1831, 2017.

Emilio Iasiello, Emilio. Cyber attack: A dull tool to shape foreign policy. In *2013 5th International Conference on Cyber Conflict (CYCON 2013)*, pages 1–18. IEEE, 2013.

Mao Jianhui, Mao. Event driven monitoring of cyber-physical systems based on hybrid automata. *National University of Defense Technology Changsha*, 2011. https://doi.org/10.3390/app131910603

Merike Kaeo, Merike. Cyber attacks on estonia: Short synopsis. *Double Shot Security*, 2007. www.doubleshotsecurity.com/pdf/NANOG eesti.pdf (accessed 18 July 2009).

Stamatis Karnouskos, Stamatis. Stuxnet worm impact on industrial cyber-physical system security. In *IECON 2011-37th Annual Conference of the IEEE Industrial Electronics Society*, pages 4490–4494. IEEE, 2011.

Rafiullah Khan, Rafiullah, Sarmad Ullah Khan, Rifaqat Zaheer, and Shahid Khan. Future internet: The internet of things architecture, possible applications and key challenges. In *2012 10th International Conference on Frontiers of Information Technology*, pages 257–260. IEEE, 2012.

Rafiullah Khan, Rafiullah, Peter Maynard, Kieran McLaughlin, David Laverty, and Sakir Sezer. Threat analysis of blackenergy malware for synchrophasor based real-time control and monitoring in smart grid. In *4th International Symposium for ICS & SCADA Cyber Security Research 2016 4*, pages 53–63, 2016. https://doi.org/10.14236/ewic/ICS2016.7

Constantinos Kolias, Constantinos, Georgios Kambourakis, Angelos Stavrou, and Jeffrey Voas. DDoS in the IoT: Mirai and other botnets. *Computer*, 50(7):80–84, 2017.

Karl Koscher, Karl, Stefan Savage, Franziska Roesner, Shwetak Patel, Tadayoshi Kohno, Alexei Czeskis, Damon McCoy, Brian Kantor, Danny Anderson, Hovav Shacham, et al. Experimental security analysis of a modern automobile. In *2010 IEEE Symposium on Security and Privacy*, pages 447–462. IEEE Computer Society, 2010.

J Sathish Kumar, J. Sathish, and Dhiren R. Patel. A survey on internet of things: Security and privacy issues. *International Journal of Computer Applications*, 90(11), 2014.

Pratyush Kumar, Pratyush, Dip Goswami, Samarjit Chakraborty, Anuradha Annaswamy, Kai Lampka, and Lothar Thiele. A hybrid approach to cyber-physical systems verification. In *DAC Design Automation Conference 2012*, pages 688–696. IEEE, 2012.

Sanjeev Kumar, Sanjeev. Smurf-based distributed denial of service (DDoS) attack amplification in internet. In *Second International Conference on Internet Monitoring and Protection (ICIMP 2007)*, pages 25–25. IEEE, 2007.

A Machie, A., Jenssen Roculan, Ryan Russell, and M.V. Velzen. Nimda worm analysis. *Technical Report*, Incident Analysis, SecurityFocus, 2001.

Bipasha Mahato, Bipasha, Deepsubhra Guha Roy, and Debashis De. Distributed bandwidth selection approach for cooperative peer to peer multi-cloud platform. *Peer-to-Peer Networking and Applications*, 14(1):177–201, 2021.

Morgan Marquis-Boire, Morgan, Marion Marschalek, and Claudio Guarnieri. Big game hunting: The peculiarities in nation-state malware research. *Black Hat*, Las Vegas, NV, 2015.

David C Mazur, David C., Ryan D. Quint, and Virgilio A. Centeno. Time synchronization of automation controllers for power applications. In *2012 IEEE Industry Applications Society Annual Meeting*, pages 1–8. IEEE, 2012.

Reem Melki, Reem, Hassan N. Noura, and Ali Chehab. Lightweight multi-factor mutual authentication protocol for IoT devices. *International Journal of Information Security*, 1–16, 2019.

Elinor Mills, Elinor. Report: Hackers broke into FAA air traffic control systems, 2009. https://www.cnet.com/news/privacy/report-hackers-hackers-broke-into-faa-air-traffic-control-systems/

Umberto Morelli, Umberto, Lorenzo Nicolodi, and Silvio Ranise. An open and flexible cybersecurity training laboratory in it/ot infrastructures. In *Computer Security*, pages 140–155. Springer, 2019.

Stuart Murdoch, Stuart and Nick Leaver. Anonymity vs. trust in cyber-security collaboration. In *Proceedings of the 2nd ACM Workshop on Information Sharing and Collaborative Security*, pages 27–29, 2015. https://doi.org/10.1145/2808128.2808134

Arvind Narayanan, Arvind, and Vitaly Shmatikov. Fast dictionary attacks on passwords using time-space tradeoff. In *Proceedings of the 12th ACM Conference on Computer and Communications Security*, pages 364–372, 2005. https://doi.org/10.1145/1102120.1102168

Troy Nash, Troy. *Backdoors and holes in network perimeters*, 2005 [Online]. http://ics-cert.us-cert.gov/controlsystems.

Hassan N Noura, Hassan N., Ola Salman, Ali Chehab, and Raphaël Couturier. Distlog: A distributed logging scheme for IoT forensics. *Ad Hoc Networks*, 98:102061, 2020.

Hassan Noura, Hassan N., Ola Salman, Ali Chehab, and Raphael Couturier. Preserving data security in distributed fog computing. *Ad Hoc Networks*, 94:101937, 2019.

Kelly O'Connell, Kelly. CIA report: Cyber extortionists attacked foreign power grid, disrupting delivery. *Internet Business Law Services*, 2008. https://www.scmagazine.com/news/cia-analyst-reports-hacker-attack-on-foreign-power-grid

Panagiotis Papantonakis, Panagiotis, Dionisios Pnevmatikatos, Ioannis Papaefstathiou, and Charalampos Manifavas. Fast, FPGA-based rainbow table creation for attacking encrypted mobile communications. In *2013 23rd International Conference on Field Programmable Logic and Applications*, pages 1–6. IEEE, 2013.

Lydia Ray, Cyber-physical systems: an overview of design process, applications, and security. *Cyber Warfare and Terrorism: Concepts, Methodologies, Tools, and Applications*, pp. 128-150, 2020.

Ryan Roemer, Ryan, Erik Buchanan, Hovav Shacham, and Stefan Savage. Return-oriented programming: Systems, languages, and applications. *ACM Transactions on Information and System Security (TISSEC)*, 15(1):1–34, 2012.

Deepsubhra Guha Roy, Deepsubhra Guha, Debashis De, Md Mozammil Alam, and Samiran Chattopadhyay. Multi-cloud scenario based qos enhancing virtual resource brokering. In *2016 3rd International Conference on Recent Advances in Information Technology (RAIT)*, pages 576–581. IEEE, 2016.

Deepsubhra Guha Roy, Deepsubhra Guha, Madhurima Das, and Debashis De. Cohort assembly: A load balancing grouping approach for traditional wi-fi infrastructure using edge cloud. In *Methodologies and Application Issues of Contemporary Computing Framework*, pages 93–108. Springer, 2018a.

Deepsubhra Guha Roy, Deepsubhra Guha, Puja Das, Debashis De, and Rajkumar Buyya. Qos-aware secure transaction framework for internet of things using blockchain mechanism. *Journal of Network and Computer Applications*, 144:59–78, 2019c.

Deepsubhra Guha Roy, Deepsubhra Guha, Ahona Ghosh, Bipasha Mahato, and Debashis De. Qos-aware task offloading using self-organized distributed cloudlet for mobile cloud computing. In *International Conference on Computational Intelligence, Communications, and Business Analytics*, pages 410–424. Springer, 2018b.

Deepsubhra Guha Roy, Deepsubhra Guha, Bipasha Mahato, and Debashis De. A competitive hedonic consumption estimation for iot service distribution. In *2019 URSI Asia-Pacific Radio Science Conference (AP-RASC)*, pages 1–4. IEEE, 2019a.

Deepsubhra Guha Roy, Deepsubhra Guha, Bipasha Mahato, Debashis De, and Rajkumar Buyya. Application-aware end-to-end delay and message loss estimation in internet of things (IoT)—mqtt-sn protocols. *Future Generation Computer Systems*, 89:300–316, 2018c.

Deepsubhra Guha Roy, Deepsubhra Guha, Bipasha Mahato, Ahona Ghosh, and Debashis De. Service aware resource management into cloudlets for data offloading towards IoT. *Microsystem Technologies*, 1–15, 2019b.

Michael Rushanan, Michael, Aviel D. Rubin, Denis Foo Kune, and Colleen M. Swanson. Sok: Security and privacy in implantable medical devices and body area networks. In *2014 IEEE Symposium on Security and Privacy*, pages 524–539. IEEE, 2014.

A Saqib, A., Raja Waseem Anwar, Omar Khadeer Hussain, Mudassar Ahmad, Md Asri Ngadi, Mohd Murtadha Mohamad, Zohair Malki, C. Noraini, Bokolo Anthony Jnr, R.N.H. Nor, et al. Cyber security for cyber physcial systems: A trust-based approach. *Journal of Theoretical and Applied Information Technology*, 71(2):144–152, 2015.

Norman Scott, Norman and Hongda Chen. Nanoscale science and engineering for agriculture and food systems. *Industrial Biotechnology*, 9(1):17–18, 2013.

Joseph Seering, Joseph, Juan Pablo Flores, Saiph Savage, and Jessica Hammer. The social roles of bots: evaluating impact of bots on discussions in online communities. *Proceedings of the ACM on Human-Computer Interaction*, 2(CSCW):1–29, 2018.

Abraham Serhane, Abraham, Mohamad Raad, Raad Raad, and Willy Susilo. Plc code-level vulnerabilities. In *2018 International Conference on Computer and Applications (ICCA)*, pages 348–352. IEEE, 2018.

Kallisthenis I Sgouras, Kallisthenis I., Avraam N. Kyriakidis, and Dimitris P. Labridis. Short-term risk assessment of botnet attacks on advanced metering infrastructure. *IET Cyber-Physical Systems: Theory & Applications*, 2(3):143–151, 2017.

Jill Slay, Jill, and Michael Miller. Lessons learned from the maroochy water breach. In *International Conference on Critical Infrastructure Protection*, pages 73–82. Springer, 2007.

Teodor Sommestad, Teodor, Göran N. Ericsson, and Jakob Nordlander. Scada system cyber security—a comparison of standards. In *IEEE PES General Meeting*, pages 1–8. IEEE, 2010.

Keith A Stouffer, Keith A., Joseph A. Falco, and Karen A. Scarfone. Sp 800–82. guide to industrial control systems (ICS) security: Supervisory control and data acquisition (SCADA) systems, distributed control systems (DCS), and other control system configurations such as programmable logic controllers (PLC), 2011. https://nvlpubs.nist.gov/nistpubs/SpecialPublications/NIST.SP.800-82r3.pdf

YingTan, Ying, Mehmet C. Vuran, Steve Goddard, Yue Yu, Miao Song, and Shangping Ren. A concept lattice-based event model for cyber-physical systems. In *Proceedings of the 1st ACM/IEEE International Conference on Cyber-physical Systems*, pages 50–60, 2010. https://doi.org/10.1145/1795194.1795202

Stephen R Vogel, Stephen R., and Steven Jeffrey Zack. Method and apparatus providing remote reprogramming of programmable logic devices using embedded JTAG physical layer and protocol. *US Patent 7,155,711*, December 26 2006.

Kaiyu Wan, Kaiyu, K.L. Man, and D. Hughes. Specification, analyzing challenges and approaches for cyber-physical systems (CPS). *Engineering Letters*, 18(3), 2010.

Wenye Wang, Wenye and Zhuo Lu. Cyber security in the smart grid: Survey and challenges. *Computer Networks*, 57(5):1344–1371, 2013.

Joseph Weiss, Joseph. *Protecting Industrial Control Systems from Electronic Threats*. Momentum Press, 2010.

Miao Wu, Miao, Ting-Lie Lu, Jing Sun, Hui-Ying Du. Research on the architecture of Internet of things. In *Advanced Computer Theory and Engineering (ICACTE)*, pages 484–487, 2010. doi: 10.1109/ICACTE.2010.5579493.

Abel Yeboah-Ofori, Abel, J. Abdulai, and Ferdinand Katsriku. Cybercrime and risks for cyber physical systems. *International Journal of Cyber-Security and Digital Forensics (IJCSDF)*, 8(1):43–57, 2019.

Bonnie Zhu, Bonnie, Anthony Joseph, and Shankar Sastry. A taxonomy of cyber attacks on scada systems. In *2011 International Conference on Internet of Things and 4th International Conference on Cyber, Physical and Social Computing*, pages 380–388. IEEE, 2011.

Chapter 9

Securing Financial Services with Federated Learning and Blockchain

Pushpita Chatterjee, Debashis Das, and Danda B. Rawat

Chapter Contents

9.1   Introduction

Financial services are crucial for economic growth as they facilitate economic activities by providing individuals, businesses, and governments with the necessary tools and resources to manage money. These services mobilize savings and investments, enabling the flow of funds between borrowers and lenders and supporting businesses to invest, expand, and create jobs. Financial services promote financial inclusion by providing access to credit and banking services, empowering individuals to manage their finances effectively. In the digital world, financial services play a vital role in driving economic development and ensuring the efficient allocation of resources.

Financial institutions handle sensitive customer information, such as bank account details, social security numbers, and transaction histories. Ensuring the security of this information is crucial to protect customers from identity theft, fraud, and unauthorized access to their funds [71]. Robust security measures, including encryption, firewalls, and access controls, are essential to safeguard customer data. The financial industry is a prime target for fraudsters and cybercriminals due to the potential financial gains involved. Cybercriminals

DOI: 10.1201/9781003376712-9

have the ability to target a wide range of businesses, and their selection process depends on the potential financial gain or the magnitude of impact that a particular target may offer [63]. Hackers may attempt to breach systems, steal sensitive information, or launch cyber-attacks to disrupt services or gain unauthorized access to accounts.

Security is crucial for financial services to protect customer information, prevent fraud and cyberattacks, maintain trust and reputation, and comply with regulatory requirements. Robust security measures are necessary to ensure the integrity, confidentiality, and availability of financial systems and data. Security breaches in the financial services sector can have severe consequences for both the institution and its customers. A security breach can damage the reputation of a financial institution, erode customer trust, and lead to financial losses [13]. Implementing strong security measures, such as multi factor authentication, intrusion detection systems, and regular security audits, can prevent such threats and protect financial systems.

Blockchain technology [27], known for its decentralized and immutable nature, can provide secure and transparent transactions in financial services [78]. It enables the creation of tamper-proof, distributed ledgers that record and validate transactions, eliminating the need for intermediaries and reducing the risk of fraud. Figure 9.1 shows that the use of Blockchain in financial sectors is greater than in other sectors [60]. By the end of the year 2028, it is anticipated that the value of the financial Blockchain industry will have increased to 36.04 billion dollars [20]. Blockchain can facilitate faster and more efficient cross-border payments, streamline trade finance processes, and improve the transparency and traceability

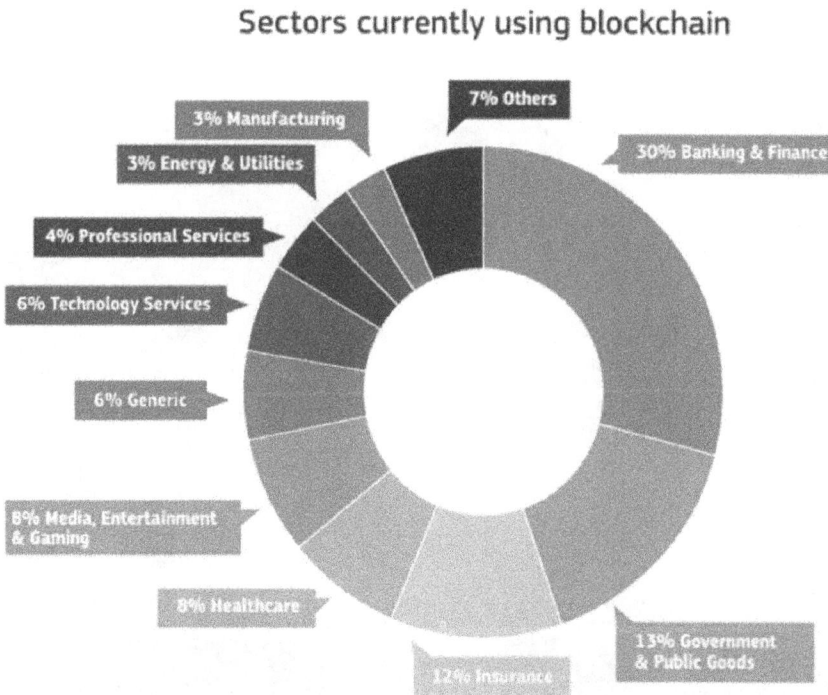

Figure 9.1 Usage of Blockchain in financial sectors rather than other sectors [60].

of supply chain financing. Federated learning, on the other hand, is a privacy-preserving approach to machine learning that allows multiple entities to collaboratively train a shared model without sharing their raw data. In the context of financial services, federated learning can enable banks and financial institutions to collectively improve their risk models, fraud detection algorithms, and customer analytics without compromising the privacy of sensitive customer data [14].

The combination of Blockchain and federated learning can further enhance financial services by providing secure and privacy-preserving data sharing and collaboration. For example, Blockchain can be used to create a decentralized marketplace where financial institutions can securely exchange data or model updates for collaborative analysis. Federated learning can ensure that the actual data remains local to the participating institutions while sharing aggregated model updates for mutual benefit. Combining these has the potential to improve federated learning's transparency, trustworthiness, and, most importantly, decentralization [7].

9.1.1 Motivation

The financial sector has benefited from several technological improvements and integrations. It still works in a centralized way, with financial institutions and governments at the center of the model. It has caused a lot of changes in how companies are set up and how they do business, which have given the financial technology sector a huge chance. Because of this, both new businesses and companies that have been around for a while and focus on making financial applications are interested in finding out if Blockchain is necessary for financial services or not. According to Figure 9.2, the use of Blockchain in financial services is greater than in other sectors. A common factor that contributes to the complexity of

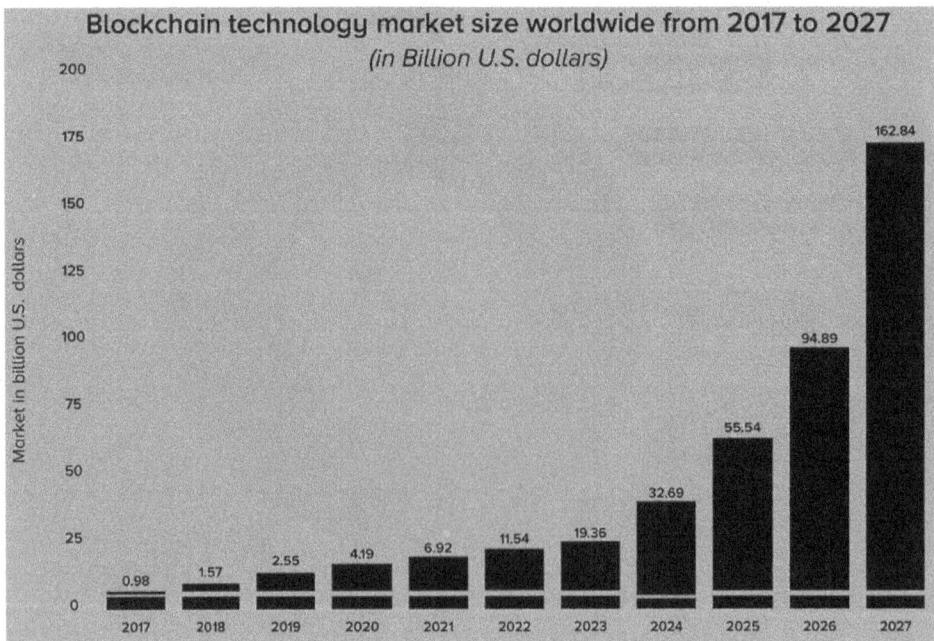

Figure 9.2 Expected Blockchain usage in ten years [21].

risk is the fact that it is impossible to completely remove or guard against all risks, regardless of how sophisticated your systems are. This is where the process of risk management comes in. Risk management is a regular, ongoing process in which the right professionals look at risks from time to time to reduce the chances that certain threats will come true. Companies that provide financial services in today's market not only have a hard time luring in new clients, but they also have a hard time doing the same with prospective workers. It may be hard to find the right people to fill new positions in information technology (IT) because of several factors, the most important of which is that millennials don't like long-term jobs. The provision of essential financial services is essential to the operation of any economy. Without them, those who have money to save may have difficulty locating others who need to borrow money, and vice versa. Without financial services, people might not buy as many goods or services because they would be so worried about saving money to protect themselves from possible losses.

Contribution of the Work: This chapter gives an overview of how Blockchain and federated learning are used in the financial services industry. The goal of this research is to look at the benefits of Blockchain and FL to get useful information that can be used to improve external statistics and make policy decisions in the financial sector. This chapter talks about and analyzes how financial activities face weaknesses and problems. Besides, it is discussed how financial services can be used and organized using decentralized technology. However, the main contribution of this chapter is presented as follows:

- We give an overview of financial services by talking about how important they are and showing how they can be used in different situations.
- We provide recent financial services vulnerabilities and related threats that have occurred. Possible risks in financial services are also discussed in this chapter.
- We demonstrate a few problems that exist in financial services and how Blockchain can be used to solve them. The benefits of Blockchain in this sector are also discussed in detail.
- We present several Blockchains and federated learning applications in different fields. Parallelly, we give some ideas for how Blockchain and FL can be used in financial services and explain what these technologies signify.

The remainder of this chapter is organized as follows. Figure 9.3 shows the overall roadmap of this chapter. Section 9.2 gives the existing use cases of financial services. In Section 9.3, existing vulnerabilities, threats, and risks in financial services are provided in detail. Section 9.4 depicts the usage of Blockchain in financial services. Applications of Blockchain and federated learning in financial services are stated in Section 9.5. Section 9.6 introduces some future research aspects in financial services. Finally, the chapter is concluded in Section 9.7.

9.2 Financial Services Use Cases

The Banking, Financial Services, and Insurance (BFSI) [66] sector is the most vulnerable to uncertainty because it depends on global trends, changing laws, and customer demographics. It is a metric that has a great deal of significance for any sector, particularly the BFSI sector. Customers often give their banks and insurance companies their most private information, like information about their health or finances. It could have consequences for both the bank and the insurance company. Data and analytics are being

Figure 9.3 Roadmap of the survey.

used by the most important institutions in this field to change the rules of the game. They are gathering additional information from sources such as telecom providers, merchants, and social media to improve the knowledge that they already have about their customers. Because they have such a comprehensive perspective on their consumers, they are in a position to increase revenue, reduce risk, cut opportunity costs, and improve operational efficiency.

Sales and Revenue Analysis: Examining the operating procedures assists financial institutions in lowering their continuing expenses. If you know the sales trends for a certain consumer, you may be able to make things easier to repeat. Sales is a key activity, and having Business Intelligence (BI) tools may aid in defining benchmarks like the number of net new customers and the lucrative sector among current customers. These are just two examples of how having these tools can be beneficial.

Sales Performance Analysis: The report is all-encompassing and includes data on employee productivity. Any employee who works with customers, like salespeople, account managers, and tellers, can benefit from the information it can give because it can help them find ways to improve and, in the end, give better service to their customers. This evaluation could be used to check the viability of new financial products or services and make strategic changes that are in line with the institution's long-term goals.

Branch/Online Sales Analysis: It may assist financial institutions in formulating the most effective channel strategy. Multiple channels are now available for customers to use when communicating with their banks. Their trips through these channels are very complicated. They often start in one channel, go through different stages of the process in another channel, and then end up in a different channel. By collecting real-time data and using analytics to learn more about the buyer's journey, financial institutions may be able to use this to give customers a truly seamless multichannel experience. In addition to this, it assists them in maintaining an awareness of their rivals.

Lending, Payment, and Transaction Analysis: Banks can use their customers' transaction history to recommend products and services that are relevant to those customers. It leads to improved conversion rates as well as increased levels of client satisfaction. The following information about customers may be analyzed more effectively with the use of banking analytics: Existing clients of a bank can ask to look at their transaction history, which could include information about deposits, withdrawals, or payments. Bankers can help their customers take advantage of good deals on their credit or debit cards or other new financial products by getting to know how they spend their money and encouraging them to do so. Using this information to send timely spending alerts and payment reminders can improve the customer experience. It can be done to improve the customer experience. By looking at their clients' transaction histories and looking for patterns, banks can find transactions that might be fraudulent and take steps to stop them. Analytics for banking include data-driven methods like digital credit evaluation, improved early-warning systems, next-generation stress testing, and analytics for collecting debt. These techniques are used to protect clients against fraud at financial institutions.

Credit Risk and Exposure Analysis: An analysis of a client's credit risk and exposure might shed light on whether or not a customer has a history of defaulting on their payments in the past. These consumers' credit profiles highlight their assets and customer behavioral data, such as past-due bills, loans or borrowings, and earnings, that may be used to calculate each customer's credit score.

Market and Portfolio Analysis: It is highly important to do market and portfolio research to recruit new clients and keep the ones you already have. An analytics system could look at a client's current portfolio to suggest new investment options to the client and help the client's portfolio managers keep a steady return. Again, doing a market study is of the utmost importance when it comes to building a portfolio that will be successful regardless of the state of the economy.

Liquidity Risk Management: Every single banking procedure has the potential to become more efficient and streamlined. Using advanced analytics, financial institutions may be able to do things like answer questions from regulators faster and more accurately and give teams more information. It helps with decision-making that is augmented by analytics. The compliance and regulatory standards that banks must meet are quite severe. This has a big effect on how poor the impairment risk they face is. Know-your-customer (KYC) analysis is vital, not only to achieve compliance with the legal requirements but also as a method of mitigating risk. Anti-money laundering (AML) analysts can more effectively detect and monitor problematic account holders with the use of these BI technologies.

Analyzing and Planning Finances: Finance is the core of every company, just as it is with every other kind of organization. In the case of banks, this issue is even more important because bank employees are responsible not only for running the day-to-day business of the bank but also for meeting the different financial needs of customers. An analytical system may find the following use cases when it comes to a bank's finances: Banks need to have their cash on hand to handle payments well and follow all of the rules set by regulators. By looking at how much they spent in the past, they can make a budget that works for them. They also take into account certain factors that could make their financial needs go up or down. This could lead to finding a clear set of important success criteria that turn short-term savings into long-term, sustainable improvements and the best way to manage costs. Business intelligence (BI) tools could make financial planning and analysis (FP&A) easier and make it easier to report to key stakeholders in a useful way. By making the necessary reports automatically and regularly, these systems could cut the amount of work needed for financial reporting by a large amount. In addition to this, they help speed up the transmission of information and ensure that decision-makers are kept up-to-date on the state of the bank's finances.

Management, Marketing, and Production: There are instances when a new strategy for approaching an established consumer is necessary. Banks need to give their current customers suggestions for new and better products, and this information should be given to these customers at the right time. When you look at the company's current customers, you can see which marketing methods have worked best in the past. You can then use these methods to bring in new customers. Business intelligence systems can be used with transactional and trade analytics to make more complete and richer profiles of customers. This, in turn, can increase the acquisition and retention of consumers as well as cross- and upselling opportunities.

Customer Portfolio and Segmentation: It is another significant use of analytics systems in financial services. It is necessary to correctly segment clients to successfully market to them. Consumers who are searching for a house loan or a vehicle loan are an example of one kind of customer segmentation used by financial services organizations. Another example would be customers who are specifically interested in a checking account or a money market account. Conversion can happen when the customer relationship manager makes an offer or calls the customer about something important to them. In the same way, a new offer may be aimed at a smaller group of people based on their credit scores.

Churn Prediction and Value Modeling: Predicting a client's likelihood to churn and estimating their lifetime worth as a client are two areas that have gained major significance for financial institutions like banks and insurance companies in recent years. It takes a massive expenditure to compete with the thousands of businesses that are fighting for consumers' attention and physical space. It is of the utmost importance to make sure that you are not leaving any value on the table after you have successfully onboarded the consumer. The process of mapping the customer journey to observe their behavior helps in understanding any requirements that the customer may have and also assists in up-selling.

Analysis of a Marketing Campaign: An analysis of a marketing campaign gives a summary of the different channels that work for a bank and finds the best way to spend money on all of them. The leading banks use the information from the transaction data of credit cards (from both their terminals and those of other banks) to develop offers that provide customers with an incentive to make regular purchases from one of the bank's merchants. These offers can be found on the websites of the leading banks.

9.3 Vulnerabilities, Threats, and Risks in Financial Services

Financial service businesses produce more data after becoming digital. Every time you make a financial transaction or contact someone, data is created and shared using multiple applications. Cybercriminals can use this information to their advantage. They may sell, utilize, or threaten to dump it. In 2022, financial institutions were most at risk from ransomware, phishing, online application, vulnerability exploitation attacks, denial of service (DoS) attacks, insider threats, nation-state and state-sponsored threat actors, and Advanced Persistent Threat (APT) groups [31]. Figure 9.4 shows vulnerabilities, threats, and risks in financial services. Trends make financial services organizations more vulnerable to cyberattacks. Thus, firms must understand their current dangers and build effective defenses against them.

9.3.1 Vulnerabilities

Organizations in the financial industry confront security risks from both internal and external sources regularly because they are high-value targets for hackers [28]. Threat actors use banking websites or virtual private networks (VPNs) to get into online banking systems to steal account information, cause trouble, or test how far they can get into a network. Internal threats often come from unhappy employees, weak third-party vendors, and human mistakes caused by phishing emails or other forms of social engineering. Weaknesses in external and internal security let sensitive financial information, client data, account balances and transactions get out. It hurts customer confidence and causes business problems [31]. Today's financial services firms need sophisticated cybersecurity solutions that can manage the growing demands of keeping customer and financial data safe, limiting attack risk, and complying with regulatory regulations.

Actively Evaluating Cybersecurity: Unknown weaknesses in cybersecurity and compliance cannot be addressed by financial organizations. In addition, failure to address these vulnerabilities may have significant repercussions. If left neglected until an event happens, institutions are compelled to use a reactive response, which may result in business

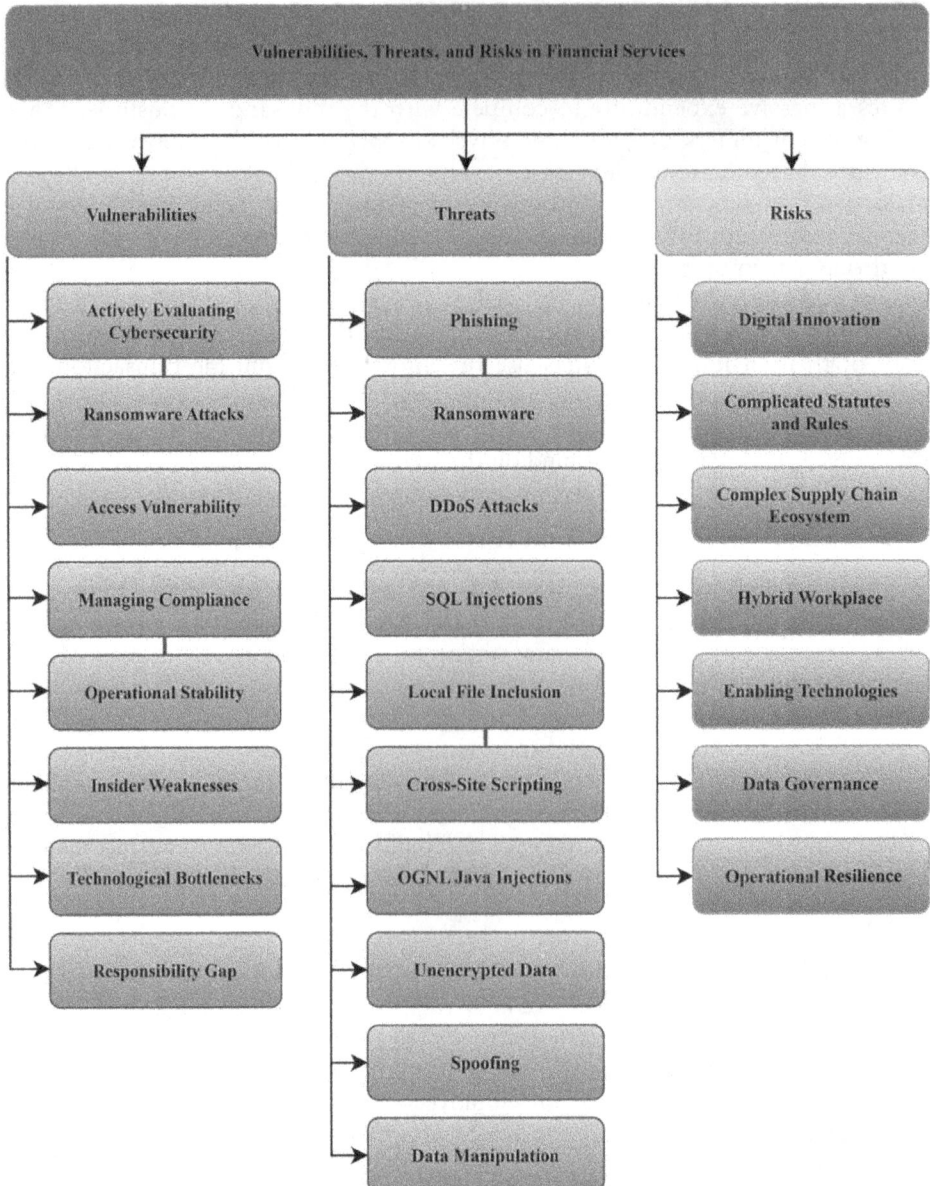

Figure 9.4 Vulnerabilities, threats, and risks in financial services.

interruptions and eroded client trust. Instead, financial service companies should take a proactive approach. Financial service organizations may conduct an initial evaluation of current vulnerabilities to discuss with a managed service provider (MSP).

Ransomware Attacks: There is an exponential increase in the number of potential targets for ransomware assaults as the globe continues to become more digitally linked. The term "ransomware" refers to an attack in which the perpetrators employ malware to get access to your business's systems or data and then keep that data hostage until the firm pays a ransom. The aftermath of these assaults has been utterly catastrophic. In addition to the cost of the ransom, there may be additional expenditures related to damage management, such as legal fees and other expenses. There is also the possibility of losing data.

Access Vulnerability: Sensitive data can be left exposed and subject to attack if there are flaws in the different levels of information access. Integration of cybersecurity measures is essential throughout all departments of a business and at each level of access. Criminals online will attempt to take advantage of whatever vulnerabilities they may find, regardless of the organizational hierarchy of the company they are targeting.

Managing Compliance: The advancement of information technology has made the financial services industry's job more difficult in terms of complying with regulations. The financial services industry in the United States is one of the most heavily regulated corporate sectors in the world. On the other hand, merely complying with the rules may not be enough anymore. Instead, aggressively managing compliance risk and increasing compliance overall are essential for gaining the trust of customers and avoiding expensive fines.

Operational Stability: A backup and disaster recovery solution that is proactive and dynamic is necessary for avoiding disruptions to corporate operations and the loss of crucial data, either of which might result in a compliance violation. It is common for off-the-shelf onsite backup systems to be unable to provide the degree of performance necessary to satisfy the requirements of finance and investment businesses. It is essential to come up with a solution in advance of an outage to guarantee a speedy recovery and reduce the amount of time customers are without service.

Insider Weaknesses: Insider vulnerabilities are a source of concern in the banking and financial industries when it comes to cybersecurity. It occurs when people who work inside a bank or other financial institution do something that puts the company at risk of being attacked. The 2019 IBM X-Force Intelligence Index [72] found that phishing emails were used in almost two-thirds (29%) of the attacks that were looked at. Whaling attacks, often called "corporate email compromise scams," were to blame for 45% of these problems. In these incidents, hackers try to break into the email accounts of important organization members, like the CEO, to get the company to reveal private information. Another common thing that might happen is that systems and servers are set up incorrectly.

Technological Bottlenecks: Websites and apps about banking and money make the architecture of the network as a whole more vulnerable. Researchers discovered that they were more likely to be hacked into banking and finance systems. Cross-site scripting (XSS) attacks [72], which allow attackers to run malicious code on a website or app, could happen to 80% of the people who were tested. Then, the bad script could change the site's content by getting to the user's cookies and other sensitive information. Users are more likely to mistrust websites and programs that have vulnerabilities like these. So, if the businesses want to stay competitive, they should look into what steps they can take to protect their websites and apps.

Responsibility Gap: Even though the global financial system is becoming more dependent on digital infrastructure, it is not clear who is responsible for protecting it from cyberattacks. It is due, in part, to the rapid pace at which the environment is changing. If people don't work together, the global financial system will continue to get worse as more innovation, more competition, and the pandemics speed up the pace of the digital revolution and make it more dangerous. Although many threat actors are motivated by a desire to make money, the number of attacks that are solely disruptive and destructive has been increasing; additionally, those who learn how to steal learn about the financial system's networks and operations, allowing them to release more obstructive or dangerous future attacks [61]. Even though the system is generally well-developed and well-regulated, this sudden change, like the risks it poses, puts a strain on its ability to respond.

9.3.2 Threats

The worst catastrophes have harmed financial information, particularly accounts, calculations, and transactions. Such assaults, which may undermine trust, now have some technological solutions. VMware reported 238% more financial institution cyberattacks in the first half of 2020 [49]. IBM and the Ponemon Institute estimate a financial data breach would cost $5.72 million in 2021 [49]. It's global. The increased frequency of assaults on targets of opportunity in low- and lower-middle-income countries is less reported than cyberattacks in high-income countries. Financial inclusion has been the biggest driver of digital banking services like mobile payment systems. Digital banking services expand financial inclusiveness but provide hackers with more targets.

Phishing: Phishing is social engineering that deceives individuals into sharing their login credentials to enter a private network. Email phishing, when victims get official-looking emails, is the most common. Visiting a phishing email's dangerous links or attachments might install malware or launch a bogus website that steals login credentials. In the first half of 2021, bank phishing attacks rose by 22%. Financial app assaults increased by 38% at the same time. Akamai's 2019 State of the Internet report found that over 50% of phishing attempts targeted financial services [49]. Phishing tactics are evolving to exploit modern worries. These troubling trends rank phishing among the banking sector's top cybersecurity dangers.

Ransomware: Ransomware also threatens financial institutions. Ransomware encrypts computers, locking victims out [49]. Only a ransom can fix the harm. Due to strict rules requiring financial institutions to be resilient to cyberattacks and data breaches, these extortion methods work effectively against them. Ransomware attacks are now data breaches, which might affect regulatory compliance requirements. Ransomware gangs target financial businesses because of their customer data. Due to the danger of data exposure on the dark web and reputational damage, many financial services companies accept extortion demands.

DDoS Attacks: The year 2020 witnessed the highest DDoS attacks on financial institutions. DDoS attacks are a prevalent cyber threat to financial services since they may target consumer accounts, payment gateways, and banks' IT systems. Due to this, the impact of DDoS attacks on financial firms is amplified. Cybercriminals may use the ensuing confusion in one of two ways. Password login attacks and DoS attacks were the two main online dangers to payment systems in 2020. In comparison to the same period in 2020, multi-vector DDoS attacks had increased by 80% in 2021 [49]. These DDoS attacks combine several campaigns to swamp security personnel.

SQL Injections: A vulnerability in a WordPress plugin that enabled Time-Based Blind SQL injections (SQLi) was found in March of 2021 [37]. This vulnerability was detected. There was a possibility that 600,000 customers were affected by this issue. Through the use of a technique known as Time-Based Blind SQLi, the vulnerability made it possible for any site visitor to access sensitive data stored in a website's database. Because the SQL query was executed inside the function object for the pages", this meant that any site visitor, even those who did not have a login, may trigger the execution of this SQL query. It would thus be possible for a hostile actor to give harmful values for either the ID or type parameters.

Local File Inclusion: A vulnerability known as Local File Inclusion (LFI) was discovered in August 2021 for a version of BIQS [49] software used by driving schools for billing customers. When a certain payload is sent to download/index.php in older versions of BIQS IT Biqs-drive than v1.83, a local file inclusion (LFI) vulnerability is present. This vulnerability may be exploited to take control of the affected system. Because of this, the attacker can access arbitrary files stored on the server using the permissions of the web–user configuration.

Cross-Site Scripting: Trend Micro revealed the details of e-commerce website cross-site scripting (XSS) attacks on April 28, 2021. EC-CUBE-built websites have also had XSS instances confirmed by JPCERT/CC (an open-source CMS for e-commerce websites). Any e-commerce website having an XSS vulnerability on its administrator page is targeted by this attack. This attack campaign continued on July 1, 2021. In order forms on targeted e-commerce websites, attackers insert malicious scripts to make purchases. XSS attacks on the administrator's page steal credentials and install Simple WebShell on the website. Attackers then utilize WebShell and JavaScript on the website to harvest and save user data. Monitoring the WebShell may allow the attackers to obtain the stolen data. During the attack, the attackers embed Adminer [2] on the e-commerce website. This is a GUI-based database content analysis tool. It supports MySQL, PostgreSQL, SQLite, MS SQL, Oracle, SimpleDB, Elasticsearch, and MongoDB. Attackers presumably accessed database information using this approach.

OGNL Java Injections: In August 2021, OGNL flaws allowed hostile actors to inject code into Atlassian Confluence servers [25]. OGNL injection vulnerabilities allow unauthenticated users to execute arbitrary codes on Confluence Server or Data Center instances. Previous versions of the Confluence Server and Data Center were affected by this problem. The vulnerability is actively abused in nature. Unauthenticated users may exploit it regardless of settings.

Unencrypted Data: When data is left unencrypted [45], fraudsters or hackers may immediately change it, causing major problems for banks. Online and financial institution data must be jumbled. It prevents attackers from using stolen data.

Spoofing: It is one of the most recent instances of a cyber threat that businesses in the financial sector need to be prepared for. The URL of a bank's website will be impersonated by hackers, who will replace it with a website that is connected to the actual one and operates in the same manner (cite 17). When a customer uses a fraudulent website and inputs his login information, the hackers will grab the customer's credentials and utilize them in the future.

Data Manipulation: One of the most common misunderstandings about cyber assaults is the belief that people are only concerned about the theft of data. It isn't always the case, though, because hackers are using data manipulation attacks more and more. Cybercriminals are always developing new methods of attack. Attacks involving data manipulation happen when a bad actor gains access to a trusted system and then makes changes to the data without being caught to help themselves [45]. One example of this would be if an employee changed information about customers. Likely, it won't be found out because the transactions will look like they were done legally. It will cause future data to be stored incorrectly. The more time that goes by before the manipulation is discovered, the more damage it will do.

9.3.3   Risks

The financial sector is getting more and more exposed to "cyber risk," which is the risk of losing money because of how much they depend on computers and digital technology. Cyber-related events, especially cyberattacks, are always at the top of polls that measure the financial stability of the United States and the rest of the world. Cyber risk, like other financial vulnerabilities, raises macroprudential issues. Similar to other financial problems, a lot of technological attention has been paid to cyber resilience, but it is still very early to measure the effects that cyber risk might have on the financial system. If you want to be strong against cyberattacks, you need to know about the problems that make the cyber risks the financial sector faces even higher. It is important to find a way to solve them all at once, as these problems are linked to each other.

Digital Innovation   In financial institutions (FIs), new technologies are being used, such as cloud computing, artificial intelligence, and digital service delivery. Most FIs are improving their data processing, fraud detection, and financial analytics by using software that is hosted in the cloud [50]. Meanwhile, the COVID-19 epidemic furthered the process of transferring the industry's IT infrastructure (digital transformation), which resulted in the proliferation of virtual banks and financial services. Because of digital transformation, businesses today run an increasing number of brand-new apps, devices, and infrastructure components, all of which expand the attack surface. A surge in cybersecurity threats for financial institutions is caused by all of the issues together. Even if the rise of new technologies in the financial sector has a major impact on industrial risk management, these technologies could help risk management by improving cybersecurity and compliance controls.

Complicated Statutes and Rules: As financial institutions use more technology and data to help their customers, regulations must change to keep up. State, federal, and international authorities have established several new restrictions for their industries in reaction to the growth in cyberattacks on financial services organizations. Data protection, privacy, and cybersecurity legislation for financial institutions (FIs) are tightening in various nations. Compliance may be time-consuming and expensive, but it's in everyone's best interest. According to BITS' technology division, chief information security officers spend 40% of their time addressing regulatory agency criteria [35]. Because of the regulatory environment's complexity, enforcement is tighter, raising regulatory costs and penalties. In August 2020, the US government fined Capital One $80 million for failing to find and deal with cyber risk, which led to a massive data breach in 2019 [69]. Capital One resolved a class-action lawsuit in late December 2021 over a 2019 Amazon Web Services cloud network intrusion that stole 100 million customers' data [8]. The settlement was for 190 million dollars.

Complex Supply Chain Ecosystem: Most financial firms outsource their digital duties. Third-party service providers may be vulnerable even if the FI's internal security is strong. Threat actors are targeting software businesses and sending malware to supply chain customers through legitimate downloads and upgrades. Threat actors gain backdoor access to client networks via these attacks on software distribution platforms. Recent assaults include the

SolarWinds breach for supply-chain assault [81]. Attackers infiltrated SolarWinds' network and planted malware in their management software to target thousands of banks and government entities. The SolarWinds breach shows how susceptible the financial services sector is to cyberattacks and disruptions since it depends on third-party suppliers and service providers with little or no cybersecurity oversight. Third-party cybersecurity vulnerabilities will grow as the government prioritizes business continuity and operational resilience.

Hybrid Workplace: COVID-19 has sped up recent changes in the way people work, like the hybrid workspace, which combines people who work in the office and those who work from home. It will increase the risk that businesses face. As we move into the five year of the pandemic, more and more people are using technologies like remote work, hybrid workforces, and software that is hosted in the cloud. Businesses had no choice but to quickly adopt the new technologies that gave them remote access, better communication, and more ways to work together. Because of this, hybrid working settings make IT systems more complicated, increase the number of ways to attack them, and create new cyber risks and threats.

Enabling Technologies: According to some estimates, the pandemic sped up the transition to digital technology by as much as three years. Enabling technologies, such as application programming interfaces, Big data analytics, artificial intelligence, biometrics, cloud computing (particularly outsourcing to the cloud), and distributed ledger (Blockchain) technology, makes it feasible for digital transformation to occur. Companies and their boards of directors need to be able to make sure that new technologies are adopted safely so that the benefits can be gained and the risks that come with trying new things can be managed proactively. This will help businesses get the most out of their innovative activities and reduce the risks that come with them.

Data Governance; The importance of having a solid strategy for data governance is only going to grow in the coming years. Companies need to realize that data is a key strategic asset before they can come up with a company-wide plan for collecting, managing, storing, protecting, retrieving, and destroying data. To put it another way, develop a strategy for data governance that is tailored to your organization. If data governance works, it will have many benefits, such as making it easier to see risks in a hybrid work environment, being able to meet the recently agreed-upon requirements for reporting climate risk, and making it easier to keep track of records.

Operational Resilience; Cybersecurity is a major problem for businesses operating in the financial industry. In a September 2021 Conference of State Bank Supervisors (CSBS) study, more than 80% of bankers regarded cybersecurity risk as "very significant" as the top internal risk [1]. This number is more than twice any other operational risk category and greater than the 60% recorded in the year 2020. This risk aversion may be attributed to a great number of different factors. For example, worries about cybersecurity can hurt both the way a company works and its reputation. If a financial institution is hit by a cyberattack, its ability to do business could be hurt or completely stopped. It is called operational risk. In addition, as a result of the hack, consumers can lose faith in the company and want to conduct their business elsewhere (reputational risk).

9.4 Blockchain for Financial Services and Banking Industries

Blockchain can make banking and lending easier by reducing the risk of the other party and the time it takes to issue and settle the money. Authenticated paperwork and KYC/AML data reduce operational risks and enable real-time financial document verification. Without a bank or financial services provider, a Blockchain lets people transmit money securely [24, 34]. Blockchain technology is known as "distributed ledger technology" in the financial services business [38]. Table 9.1 shows the existing methods of using Blockchain in financial services. Since all transactions on a Blockchain are saved in a shared database, it could make banking more open. Because of this openness, problems like fraud might be found and fixed, which could make the risk for financial institutions lower [9]. Figure 9.5 shows the usage of Blockchain in several cases in financial sectors.

9.4.1 Addressable Challenges in Financial Services

Financial services often have problems, like not reaching their goals, taking a long time to raise money, and losing more and more money. These problems are often caused by inadequate management. The following is a list of challenges that Blockchain technology has the potential to solve in the financial technology industry [39,73]:

Centralized System: Even though financial services solutions make things seem easier, the real power is still in the hands of third parties [6,41]. Higher-ups are still the only ones who can approve transactions, so users are still waiting for confirmation that they can move forward with their transactions. Because of the introduction of Blockchain technology, this is the first problem in the financial services industry that could be addressed.

Trust Issues: When consumers take any action inside financial services apps, they are not aware of what is occurring on the other side of the transaction [3]. This leads to a great deal of uncertainty as well as an increase in the fear of having one's identity stolen, which eventually results in a decrease in faith in the process. Because the Blockchain is open and can't be changed, these Blockchain application development services can solve this problem in the field of financial technology.

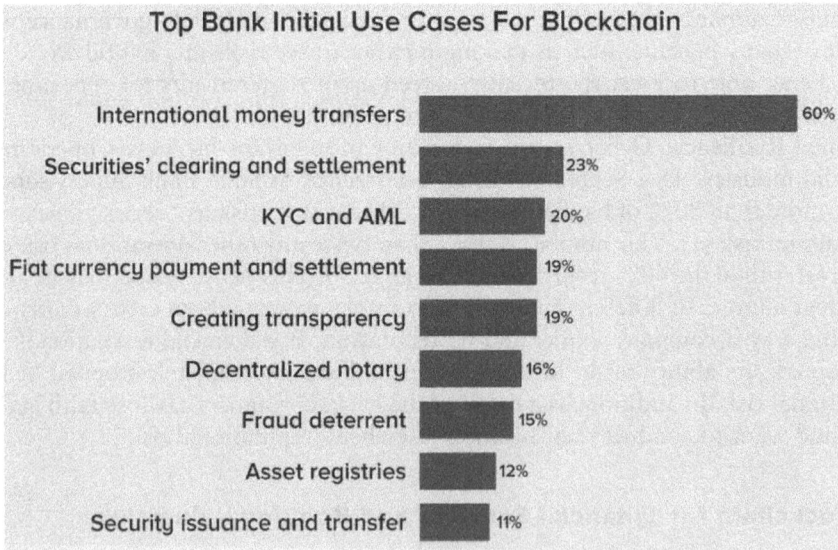

Top Bank Initial Use Cases For Blockchain

Use Case	Percentage
International money transfers	60%
Securities' clearing and settlement	23%
KYC and AML	20%
Fiat currency payment and settlement	19%
Creating transparency	19%
Decentralized notary	16%
Fraud deterrent	15%
Asset registries	12%
Security issuance and transfer	11%

Figure 9.5 Usage of Blockchain in different scenarios of financial services [21].

Table 9.1 Existing Blockchain-Based Methods for Financial Services

Ref	Proposed Method	Description	Pros	Cons
Liu et al. [54]	Hybrid chain model combining PANDA and X-alliance	Hybrid chain model may handle each account's transaction in parallel, asynchronized from other adjacent accounts in the network. It provides efficient data storage and authorization control and ownership of change-tracking data.	• High performance. • Lower protection cost.	• Performance accuracy is low. • Increased sample data. • Smart contract adoption issue.
Lorenz et al., [55]	Money-laundering detection	Active learning approach that matches a fully supervised baseline with 5% of labels.	• Detects money laundering. • Simulates a real-world situation with few analysts for manual labeling.	• Lack of proper Anti-Money Laundering (AML) processing in all financial services.
Chen et al. [17]	Blockchain-enabled Financial Surveillance Systems	Research on Blockchain credit information preservation and supervision, post-loan management, and time-based financial supervision chain.	• Safeguarding financial and user data. • Cost reduction in auditing. • Provides repayment ability function.	• User identity verification. • Unauthorized persons can read financial data.
Chen et al. [15]	Online P2P lending	Used machine learning and neural networks to forecast online lending credit risk.	• Improves the efficiency of online lending. • Reduced credit risks. • Increased trust in the models	• Time consuming. • Prediction accuracy is lower.
Kherbouche et al. [47]	Combination of UML-AD and Blockchain	Developed a Blockchain UML profile, activity diagram, and Petri net verification to ensure a distributed computing system with modifiable information.	• Integrity maintenance for financial organizations. • Facilitates insurance claiming.	• Lack of financial business policies. • Patterns, collection is minimal.
Ma et al. [58]	Blockchain-based supply chain finance	Hyperledger Fabric privacy security and supply chain financing.	• Designed privacy protection mechanism.	• Lack of data privacy.

(Continued)

Table 9.1 (Continued)

Reference	Technique	Description	Benefits	Limitations
Du et al. [32]	Blockchain-enabled supply chain financing	The suggested supply chain finance platform overcomes the issue of non-trust among supply chain players and uses homomorphic encryption on the Blockchain to safeguard confidential information in financial services.	• Increase capital flow and data flow efficiency. • Protect user privacy and data privacy.	• Inadequate for massive data processing. • Higher computational cost.
Wu and Duan [76]	Blockchain for small and microenterprises	Proposed that Xi'an's tiny and micro firms employ Blockchain technology to solve their funding problems.	• Enhances the asymmetry of information. • Additional financing assistance for small-and medium-sized enterprises.	• Weak incentive policies. • Privacy breach.
Camino et al. [10]	Financial Transaction Suspicions	A methodology for doing financial data analysis that does not involve the use of ground truth.	• Detection of expected anomalies. • Exploratory data analytics for financial data sources.	• Vast and challenging to comprehend.
Ye Guo and Chen Liang [40]	Blockchain-enabled regulatory sandbox	Proposed a "regulatory sandbox" and industry norms immediately.	• Improves banking efficiency. • Increases security in banking industries. • Solves regulatory problems.	• Lack of authenticity. • Lack of reliability.
Chod et al. [22]	b_verify an open-source software protocol	Blockchain technology's transparency into a firm's supply chain allows it to get attractive financing conditions at reduced signaling costs.	• Security of financing conditions. • Less signaling cost.	• It is complex for humans to read cryptographic verification proof.
Li et al. [51]	Fabric-SCF	A Blockchain-based secure storage solution using distributed consensus to protect, trace, and immutably store supply chain financial information.	• Provides dynamic, fine-grained access control. • Secure storage management.	• Complex business logic is required instead of simple SCF logic. • Throughput verification is required.
Li et al. [52]	Fabric-Chain & Chain	Proposed a Blockchain and bloom-based storage mechanism for the electronic document data utilized in the supply chain finance business process.	• Improved access efficiency. • Increased retrieval efficiency.	• Lack of portable business logic. • Lack of data updating.

Less Efficient Methods : One further reason why the financial technology industry requires Blockchain is that the presence of many different third parties often causes the procedures to be delayed. This, in the end, leads to poorer rates of customer satisfaction and increased levels of volatility in the commercial sector as it generates a lot of data [42].

Higher Operational Cost: In the financial technology industry, time is money. As a result, Blockchain technology has once again shown itself to be one of the financial services innovations that have the potential to lower costs by almost half. This is because it reduces the dependence on many individuals, makes the process public, and shortens the required time.

9.4.2 Blockchain in Financial Services

When talking about how the Blockchain technology has changed the financial technology industry, it's best to focus on the most important parts of the economy to better understand and analyze the changes. The following is a list of Blockchain use cases for financial services [33].

P2P Payments : Bank clearing and settlement regions have concerns about costly bureaucracy and unclear expertise. These concerns are present in most financial arrangements and cause concern. Old and hierarchical financial systems produce these gaps. Decentralized consensus methods can close them quickly. Blockchain technology helps financial services. Decentralized ledger technology will enable mobile banking for those without bank accounts. A Blockchain mobile app development business may simplify cross-checking data between companies engaged in international payment transactions. Blockchain technology allows several checks at once.

Financial Trading: Documents are still being sent or faxed to confirm information that is necessary for trade financing, which means that paperwork is being sent across the world to verify the information. To buy stocks or shares, you still have to go through the complicated and time-consuming steps of brokerage, exchanges, clearing, and settlement. The settlement process takes three days on average, but it may take longer if it occurs over the weekend. This is because every trader is required to keep databases for all of the transaction-based documents, and they must routinely check these databases against each other to ensure that they are accurate. The application of Blockchain technology to the provision of financial services in this sector has the potential to free traders from the need to do time-consuming checks on counter-parties while also improving the efficiency of the whole life cycle [30,42]. This not only speeds up the settlement process but also makes transactions more accurate and reduces the risks involved [44].

Crypto Lending: Thanks to crypto lending, the financial world now has a new, easy, and transparent way to lend money. The lenders will give the borrowers the assets they need for the loan at a rate of interest that was agreed upon ahead of time. Borrowers will be able to keep their crypto assets as collateral for a loan based on fiat currency or stablecoins. It is also true when read backward. When borrowers need to borrow crypto assets, they will occasionally use their stable coins or traditional cash as collateral.

Regulatory Compliance: For the second time, this is one of the most consequential uses of Blockchain in the financial sector. Since it is expected that the global need for regulatory services will expand in the next few years, financial services businesses are integrating Blockchain technology to improve regulatory compliance [65]. They expect this technology to record the actions of all parties involved in every verified transaction, eliminating the need for regulators to verify the records' veracity. Technology is also allowing scholars to return to the original documents rather than relying on the many copies that have been produced. Errors are less likely to occur, the integrity of records for financial reporting and audits is being preserved, and the time and resources spent on auditing and accounting are being drastically reduced thanks to the Blockchain's promise of immutability [4,62].

Digital Identity: The number of accounts that have been made with fake information keeps going up. Even though banks do have stringent Know Your Customer and Anti-Money Laundering inspections, these measures are not failsafe. A digital identification system may benefit from using Blockchain technology. The customers only need to go through the validation process once, and then they may use their credentials to conduct transactions in any part of the world. On this front, Blockchain may also aid financial users in the following ways: 1) Managing personal information; 2) communicating personal information to other parties while minimizing security concerns; and 3) digitally signing legal documents, such as claims and transactions [75, 77].

Auditing: It is a procedure that checks the finances and brings to light any discrepancies that may exist. The procedure is not only difficult to understand, but it also moves quite slowly. Blockchain technology, on the other hand, makes the procedure simpler. Because of this technology, you can ask the Blockchain application development firm with whom you are paired to add the record straight to the ledger, making it possible to see and update data in a time-efficient manner [43].

New Crowdfunding Models: The concept of "crowdfunding" refers to a method of supporting a project by soliciting contributions from a large number of individuals, often via the Internet. ICOs, IEOs, and other mechanisms may make the fundraising process using Blockchain technology more open and efficient, as opposed to more traditional methods of financing. However, it is highly recommended to have a clear understanding of what all financial services organizations that use Blockchain technology are doing with it.

9.4.3 Benefits of Blockchain in Finance

Blockchain technology has made it possible to create inclusive, open, and safe corporate networks. These networks make it possible to issue digital security credentials in a shorter amount of time, at lower unit prices, and with a higher degree of customization [11, 67]. Over the past few years, the use of Blockchain technology in the financial sector has grown, which has shown the following benefits [23, 29, 80].

Transparency: Protocols, standards that everyone agrees on, and common procedures are all used in Blockchain technology. Together, these things serve as a single source of growth for all members of the network. It makes the data more reliable and improves the customer experience by making processing faster.

Security: In the financial sector, Blockchain technology has made it possible to use secure application code that is designed to be impossible to change by hostile or third-party actors [18]. It makes it practically impossible to modify or hack the system.

Trust: The immutable and clear ledger makes it easier for the different people in a business network to work together on data management and get along. The Blockchain is a distributed ledger technology that enables the safe recording, management, storage, and transmission of transactions across a wide variety of industries.

Privacy: When Blockchain technology is used in the financial industry, it protects data privacy at all levels of software stacks in a way that is unmatched in the industry. This makes it possible for businesses to share data selectively within their networks. This increases confidence and openness while preserving anonymity and privacy at the same time.

Programmability: It makes it possible to design and run smart contracts, which are pieces of software that automate business logic and can't be changed. This makes them easier to program, more efficient, and more trustworthy.

Scalability and High Performance: In the financial sector, Blockchain technology is used through hybrid and private networks that were built to handle hundreds of transactions per second. It provides enterprises with significant resilience and worldwide reach by completely supporting interoperability across the public and private sectors.

9.5   Blockchain-Enabled Federated Learning for Financial Services

With a central server, FL has reached a point where it can't do anymore. At the same time, new risks are showing up. There are a lot of things going on, but the most important ones are centralized processing, making up data, and not having any incentives. Businesses and academics are paying a lot of attention to Blockchain-enabled distributed ledger technology in the hopes that it will speed up the use of FL. There have been a lot of creative answers made to meet the needs of a wide range of situations that are always getting more complicated. The functionality of FL that is allowed by Blockchain offers both ideas and approaches to increasing the functionality of FL from a variety of points of view.

9.5.1   Federated Learning

Federated learning (FL) is an exciting new decentralized DL technique that lets users update their models together without having to share their data. FL is changing how mathematical modeling and analysis are done in the business world. This makes it possible for more and more industries to build distributed machine learning models that protect privacy and are safe. Nonetheless, the properties that are intrinsic to FL have resulted in several issues, including those about the security of personal information, the expense of communication, the heterogeneity of the systems, and the unreliability of model upload during real-world operation. It's interesting to think about how adding Blockchain technology could improve FL's security and performance and increase the number of things that can be done with it. FL is a strategy for training artificial intelligence systems using data that is kept confidential. It lets centralized AI systems learn from data, which is often personal, without the data's actual content being shared or made public. Instead, only the lessons that can be gleaned from the data's structure are used.

9.5.2   Blockchain-Enabled FL for Financial Services

The conventional architecture of federated learning is based on centralization. However, a model that depends on a trustworthy central server has several security holes. If the central server is attacked, the whole federated learning process could be ruined. This is because the global model update wouldn't be able to happen, training results might not be accurate, and the federation would be vulnerable to threats while collecting and updating model parameters. Joint sharing modeling of multi party data is made possible by the use of Blockchain technology, which is integrated with federated learning to meet specific data-sharing requirements. Table 9.2 provides the existing Blockchain-enabled FL methodologies designed and developed by several researchers. These methods can be used in financial services to improve service quality. Because the Blockchain has a strong way to store certificates, it is possible to

Table 9.2 Existing Blockchain-Enabled FL Methods for Data Sharing

Ref	Proposed Method	Description	Pros	Cons
Feng et al. [36]	Credibility-based RAFT-efficient consensus method	Used Consortium Blockchain and federated learning to enterprise data sharing and monitor the process.	• High-enterprise application efficiency. • Enterprise data exchange without compromising privacy.	• It is unable to defend itself against Byzantine attacks.
Zhao et al. [79]	Reputation-based FL method	This strategy trains a machine-learning model using consumer data. Manufacturers can then estimate client needs and expenditures.	• Introduced a reward system. • Entices clients to the crowd-sourced FL activity. • Provides privacy and test accuracy.	• Deterministically optimum local–global balance for greater test accuracy.
Lu et al. [56]	Privacy-preserving distributed data sharing	This approach involves training a machine-learning model with consumer data. This enables manufacturers to predict client requirements and spending patterns.	• Improved computational resource efficiency. • Improved data exchange efficiency. • Secure information exchange with higher accuracy.	• Data utility. • Inefficient data-sharing method, if the number of devices is limited.
Bandara et al. [5]	Bassa-ML	An integrated federated learning platform that is built on Blockchain technology and Model Cards.	• More openness and transparency. • Bringing audit ability to the process of federated learning. • Enhanced trust for the models.	• Data utility.
Li et al. [53]	BFLC	Blockchain is used by the framework in place of a centralized server for both the storing of global model data and the communication of local model update information.	• Cuts down on the amount of computing needed for consensus. • Reduces the number of harmful assaults.	• Scalability of the BFLC. • Storage is expensive because Blockchain stores all information.
Kim et al. [48]	BlockFL	Architecture for distributed learning that allows for the secure and transparent sharing and verification of localized modifications to learning models.	• Incentive-based method. • Federates additional devices with more training samples.	• Uncontrolled data quality.

Reference	Name	Description	Advantages	Limitations
Ramanan et al. [68]	BAFFLE	Decomposing the global parameter space into discrete pieces and then using a score and bid technique is how BAFFLE increases computing efficiency.	• Updates many sections simultaneously. • A low cost for the computation.	• No method for data quality control.
Shayan et al. [70]	Biscotti	A completely decentralized peer-to-peer (P2P) method for multi party machine learning, as well as a client-to-client machine learning procedure that protects users' privacy.	• The PoF consensus algorithm was introduced. • Scalable, resistant to errors, and able to withstand recognized forms of assault.	• Scalability issues for the large model. • Privacy issues. • Stake limitations.
Chen et al. [16]	Learning chain	Decentralized privacy-preserving and secure machine learning system using a broad (linear or nonlinear) deep-learning approach with no centralized trusted servers.	• A decentralized framework. • Protection against attacks performed by Byzantine.	• Prone to being influenced by biases.
Lyu et al. [57]	FPPDL	A local credibility mutual assessment method for fairness and a three-layer onion-style encryption technique for accuracy and privacy.	• Privacy-preserved. • Fairness. • Accuracy.	• Many encryption methods may impede node processing. • Fairness in non-IID setting.
Ma et al. [59]	BLADE-FL	Prevents dishonest customers from tainting the learning process and further gives clients a learning environment that is self-motivating and dependable.	• Privacy preserved. • Lazy client protection.	• Computing capability. • Training data size. • Transmitting diversity.

provide controllable data traceability. For federated learning, both making data visible and making it available are possible. the FL functions are somewhat complementary with each other If they can be connected, it will not only be possible to improve the way data is shared, but it will also be possible to keep data modeling private [26].

Data Protection: Feng and Chen [36] proposed a Blockchain-enabled FL data-preserving strategy, which can be used by a financial business. Each business node that joins the consortium chain must pass a certain qualification test. Business nodes that take part in federated learning within the consortium chain must also verify their strong reputation value, but only if it is higher than a certain threshold. In federated learning, only nodes with a certain minimum level of reputation can be chosen to take part. They divided federated learning nodes into monitoring and training groups. The training node group does local model training rounds. The monitoring node group determines the training node's reputation and global parameters. The top reputation-valued nodes in this cycle of federated learning are the monitoring nodes. This process is done to ensure that malicious nodes will be excluded in the learning cycle.

Data Traceability: Federated computing ensures that the original data never leaves the local node; that only the gradient information of the model update needs to be uploaded; and that private data is protected while taking advantage of the Blockchain's immutability, certificate storage function strength, and controllable traceability. Based on this concept, it is possible to train complicated joint models. Therefore, they serve a similar purpose but in different ways. When used in tandem, they may not only make data exchange a reality but also keep personal information secure.

Profit Sharing: Because nodes in a community can use the model whenever they want without making any changes, a strong incentive is needed to get nodes to train models. Li et al. [53] suggested that a system called "profit sharing by contribution" be put in place as a way to solve this problem. Following the aggregation of each round, the managers will then give incentives to the appropriate nodes based on the scores of the updates that they have provided. As a result, giving updates regularly may bring more benefits, and the constantly updated global model may encourage other nodes to join the network. It is important to research this incentive mechanism since it can easily be adapted to a variety of applications in the real world and has great scalability.

Storage Optimization: When it comes to applications that take place in the real world, storage overhead is a critical component that helps determine the hardware requirements for the training devices. Historical models and updates may help with recovery after a disaster, but they take up a huge amount of space. Here, Li et al. [53] gave a simple and workable plan for reducing the storage overhead: Nodes that don't have enough space can get rid of old blocks locally, and they only need to keep the most recent model and updates for the current round. In this way, the problem of not having enough storage space can be solved, and the core nodes can keep their ability to recover and verify data. Having said that, it is also easy to see where this approach falls short. With each transaction that is deleted, the legitimacy of the Blockchain suffers. It is possible that each node will not adopt this technique due to the high level of mutual mistrust that exists among this group. So, the best thing to do might be to store your data with a reputable and trustworthy third party. The only information that the Blockchain stores are the network addresses of the locations of each model or updated file, as well as records of any modifications that were made to those files. For other nodes to access the model and its changes, the centralized store must interface with those nodes. This central storage will be in charge of making backups in case of a disaster and offering services for storing files in different places.

Safeguarding Payment Networks: One of the best things about Blockchain technology is that it can be used as a payment network that is not limited by national borders. As a decentralized answer to the problem of making payments without friction and with low transaction costs, many different Blockchain protocols have been made. Centralized financial institutions are known for their high fees and painfully slow processing times, which led to the creation of these alternatives. Even though this use of Blockchain technology has a lot of potential, it is still hard to use it widely because of security concerns [74]. Theft and fraud are all too frequent in Blockchain transactions since all that is required to complete them is a set of public and private keys. On the other hand, advanced machine learning can easily find unusual account activity, which then calls for human help. Both the companies that supply financial services and the people who use them are protected by this extra security measure [64].

Effective Financial Services: Financial institutions that use FL and Blockchain technologies often have the goal of increasing both the speed at which they provide their services and the quality of those services. In the same way that any other company would, these establishments have the incentive to find ways to save expenses and, as a result, generate greater value. In a survey that shows this trend, A survey in [19] found that 57% of firms see cost savings as the main benefit of joining consortium Blockchain networks. Those that supply financial services may deliver more value to their consumers while also optimizing the returns on their investments if they use these two technologies to drive business operations.

Controlled Finance Automation: The trend toward more automation cannot be denied, but if it is allowed to continue unchecked, it may result in undesirable consequences. Businesses will always lose control of their operations over time if there are no limits on how they can use automated processes. As a result of this, the tasks of automation need to be carried out in conjunction with built-in checks and balances. Using FL and smart contracts backed by Blockchain technology could help make this dynamic happen. Smart contacts make it possible to automate procedures, while machine learning can look for problems and only call for human help when it's necessary. Because of this very important infrastructure, all financial transactions would be completely safe, open, and efficient.

9.6 Future Scopes

Financial companies have always been concerned about cybersecurity breaches, but cyber-attacks now pose considerably greater hazards to their operations and reputations. They must know the biggest threats first. It will help them prioritize cybersecurity activities and maximize ROI. It will also aid cybersecurity strategy development. Ransomware, phishing, online application, vulnerability exploitation assaults, denial of service (DoS) attacks [46], insider threats [12], nation-state attacks, and state-sponsored threats will be the biggest dangers to financial institutions in 2022. Financial institutions must upgrade their security to protect against ever-changing threats. Despite spending millions and adding several levels of protection to their infrastructure, they still don't know how to employ security measures. To remain ahead of threat actors, they should test their IT security infrastructure against real-world threats often. IT and security executives must verify security measures to assess firms' cybersecurity posture and cyber resilience and show that they prevent intrusions.

Effectiveness of Transmission: The storage and synchronization of the Blockchain need a significant amount of hardware resources, including not only the space on the hard drive but also the bandwidth on the network. Therefore, the best way to lessen the amount of transmission that is needed while still preserving the reliability of model training is a subject deserving of further investigation.

The Scene in Public: The authentication processes are handled by the alliance Blockchain system, although this does have the side effect of making it more difficult to join the training community. How to create a public community by using a Proof-of-Work (PoW) mechanism is another interesting problem, which is finding a way to work together while protecting yourself from attacks from hostile nodes in financial services.

Lightweight Training: A lot of the Internet of Things (IoT) devices that financial clients use don't have enough hardware features to train a deep neural network well. Because of this, the question of how to make training models easier (e.g., by using edge servers) while still protecting users' privacy is an important one that deserves more research.

Security Control Validation: Even though it's good to see financial institutions work toward high levels of cyber maturity, we strongly suggest doing an objective evaluation of these assumptions and maturity levels and fixing any gaps between what was expected and what was found in an assessment. Banks and other financial institutions, as well as other businesses, must constantly test the efficacy of their security policies against real-world attacks to stay one step ahead of threat actors. The term "security control validation" refers to an approach that is centered on potential threats and that enables businesses to evaluate and analyze their cybersecurity posture and their overall cyber resilience. It also checks to see if the security controls are working well enough to stop cyberattacks.

Automation: As long as AI keeps getting good investments, FL seems to be in a great position to grow a lot. Because of the adaptability of the technology, it is expected that it will make its way into a rising number of different businesses across a wide variety of use cases. The financial services industry, in particular, is in a great position to get a lot of value out of the combination of Blockchain and FL technologies. These technologies work together to make big changes in the financial sector by making it safer, making it run better, and giving people more control over automation. Figure 9.6 shows the potential outcomes of using Blockchain technology in the financial sector.

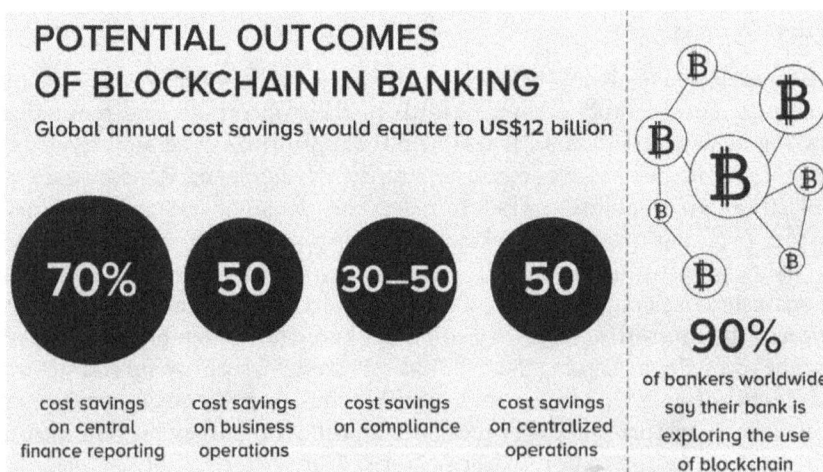

Figure 9.6 The potential outcomes of using Blockchain technology in the financial sector [21].

Next-Generation Marketing: However, to make use of this potential, it is essential to keep in mind why Blockchain technology was developed in the first place. Also, for FL learning systems to keep being useful, they need to be fed with good enough data. Although many questions remain unanswered, these technologies are not going away, either together or separately. In a financial market that is always changing, Blockchain and FL integrations might be the next drivers of disruptive change across the finance sector.

However, it's important to note that both Blockchain and FL face challenges and limitations. Blockchain's scalability, energy consumption, and regulatory considerations are areas that require further investigation. FL may face issues related to model synchronization, privacy leakage, and communication overhead. Overcoming these challenges and ensuring appropriate governance frameworks will be essential for the successful implementation of Blockchain and federated learning in future financial services.

9.7   Conclusion

FL is attracting attention for its privacy-enhancing and scalable financial services and applications. We conducted a state-of-the-art assessment and comprehensive survey based on recent research to examine how Blockchain and FL may improve financial services. This study was motivated by the lack of a comprehensive FL and Blockchain survey in financial services. We first discussed FL and Blockchain technologies and their combination to bridge this gap. Then, we extended our study by providing measures of how FL and Blockchain can be used in financial services to protect financial data, make decentralized storage more efficient, keep payment networks safe, automate tasks, protect privacy, and keep things safe. Lastly, we talked about some of the problems with research and gave notions for future initiatives that will bring more attention to this new field and encourage more research to realize FL and Blockchain.

Bibliography

[1] CSBS 2022 national survey of community banks, findings from the 2022 CSBS national survey of community banks presented at the 10th annual community banking research conference, 2022. https://www.picussecurity.com/resource/blog/six-stages-of-dealing-with-a-global-security-incident [Online; Accessed on Oct. 24, 2022].

[2] Adminer. *Database Management in a Single PHP File*, 2021. https://www.adminer.org/en/ [Online; Accessed on Nov. 17, 2022].

[3] David Andolfatto. Assessing the impact of central bank digital currency on private banks. *The Economic Journal*, 131(634):525–540, 2021.

[4] Raphael Auer. *Embedded Supervision: How to Build Regulation into Blockchain Finance*, 2019. https://ssrn.com/abstract=3486246

[5] Eranga Bandara, Sachin Shetty, Abdul Rahman, Ravi Mukkamala, Juan Zhao, and Xueping Liang. Bassa-ml—a blockchain and model card integrated federated learning provenance platform. In *2022 IEEE 19th Annual Consumer Communications & Networking Conference (CCNC)*, pages 753–759, 2022. doi: 10.1109/CCNC49033.2022.9700513.

[6] Christian Barontini and Henry Holden. Proceeding with caution-a survey on central bank digital currency. *Proceeding with Caution-A Survey on Central Bank Digital Currency (January 8, 2019)*. BIS Paper, (101), 2019.

[7] Laveen Bhatia and Saeed Samet. Decentralized federated learning: A comprehensive survey and a new blockchain-based data evaluation scheme. In *2022 Fourth International Conference on Blockchain Computing and Applications (BCCA)*, pages 289–296, 2022. doi: 10.1109/BCCA55292.2022.9922390.

[8] Amanda Bronstad. *Capital One Reaches $190m Settlement Over 2019 Data Breach*, 2021. https://www.law.com/2021/12/21/capital-one-settles-lawsuits-over-2019-data-breach/ [Online; Accessed on May 19, 2022].

[9] Volker Brulhl. Virtual currencies, distributed ledgers and the future of financial services. *Intereconomics*, 52(6):370–378, 2017.

[10] Ramiro Daniel Camino, Radu State, Leandro Montero, and Petko Valtchev. Finding suspicious activities in financial transactions and distributed ledgers. In *2017 IEEE International Conference on Data Mining Workshops (ICDMW)*, pages 787–796, 2017. doi: 10.1109/ICDMW.2017.109.

[11] Caytas, Joanna, Developing Blockchain Real-Time Clearing and Settlement in the EU, U.S., and Globally (June 22, 2016). *Columbia Journal of European Law: Preliminary Reference (June 22, 2016)*, Available at SSRN: https://ssrn.com/abstract=2807675

[12] Sarathiel Chaipa, Ernest Ketcha Ngassam, and Singh Shawren. Towards a new taxonomy of insider threats. In *2022 IST-Africa Conference (IST-Africa)*, pages 1–10, 2022. doi: 10.23919/IST-Africa56635.2022.9845581.

[13] Pushpita Chatterjee, Debashis Das, and Danda Rawat. *Securing Financial Transactions: Exploring the Role of Federated Learning and Blockchain in Credit Card Fraud Detection*, 2023. ISSN 0167-739X, https://doi.org/10.1016/j.future.2024.04.057.

[14] Pushpita Chatterjee, Debashis Das, and Danda Rawat. *Use of Federated Learning and Blockchain Towards Securing Financial Services*, 2023. doi: 10.1109/TCE.2023.3339702.

[15] Chen, Shuhui, Qing Wang and Shu-an Liu. "Credit Risk Prediction in Peer-to-Peer Lending with Ensemble Learning Framework." *2019 Chinese Control And Decision Conference (CCDC)* (2019): 4373-4377.

[16] Xuhui Chen, Jinlong Ji, Changqing Luo, Weixian Liao, and Pan Li. When machine learning meets blockchain: A decentralized, privacy-preserving and secure design. In *2018 IEEE International Conference on Big Data (Big Data)*, pages 1178–1187, 2018. doi: 10.1109/BigData.2018.8622598.

[17] Yi-Hui Chen, Li-Chin Huang, Iuon-Chang Lin, and Min-Shiang Hwang. Research on the secure financial surveillance blockchain systems. *International Journal of Network Security*, 22(4):708–716, 2020.

[18] Yi-Hui Chen, Li-Chin Huang, Iuon-Chang Lin, and Min-Shiang Hwang. Research on the secure financial surveillance blockchain systems. *International Journal of Network Security*, 22(4):708–716, 2020.

[19] Z. Chen. *What the Convergence of Blockchain and Machine Learning Means for the Future of Finance*, 2023. https://www.nasdaq.com/news-and-insights [Online; Accessed on Jan. 15, 2023].

[20] Chirag. *Blockchain in Financial Services: A Catalyst for Disruption in Finance World*, 2022. https://appinventiv.com/blog/blockchain-and-fintech/ [Online; Accessed on Jan. 13, 2023].

[21] Chirag. *Why Are Banks Adopting Blockchain Technology?*, 2023 https://appinventiv.com/blog/blockchain-in-banking/ [Online; Accessed on Jan. 18, 2023].

[22] Jiri Chod, Nikolaos Trichakis, Gerry Tsoukalas, Henry Aspegren, and Mark Weber. On the financing benefits of supply chain transparency and blockchain adoption. *Management Science*, 66(10):4378–4396, 2020.

[23] Tsan-Ming Choi. Financing product development projects in the blockchain era: Initial coin offerings versus traditional bank loans. *IEEE Transactions on Engineering Management*, 69(6):3184–3196, 2022.

[24] Luisanna Cocco, Andrea Pinna, and Michele Marchesi. Banking on blockchain: Costs savings thanks to the blockchain technology. *Future Internet*, 9(3):25, 2017.

[25] Atlassian Community. *Confluence Server and Data Center—CVE-2021–26084, Confluence Server Webwork Ognl Injection*, 2021. https://confluence.atlassian.com/doc/confluence-security-advisory-2021-08-25-1077906215.html [Online; Accessed on Dec. 17, 2022].

[26] Debashis Das, Sourav Banerjee, and Utpal Biswas. A secure vehicle theft detection framework using blockchain and smart contract. *Peer-to-Peer Networking and Applications*, 14:672–686, 2021.

[27] Debashis Das, Sourav Banerjee, Pushpita Chatterjee, Uttam Ghosh, and Utpal Biswas. A secure blockchain enabled v2v communication system using smart contracts. *IEEE Transactions on Intelligent Transportation Systems*, pages 1–10, 2022. doi: 10.1109/TITS.2022.3226626.

[28] Debashis Das, Sourav Banerjee, Kousik Dasgupta, Pushpita Chatterjee, Uttam Ghosh, and Utpal Biswas. Blockchain enabled sdn framework for security management in 5g applications. In *Proceedings of the 24th International Conference on Distributed Computing and Networking*, pages 414–419, 2023. https://doi.org/10.1145/3571306.3571445

[29] Debashis Das, Sourav Banerjee, Uttam Ghosh, Utpal Biswas, and Ali Kashif Bashir. A decentralized vehicle anti-theft system using blockchain and smart contracts. *Peer-to-Peer Networking and Applications*, 14:2775–2788, 2021.

[30] Natalia Dashkevich, Steve Counsell, and Giuseppe Destefanis. Blockchain application for central banks: A systematic mapping study. *IEEE Access*, 8:139918–139952, 2020.

[31] D. Donaldson. *Vulnerability of Financial Institutions to Cyber Crime*, 2022. https://www.iap-association.org/getattachment/Conferences/Regional-Conferences/North-America-and-Caribbean/4th-North-American-and-Caribbean-Conference/Conference-Documentation/4NACC_Jamaica_WS2B_PPT_Damien_Donaldson.pdf.aspx [Online; Accessed on Dec. 18, 2022].

[32] Mingxiao Du, Qijun Chen, Jie Xiao, Houhao Yang, and Xiaofeng Ma. Supply chain finance innovation using blockchain. *IEEE Transactions on Engineering Management*, 67(4):1045–1058, 2020.

[33] Mingxiao Du, Qijun Chen, Jie Xiao, Houhao Yang, and Xiaofeng Ma. Supply chain finance innovation using blockchain. *IEEE Transactions on Engineering Management*, 67(4):1045–1058, 2020.

[34] Olaniyi Evan s. *Blockchain Technology and the Financial Market: An Empirical Analysis*, 2018. https://mpra.ub.uni-muenchen.de/99212/2/MPRA_paper_99212.pdf

[35] F. Christopher Feeney. *Cybersecurity Regulation Harmonization: The Financial Services Roundtable—Bits*, 2017. https://www.hsgac.senate.gov/imo/media/doc/Testimony-Feeney-2017-06-21.pdf [Online; Accessed on Nov. 25, 2022].

[36] Xiaoqing Feng and Lei Chen. Data privacy protection sharing strategy based on consortium blockchain and federated learning. In *2022 International Conference on Artificial Intelligence and Computer Information Technology (AICIT)*, pages 1–4, 2022. doi: 10.1109/AICIT55386.2022.9930188.

[37] Ram Gall. *Over 600,000 Sites Impacted by WP Statistics Patch*. https://www.wordfence.com/blog/2021/05/over-600000-sites-impacted-by-wp-statistics-patch/ [Online; Accessed on Aug. 22, 2022].

[38] Teja Goud Allam, A. B. M. Mehedi Hasan, Angelika Maag, and P. W. C. Prasad. Ledger technology of blockchain and its impact on operational performance of banks: A review. In *2021 6th International Conference on Innovative Technology in Intelligent System and Industrial Applications (CITISIA)*, pages 1–10, 2021. doi: 10.1109/CITISIA53721.2021.9719886.

[39] Meriem Guerar, Alessio Merlo, Mauro Migliardi, Francesco Palmieri, and Luca Verderame. A fraud-resilient blockchain-based solution for invoice financing. *IEEE Transactions on Engineering Management*, 67(4):1086–1098, 2020.

[40] Ye Guo and Chen Liang. Blockchain application and outlook in the banking industry. *Financial Innovation*, 2(1):1–12, 2016.

[41] Xuan Han, Yong Yuan, and Fei-Yue Wang. A blockchain-based framework for central bank digital currency. In *2019 IEEE International Conference on Service Operations and Logistics, and Informatics (SOLI)*, pages 263–268. IEEE, 2019.

[42] Hossein Hassani, Xu Huang, Emmanuel Sirimal Silva, Hossein Hassani, Xu Huang, and Emmanuel Sirimal Silva. *Fusing Big Data, Blockchain, and Cryptocurrency*. Springer, 2019.

[43] Adam Hayes. Decentralized banking: Monetary technocracy in the digital age. In *Banking Beyond Banks and Money*, pages 121–131. Springer, 2016.

[44] Friedrich Holotiuk, Francesco Pisani, and Jlrrgen Moormann. The Impact of Blockchain Technology on Business Models in the Payments Industry, in Leimeister, J.M., Brenner, W. (Hrsg.): Proceedings der 13. *Internationalen Tagung Wirtschaftsinformatik (WI 2017)*, St. Gallen, S. 912-926.

[45] Intellipaat. *The Importance of Cyber Security in Banking Sector: Cyber Security in Banking*, 2022. https://intellipaat.com/blog/cyber-security-in-banking/ [Online; Accessed on Sep. 05, 2022].

[46] Clifford Kemp, Chad Calvert, and Taghi M. Khoshgoftaar. Detecting slow application-layer dos attacks with PCA. In *2021 IEEE 22nd International Conference on Information Reuse and Integration for Data Science (IRI)*, pages 176–183, 2021.doi: 10.1109/IRI51335.2021.00030.

[47] Meriem Kherbouche, Galena Pisoni, and Bálint Molnár. Model to program and blockchain approaches for business processes and workflows in finance. *Applied System Innovation*, 5(1), 2022.

[48] Hyesung Kim, Jihong Park, Mehdi Bennis, and Seong-Lyun Kim. Blockchained on-device federated learning. *IEEE Communications Letters*, 24(6):1279–1283, 2020.

[49] Edward Kost. *The 6 Biggest Cyber Threats for Financial Services in 2023*, 2023. https://www.upguard.com/blog/biggest-cyber-threats-for-financial-services [Online; Accessed on Jan. 15, 2023].

[50] Pralay Kumar Lahiri, Debashis Das, Wathiq Mansoor, Sourav Banerjee, and Pushpita Chatterjee. A trustworthy blockchain based framework for impregnable iov in edge computing. In *2020 IEEE 17th International Conference on Mobile Ad Hoc and Sensor Systems (MASS)*, pages 26–31. IEEE, 2020.

[51] Dun Li, Dezhi Han, Noel Crespi, Roberto Minerva, and Kuan-Ching Li. A blockchain-based secure storage and access control scheme for supply chain finance. *The Journal of Supercomputing*, 1–30, 2022.

[52] Dun Li, Dezhi Han, and Han Liu. Fabric-chain & chain: A blockchain-based electronic document system for supply chain finance. In *International Conference on Blockchain and Trustworthy Systems*, pages 601–608. Springer, 2020.

[53] Yuzheng Li, Chuan Chen, Nan Liu, Huawei Huang, Zibin Zheng, and Qiang Yan. A blockchain-based decentralized federated learning framework with committee consensus. *IEEE Network*, 35(1):234–241, 2021.

[54] Jingkuang Liu, Lemei Yan, and Dong Wang. A hybrid blockchain model for trusted data of supply chain finance. *Wireless Personal Communications*, 127:919–943, 2021.

[55] Joana Lorenz, Maria Inês Silva, David Oliveira Aparício, João Tiago Ascensão, and P. Bizarro. Machine learning methods to detect money laundering in the bitcoin blockchain in the presence of label scarcity. *Proceedings of the First ACM International Conference on AI in Finance*, 2020. https://doi.org/10.1145/3383455.3422549

[56] Yunlong Lu, Xiaohong Huang, Yueyue Dai, Sabita Maharjan, and Yan Zhang. Blockchain and federated learning for privacy-preserved data sharing in industrial IoT. *IEEE Transactions on Industrial Informatics*, 16(6):4177–4186, 2020.

[57] L. Lyu, Jiangshan Yu, Karthik Nandakumar, Yitong Li, Xingjun Ma, and Jiong Jin. Towards fair and decentralized privacy-preserving deep learning with blockchain. *ArXiv*, abs/1906.01167, 2019.

[58] Chaoqun Ma, Xiaolin Kong, Qiujun Lan, and Zhongding Zhou. The privacy protection mechanism of hyperledger fabric and its application in supply chain finance. *Cybersecurity*, 2(1):1–9, 2019.

[59] Chuan Ma, Jun Li, Long Shi, Ming Ding, Taotao Wang, Zhu Han, and H. Vincent Poor. When federated learning meets blockchain: A new distributed learning paradigm. *IEEE Computational Intelligence Magazine*, 17(3):26–33, 2022.

[60] Konstantinos Mantzaris. *How Can We Benefit from Blockchain Technologies?*, 2018 https://economistmk.blogspot.com/2018/03/how-can-we-benefit-from-blockchain.html [Online; Accessed on Jan. 18, 2023].

[61] Tim Maurer and Arthur Nelson. *The Global Cyber Threat, Finance & Development*, 2021. https://www.imf.org/external/pubs/ft/fandd/2021/03/pdf/global-cyber-threat-to-financial-systems-maurer.pdf [Online; Accessed on Sep. 11, 2022].

[62] Quoc Khanh Nguyen. Blockchain-a financial technology for future sustainable development. In *2016 3rd International Conference on Green Technology and Sustainable Development (GTSD)*, pages 51–54. IEEE, 2016.

[63] Jack Nicholls, Aditya Kuppa, and Nhien-An Le-Khac. Financial cybercrime: A comprehensive survey of deep learning approaches to tackle the evolving financial crime landscape. *IEEE Access*, 9:163965–163986, 2021.

[64] Mildred Chidinma Okoye and Jeremy Clark. Toward cryptocurrency lending. In *Financial Cryptography and Data Security: FC 2018 International Workshops, BITCOIN, VOTING, and WTSC, Nieuwpoort, Curaçao, March 2, 2018, Revised Selected Papers 22*, pages 367–380. Springer, 2019.

[65] Peterson K. Ozili. Blockchain finance: Questions regulators ask. In *Disruptive Innovation in Business and Finance in the Digital World*. Emerald Publishing Limited, 2019.

[66] Polestar. *Top Financial Services Banking Analytics Use Cases*, 2022. https://www.polestarllp.com/ top-financial-services-banking-analytics-use-cases [Online; Accessed on Jan. 10, 2023].

[67] Randy Priem. Distributed ledger technology for securities clearing and settlement: Benefits, risks, and regulatory implications. *Financial Innovation*, 6(1):1–25, 2020.

[68] Paritosh Ramanan and Kiyoshi Nakayama. Baffle: Blockchain based aggregator free federated learning. In *2020 IEEE International Conference on Blockchain (Blockchain)*, Rhodes Island, 2020 pp. 72-81. doi: 10.1109/Blockchain50366.2020.00017

[69] Pete Schroeder. *Capital One to Pay $80 Million Fine After Data Breach*, 2020. https://www.reuters.com/article/us-usa-banks-capital-one-fin-idUSKCN2522DA [Online; Accessed on Sep. 06, 2022].

[70] Muhammad Shayan, Clement Fung, Chris J. M. Yoon, and Ivan Beschastnikh. Biscotti: A blockchain system for private and secure federated learning. *IEEE Transactions on Parallel and Distributed Systems*, 32(7):1513–1525, 2021.

[71] Rajani Singh, Ashutosh Dhar Dwivedi, Gautam Srivastava, Pushpita Chatterjee, and Jerry Chun-Wei Lin. A privacy preserving internet of things smart healthcare financial system. *IEEE Internet of Things Journal*, 1–1, 2022.

[72] Swivelsecure. *5 Cybersecurity Weaknesses in Banking and Finance*, 2022. https://swivelsecure.com/solutions/banking-finance/5-cybersecurity-weaknesses-threats-in-banking-and-finance-industry/. [Online; Accessed on Dec. 12, 2022].

[73] Wei-Tek Tsai, Robert Blower, Yan Zhu, and Lian Yu. A system view of financial blockchains. In *2016 IEEE Symposium on Service-Oriented System Engineering (SOSE)*, pages 450–457. IEEE, 2016.

[74] Wei-Tek Tsai, Enyan Deng, Xiaoqiang Ding, and Jie Li. Application of blockchain to trade clearing. In *2018 IEEE International Conference on Software Quality, Reliability and Security Companion (QRS-C)*, pages 154–163. IEEE, 2018.

[75] Xin Wang, Xiaomin Xu, Lance Feagan, Sheng Huang, Limei Jiao, and Wei Zhao. Inter-bank payment system on enterprise blockchain platform. In *2018 IEEE 11th International Conference on Cloud Computing (CLOUD)*, pages 614–621. IEEE, 2018.

[76] Binghui Wu and Tingting Duan. Application blockchain in supply chain finance: A study on small and micro enterprises in xi'an. In *2021 2nd International Conference on Big Data Economy and Information Management (BDEIM)*, Sanya, China, 2021, pp. 479-482, doi: 10.1109/BDEIM55082.2021.00104.

[77] Tong Wu and Xiubo Liang. Exploration and practice of inter-bank application based on blockchain. In *2017 12th International Conference on Computer Science and Education (ICCSE)*, pages 219–224. IEEE, 2017.

[78] Yuanxin Zhang, Zeyu Wang, Jiaying Deng, Zaijing Gong, Ian Flood, and Yueren Wang. Framework for a blockchain-based infrastructure project financing system. *IEEE Access*, 9:141555–141570, 2021.

[79] Yang Zhao, Jun Zhao, Linshan Jiang, Rui Tan, Dusit Niyato, Zengxiang Li, Lingjuan Lyu, and Yingbo Liu. Privacy-preserving blockchain-based federated learning for IoT devices. *IEEE Internet of Things Journal*, 8(3):1817–1829, 2021.

[80] Weilin Zheng, Zibin Zheng, Xiangping Chen, Kemian Dai, Peishan Li, and Renfei Chen. Nutbaas: A blockchain-as-a-service platform. *IEEE Access*, 7:134422–134433, 2019.

[81] Sİlleyman Ozarslan. *Six Stages of Dealing with a Global Security Incident*, 2021. https://www.picussecurity.com/resource/blog/six-stages-of-dealing-with-a-global-security-incident [Online; Accessed on May 27, 2022].

Chapter 10

A Comprehensive Survey on Blockchain-Integrated Smart Grids

Uttam Ghosh, Laurent L. Njilla, Danda B. Rawat, and Charles A. Kamhoua

Chapter Contents

10.1 Introduction

A smart grid network refers to a complex cyber-physical system (CPS) that provides efficient and cost-effective management of the electric energy grid by allowing real-time monitoring, coordinating, and controlling of the system [1]. Unlike power grid, the smart grid can generate the energy seamlessly from distributed energy resources (DERs) (including nuclear, hydro, solar, wind, solar, and thermal) and transmit the energy using transmission lines to distribution center, and then distribute to the individual customers based on their usage and requirements. It can store the excessive energy in a battery for future use. By leveraging real-time data, advanced analytics and technologies, and automated control systems, the smart grid can meet the evolving demands of different customers (including residential, commercial, and industrial customers) and transform the traditional power grid into a more resilient, efficient, and reliable energy infrastructure. It incorporates self-healing capabilities

DOI: 10.1201/9781003376712-10

to quickly detect the issues, reroute power flows, and restore service to affected areas, minimizing downtime, and improving grid reliability. It utilizes automation to monitor, control, manage, and make informed decisions to optimize the grid's operation and ensuring efficient energy delivery. Overall, the smart grid is an automated, self-healing, and distributed advanced energy delivery network that provides the following features to enhance the functionality and efficiency of the power grid [2]:

- The smart grid supports two-way communication of electricity and information. In traditional power grid, the electricity flows in a one-way direction from the utility provider to the consumers. Whereas the smart grid supports bidirectional flow of the electricity. It allows for the installation of DERs (solar panels, wind turbines) at the consumer premises. These DERs can generate excess electricity and feed back into the smart grid for contributing to the overall electricity supply.
- The smart grid facilitates interaction between users and the electricity market. As mentioned above, users can generate renewable energy using DERs, store excess energy, and sell it back to the grid or participate in local energy markets. Smart meters allow bidirectional communication between users and utilities where users can access detailed information on their energy usage, receive real-time billing data, and even provide feedback about their energy preferences or concerns.

In the smart, interconnected grid, as envisaged in the U.S. Department of Energy's 2030 roadmap [3], increased renewable energy generation with a decrease in battery storage costs has led to a stronger global focus on energy storage solutions and services. The future grid, as depicted in Figure 10.1, relies upon the uninterrupted flow of information between consumers and power generation, transmission, and distribution companies. The exchange of this real-time information is susceptible to cyber tampering from hackers and malicious actors that propagate throughout the smart grid system due to the interdependencies between the cyber, physical, and communication components. The smart grid features are designed to improve the reliability, performance, and security of traditional power grids [4]. The heterogeneous nature of the smart grid makes it difficult to standardize procedures and communication paradigms. Lately, the smart grid has seen extensive attention, and the global smart grid market size was valued at USD 30.6 billion in 2020 and is expected to reach USD 162.4 billion by 2030 as shown in Figure 10.2. The rising demand for reliable and regular power supply owing to the increasing shift toward sustainable energy sources is boosting the growth of the smart grid market [5].

The smart grid network also suffers from a lot of security problems for its resource restrictions. Blockchain applications like Bitcoin and Ethereum have attained excellent achievements that are beyond anticipation. The technologies behind the Blockchain are distributed ledger, cryptography, and digital signature to ensure the security and tamper-proof architecture [6]. It is also a decentralized system where every server carries an identical copy of the entire ledger. Meanwhile, sustainable Blockchain can be utilized to enhance smart grid security and safety. It can also provide data security and confidentiality for all information stored in it. Sustainable Blockchain can be viewed as an effective method of the smart grid safety scaling by reducing energy consumption and communication cost. However, the effectiveness of the sustainable Blockchain for the smart grid safety is still uncertain. Thus, it is chosen to improve the smart grid security. Figure 10.3 presents the overview of Blockchain-enabled smart grid networks.

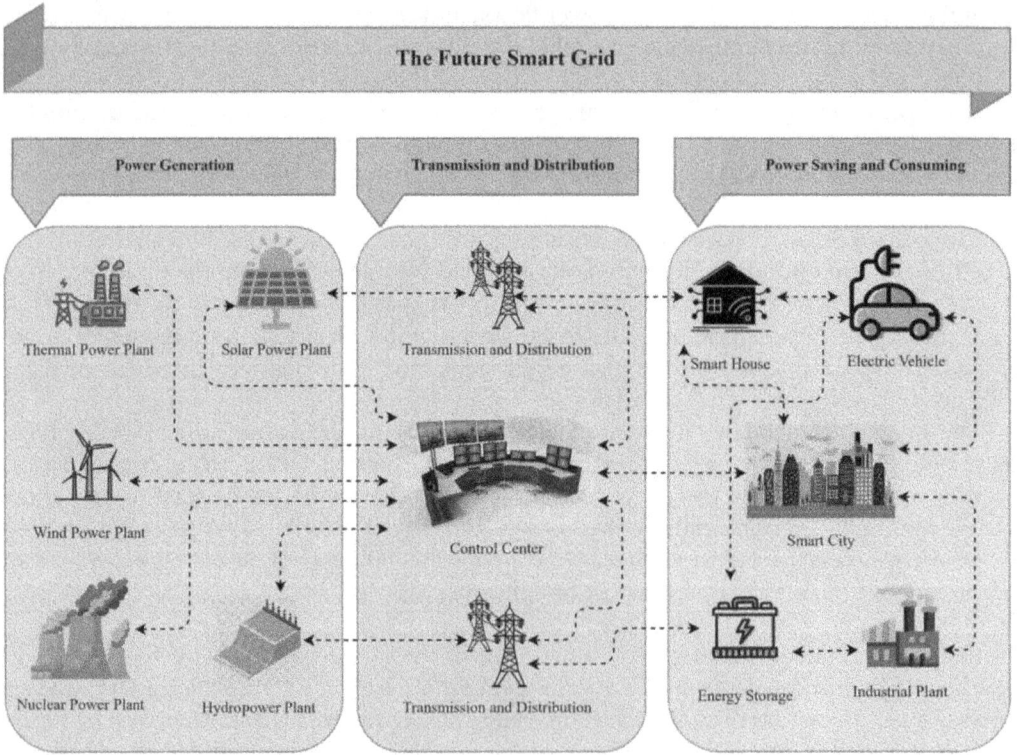

Figure 10.1 The future smart grid model.

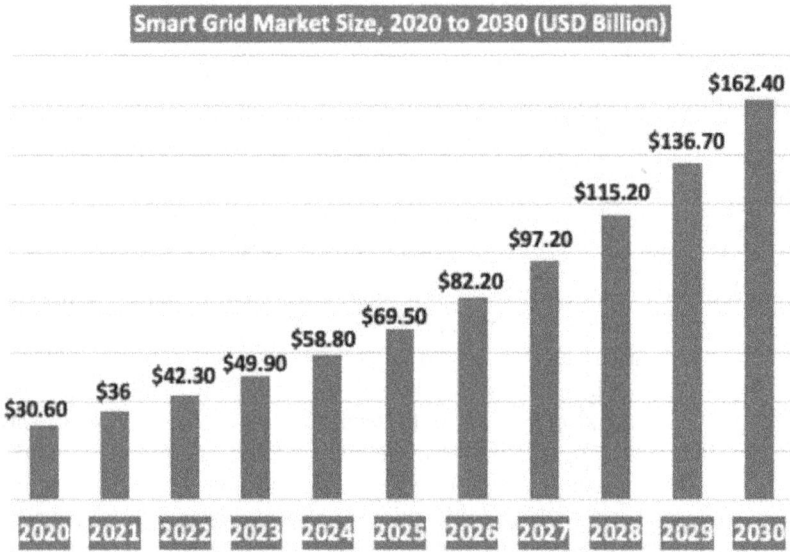

Figure 10.2 The global smart grid market size from 2020 to 2030 (USD billion).

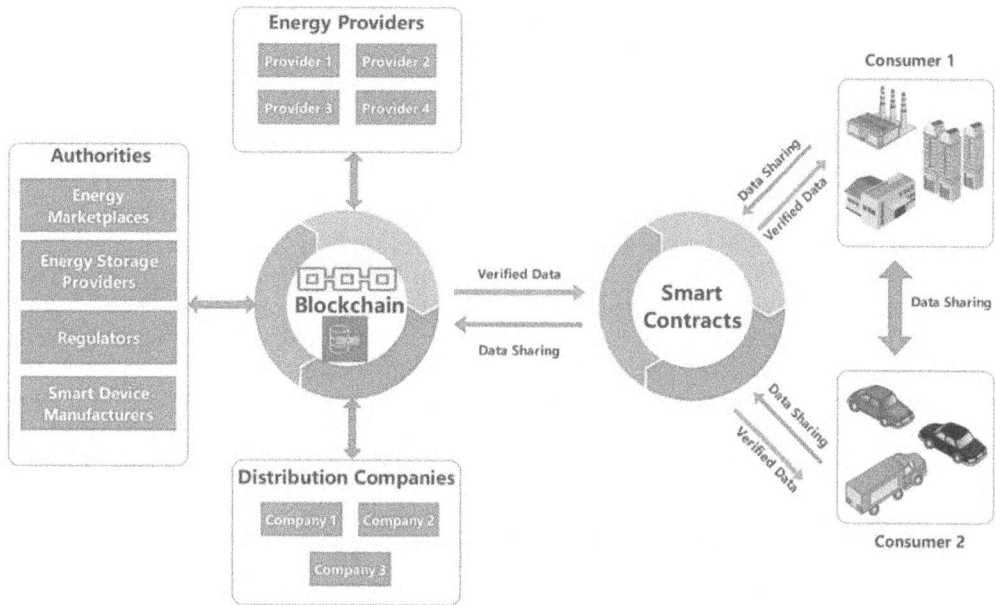

Figure 10.3 Overview of Blockchain integrated smart grid networks.

10.1.1 Motivation and Contribution

Security and privacy are key considerations in the design and operation of a smart grid system. We need to consider the following key measures related to security and privacy for the smart grid system [7, 8]: (a) Utilizing appropriate encryption techniques to protect data transmission and storage to ensure that unauthorized nodes neither can access nor can modify any information, (b) implementing access control mechanisms to restrict data and system access based on user roles and privileges, (c) implementing robust monitoring system, (d) utilizing advanced privacy-preservation techniques to protect information disclosure, (e) implementing a fault-tolerant network to protect against availability attacks, (f) providing non-repudiation to ensure that a node cannot deny or dispute actions it has performed in the past, and (g) promoting transparency and democracy among all the participating nodes.

As mentioned earlier, the smart grid combines advanced technologies, data analytics, smart devices, and communication systems to provide effective electricity distribution, supply security, minimum losses, and services including billing, bidding, and energy trading. Moreover, it needs to integrate a several number of electric vehicles, DERs, and prosumers (can produce and consume electricity). However, the traditional centralized grid system faces several challenges to effectively manage the smart devices, maintaining security and stability of the grid system. As a result, the smart grid is leading to a shift from a centralized topology to a more decentralized, distributed, and fully automated model for allowing greater interaction among all the grid devices [7].

On the other hand, Blockchain is a decentralized and distributed ledger technology that provides transparency and security while multiple participants maintain a shared and immutable tamper-proof record of transactions or data. It can be a promising alternative to the conventional centralized security systems to improve security and privacy. Blockchain can provide confidentiality, integrity, authentication, access control, transparency, and automaticity which are necessary for the smart grid system also. Table 10.1 summarizes the objectives of security and privacy in Blockchain paradigm.

Contribution: This chapter aims to present a comprehensive study of Blockchain applications to smart grids. First, it presents a detailed background, classification, working principle, characteristics, and applications of Blockchain. Then, it discusses a number of recent research works proposed in different literatures on Blockchain integration into smart grid systems for energy management, energy trading, security and privacy, microgrid management, and electric vehicle management. Finally, the chapter presents the limitations of Blockchain and future research direction of applying Blockchain in smart grids. With this chapter, the readers can have a more thorough understanding of Blockchain and smart grid architectures, Blockchain integration into smart grid systems, challenges and limitations of Blockchain, and present and future research trends in this area.

Chapter Organization: The rest of the chapter is organized as follows: Section 10.2 presents a background of Blockchain and its classification, working principle, characteristics, and

Table 10.1 Security and Privacy in Blockchain Paradigm

Objective	Blockchain Paradigm
Confidentiality	Public Blockchain is an open network to anyone for accessing the data stored on-chain. Thus, we should either avoid sharing sensitive/private data on public chain or use cryptography to encrypt the data before sharing on public chain. We need to make sure that the trusted and authorized nodes join the network.
Integrity	The immutable nature of Blockchain protects the integrity of data by design. Blockchain uses cryptographic hash function, Merkle tree, nonce, and timestamps for providing data integrity. We can detect if the data stored on-chain is modified by the attacker and also prevent decentralized access.
Authentication	Blockchain uses digital signature for the authentication where the participating nodes sign the transaction using their private key and verify the signature using the corresponding public key.
Privacy	Blockchain typically uses pseudonyms or cryptographic addresses to represent identities of the participants instead of real-world identities and provides a certain level of privacy. Zero-knowledge proofs can be employed in Blockchain to validate the correctness of a transaction or claim without disclosing the underlying data.
Availability	Blockchain operates on a distributed network nodes and does not rely on single point of failure. Further, it incorporates redundancy and data replication mechanisms to enhance the availability of data and maintain the continuity of the Blockchain network.
Transparency	Blockchain maintains an immutable distributed ledger that includes all records, transactions, events, and logs. In public Blockchain, participants can access, audit, and verify the transactions and records independently.
Access Control	Access control logic can be embedded with smart contract to define who can perform specific actions and ensure only that authorized entities can interact with the Blockchain and execute predefined operations. Blockchain can implement user authentication mechanisms including username–password combinations, multifactor authentication, digital certificates, and biometric authentication for the identity verification.
Automaticity	Blockchain can automate various actions and streamline operations within decentralized networks by leveraging smart contracts, predefined rules, event-driven triggers, and consensus mechanisms.

applications. The literature survey on Blockchain-assisted smart grids has been presented in Section 10.3. Section 10.4 presents the contribution of Blockchain into smart grids for energy management, energy trading, security and privacy, microgrid management, and electric vehicle management. The limitations of Blockchain and future research direction have been presented in Section 10.5. Finally, Section 10.6 concludes the chapter.

10.2   Preliminaries

Blockchain is a decentralized and distributed digital ledger that is used to record and store information securely and transparently. It consists of a chain of blocks, where each block contains information and a reference to the previous block [9, 10]. One of the key features of Blockchain technology is its immutability. This means that once data is recorded and confirmed in a block, it becomes very hard to alter or tamper with. Further, it does not need a third-party middleman entity, such as governments or banks, as Blockchain technology makes it especially safe to transfer money, properties, and contracts [11]. Blockchain is often considered a type of software protocol that relies on the Internet to function. It comprises various components, including a database, software applications, and a network of interconnected computers. Figure 10.4 presents the timeline for Blockchain use and deployment with coordination of the power system. In the following section, we provide the overview of Blockchain technology.

10.2.1   Types of Blockchain

Blockchain technology can be classified into public, private, and consortium types based on the underlying architecture, consensus mechanisms, and permission levels. Table 10.2 summarizes the classification of Blockchain [9, 12].

- **Public Blockchain:** Public Blockchain is an open, fully decentralized, and permissionless network where the general public can participate as a node, read and write data, validate transactions, and contribute to the consensus process. It provides a high level of decentralization, transparency, and security. However, it may have scalability limitations.
- **Private Blockchain:** Private Blockchain is a restricted network under the governance of one organization, which defines the access policies to participate in its private network.

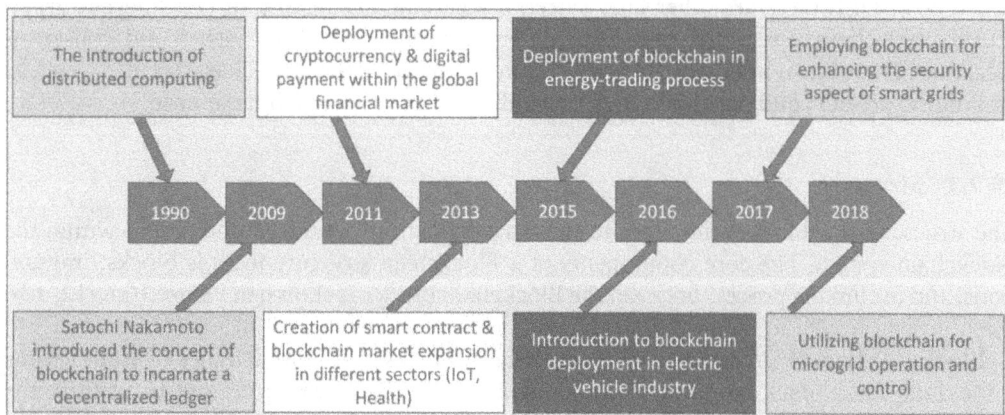

Figure 10.4 Timeline for Blockchain use and development with coordination of the power system.

Table 10.2 Classification of Blockchains

Parameter/Advantage/ Disadvantage/Use Case	Public Blockchain (Permissionless)	Consortium Blockchain	Private Blockchain (Permissioned)
Receptivity	Fully open	Available to some nodes	Open to a person/entity
Writers' access	Anyone	Certain nodes	Internally managed
Reading access	Anyone	Anyone	Public accessibility
Obscurity	More	Less	Less
Transaction speed	Low	High	Extremely high
Decentralization	Fully decentralized	Less decentralized	Less decentralized
Technology	Bitcoin, Ethereum	Quorum, MultiChain	Hyperledger Fabric, R3 Corda
Advantage	Independence Transparency Security	Access control Scalability Security	Access control Performance
Disadvantage	Performance Scalability	Transparency	Not decentralized Auditability
Use case	Cryptocurrency Document validation	Banking Supply chain research	Supply chain Asset ownership

Therefore, only authorized participants have the ability to read and access the transaction data. It offers more control, privacy, and scalability compared to public Blockchain. However, it does not support decentralization.

- **Consortium Blockchain:** Consortium Blockchain is governed and operated by a consortium or group of organizations that work together and collectively validate transactions. It has control over who can join the network and who can participate in the consensus process of the Blockchain. It provides a balance between public and private Blockchains. Consortium Blockchain maintains a certain level of decentralization.

10.2.2   Digital Ledger Technology

Distributed Ledger Technology (DLT) encompasses various types of digital ledgers that enable the recording, sharing, and verification of transactions or data in a decentralized manner, without the need for a central authority. Whereas Blockchain is a specific implementation of DLT that uses a chain of blocks to store and validate transactions. While Blockchain is a specific type of DLT, there are other DLT systems, including hashgraph, directed acyclic graph (DAG), holochain, and radix that have different architectural approaches and consensus mechanisms as shown in Figure 10.5. DLT systems can have different characteristics, such as scalability, privacy, and permissioning, depending on their design and use case.

10.2.3   Structure

The structure of a Blockchain refers to the way the data is organized and stored within the Blockchain system. The core components of a Blockchain structure include blocks, transactions, and the linking process between the Blockchain blocks as shown in Figure 10.6 [12, 13].

- **Blocks:** A Blockchain consists of a sequence of blocks that store batches of transactions or data. Each block contains a header and a body. The header is used to identify the particular block in the entire Blockchain, and it typically includes metadata such as a

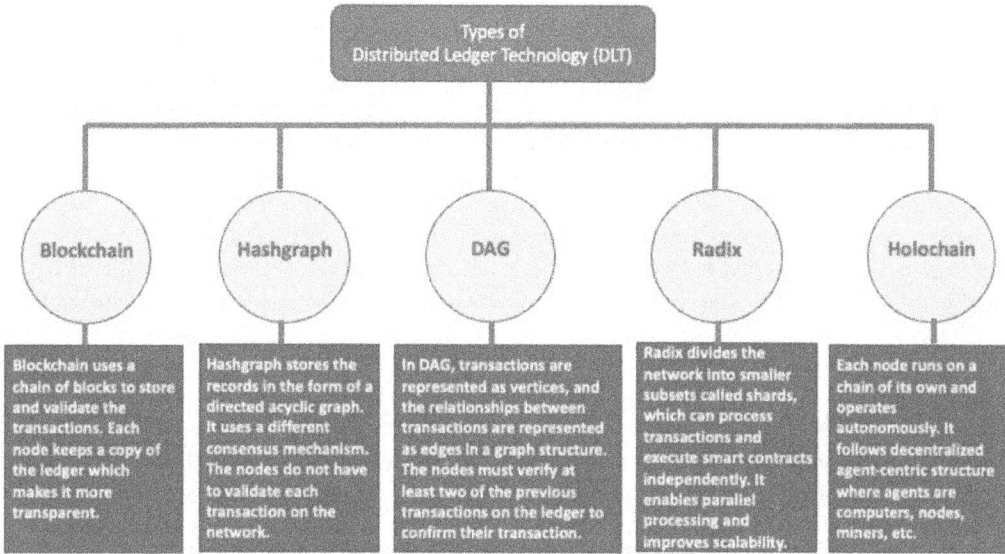

Figure 10.5 Types of Distributed Ledger Technology (DLT).

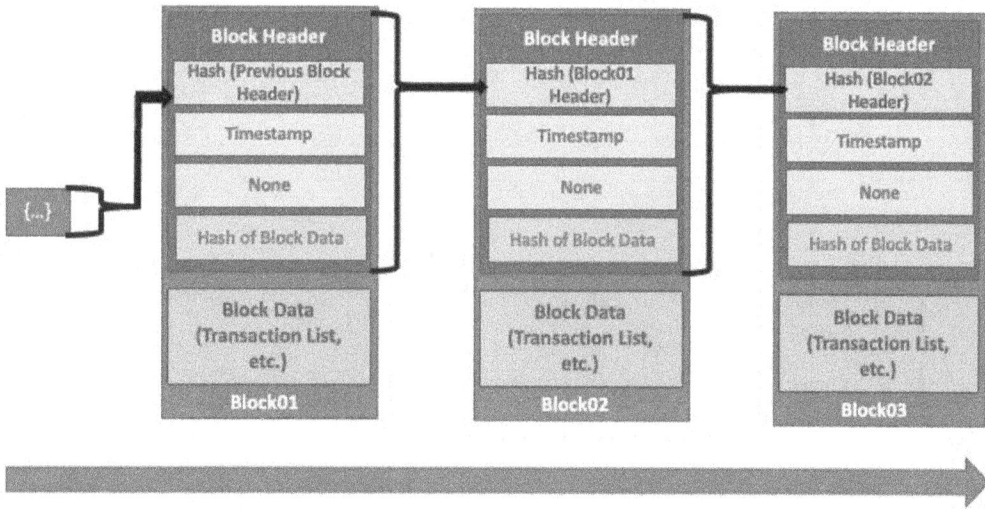

Figure 10.6 Blockchain structure.

timestamp, a nonce, a unique block identifier (hash), and a reference to the previous block. The timestamp is used to identify the document or event uniquely and indicate when the document was created. A nonce is a number that can only be used once. It is a 32-bit field that miners change as they work on new blocks in the Blockchain to ensure that each block has a unique hash. The body of the block holds a list of transactions, data, or other information that is being recorded.

- **Transactions:** In a Blockchain, transactions are the essential units of data. They represent actions or exchanges of information, which are recorded on the Blockchain. Each transaction typically includes details of the sender, receiver, amount, and additional data specific to the application or use case. Transactions are grouped together within a block and are recorded in a defined format based on the Blockchain's protocol.
- **Linking Mechanism:** Blockchain uses cryptographic hashing to ensure the integrity and continuity of the Blockchain structure. Each block includes a reference (hash) to the previous block and creates a structure like the chain. Blockchain creates a permanent and tamper-resistant record by this linking mechanism. It is computationally impractical to alter historical data as any change to a block would require modifying the subsequent blocks in the chain.

10.2.4 Working Principle

A new transaction needs to be approved, or authorized,, before it is added to a block in the chain. The decision to add a transaction to the chain is made by consensus for the public Blockchain. This means that most "nodes" (or computers in the network) must agree that the transaction is valid. The people who own the computers in the network are incentivized to verify transactions through rewards. This process is known as "proof of work" [7]. The working principle of Blockchain is shown in Figure 10.7.

10.2.5 Characteristics of Blockchain

Blockchain is a decentralized, immutable, and distributed ledger technology that provides for secure and transparent transactions for record-keeping and enabling various applications beyond cryptocurrencies, such as supply chain management, healthcare, finance, smart grid, and more. Figure 10.8 discusses the important characteristics of Blockchain.

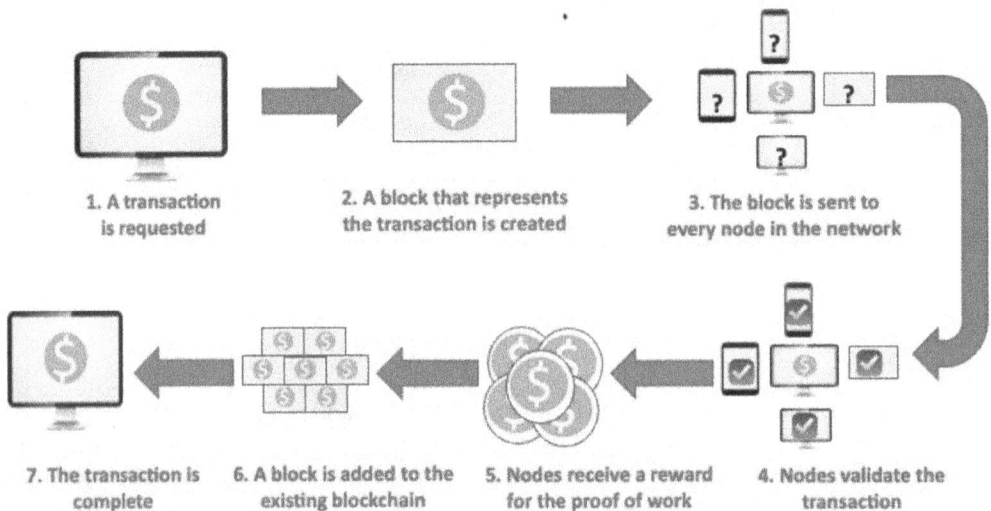

1. A transaction is requested
2. A block that represents the transaction is created
3. The block is sent to every node in the network
4. Nodes validate the transaction
5. Nodes receive a reward for the proof of work
6. A block is added to the existing blockchain
7. The transaction is complete

Figure 10.7 Working principle of Blockchain.

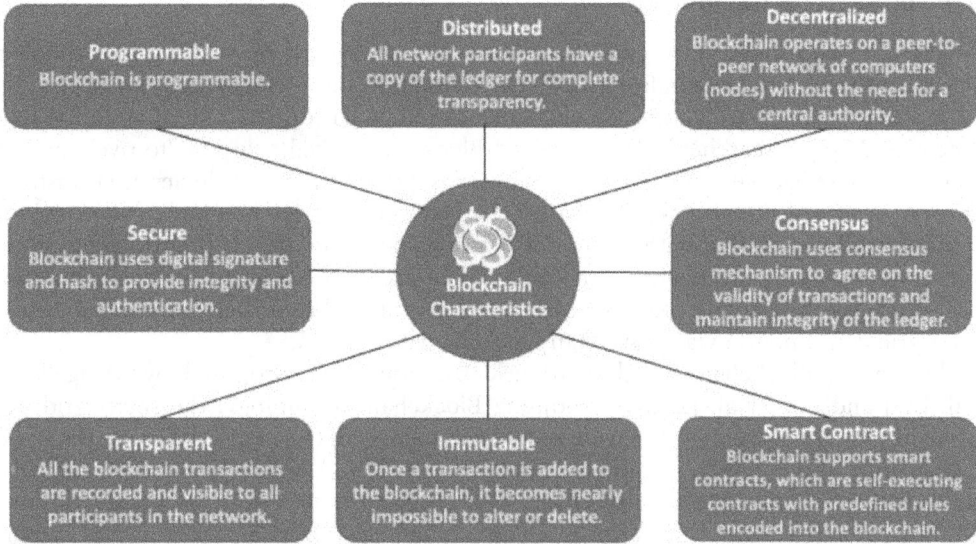

Figure 10.8 Characteristics of Blockchain.

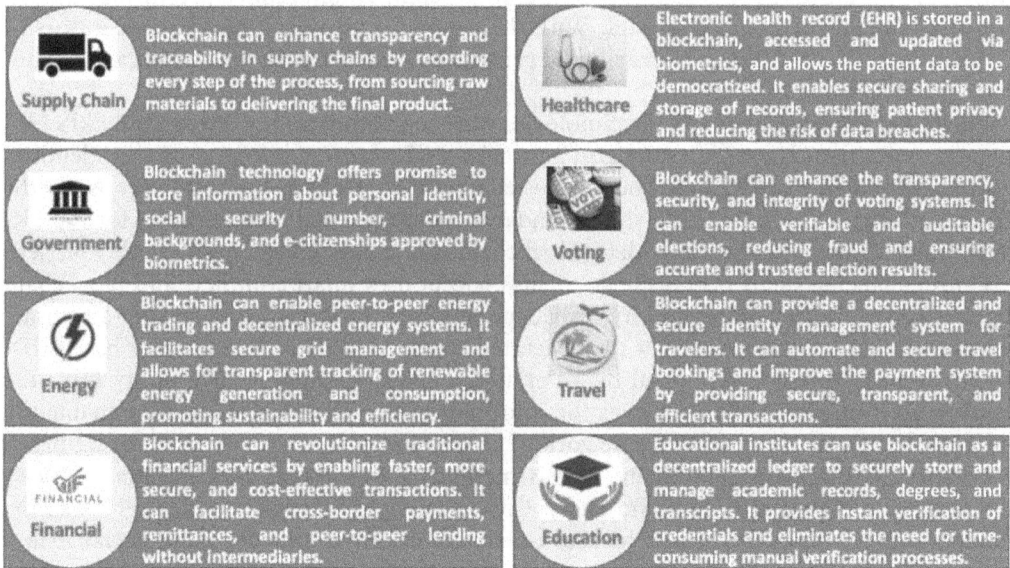

Figure 10.9 Applications of Blockchain technology.

10.2.6   Applications of Blockchain

Blockchain technology offers several potential applications and benefits across various industries including government, healthcare, energy sectors, education, travel, and financial. Figure 10.9 summarizes the applications of Blockchain technology [13–20].

10.3   Literature Review

New security and intelligence challenges are brought forward by smart grid technology [21]. Blockchain uses decentralized database technology and is transparent and open. The development of smart grids using Blockchain technology has drawn a lot of attention from academics. Many researchers have proposed Blockchain-based solutions to overcome the security and intelligence problems with smart grids. Smart grid technologies are transforming every aspect of the grid supply chain [22]. Data interchange in supply chains may be made safe and transparent with the use of Blockchain technology [23]. Various businesses from both upstream and downstream markets that are involved in the life cycle of a good or service make up a supply chain. Gao et al. [24] automated recording and management of power grid consumption via a sovereign blockchain.

Additionally, Blockchain-based smart grid solutions were proposed by Mengelkamp et al. [25] and Pop et al. [26]. By adopting Blockchain technology, the smart grid supply chain may simultaneously provide secure transaction data storage, promote the use of green energy, and enhance sustainable development [27]. A smart grid is a concept that combines several different elements. Dileep [28] defined a "smart grid" as a quick, smooth, and transparent two-way transfer of information and electricity. Smart grids will increase the availability of grid panoramic data in comparison to conventional electricity networks.

Smart technology can enhance the real-time monitoring, diagnosis, and optimization of business flows inside the grid to improve power grid operation and management. Rusitschka et al. [29] proposed adopting a cloud computing architecture to manage the real-time streams of smart grid data. With their model, customers, retailers, operators of virtual power plants for widely distributed generation, and network operators can collaborate and share information.

Public Key Infrastructure (PKI) technologies were the foundation of Metke and Ekl's [30] suggested security system for smart grid networks. PKI relates public keys to user identities via the use of digital certificates. They suggested that PKI standards be created for use by the critical infrastructure sector, and these standards would be used to set criteria for the PKI operations of energy service providers. Blockchain was suggested by Mylrea and Gourisetti [31] for enhancing smart grid resilience. This concept uses smart contracts and Blockchain as intermediaries between electricity providers and users to lower costs, expedite transactions, and enhance the security of the generated transaction data. The identity-based (ID-based) security strategy developed by Ye et al. [32] was intended to give utility firms the greatest level of privacy control over the information that is generated and to enable the Internet Communication (ICT) framework to accommodate more users. IBSC, or ID-based encryption, is the main component of this method. This method concurrently performs the tasks of a digital signature and encryption. The triple bottom line idea serves as the foundation for this essay, which focuses on economic, environmental, and social issues. The authors consider the economic, social, and environmental aspects while also taking into account the cost and effectiveness of public infrastructure, social welfare, and top-level design challenges.

10.4   Contributions of Blockchain to Smart Grid

Blockchain technology can significantly improve the critical infrastructure of the smart grid. Presently, the smart grid incorporates information and communication technologies (ICTs), sensors, and intelligent AI/ML technologies into the power grid for improving the energy efficiency and the safety of the grid. Several numbers of smart devices are deployed throughout the smart grid to efficiently control and manage energy generation, transmission, distribution, and consumption. However, traditional centralized approaches face critical challenges to effectively manage these smart devices and maintain security and stability of the smart grid. Blockchain technology can bring several benefits to the smart grids as shown in Figure 10.10. Especially, Blockchain can be integrated into smart grid systems for energy management, energy trading, security, privacy, microgrid management, and electric vehicle management [33].

We have discussed numerous Blockchain application categories and how Blockchain technology advances the smart grid in this chapter. According to several academics and business leaders, the emergence of Blockchain technology will potentially adopt advanced development and streamline the switch to the smart grid. Some smart grid systems have historically been built on decentralized technologies [34]. The electrical grid's integration of electric vehicles, energy storage technologies, and renewable energy sources has sparked extensive study of new control techniques to handle these problems [35]. Researchers from academics and industries are interested in investigating and applying Blockchain technology in smart grids because of its many compelling advantages [36]. The following categories could be used to categorize blockchain applications in the smart grid's various components [21]:

- Power Generation: The dispatching agencies now have complete real-time understanding of the total functioning state of the power system thanks to Blockchain technology. They can then create dispatching plans that would optimize their revenues thanks to this.
- Power Transmission and Distribution: By using Blockchain technology and automation, the control centers can have systemic decentralization that does away with the major drawbacks of conventional centralized systems.

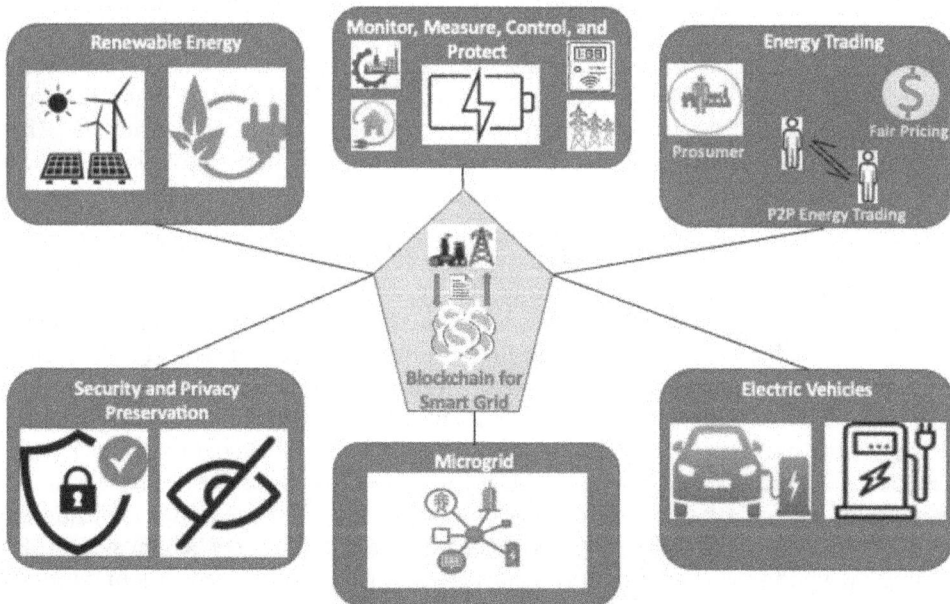

Figure 10.10 Blockchain technology in smart grid applications.

- Power Consumptions: Like the management of the generating and transmitting sides, this area could benefit from the management of energy trading between prosumers, various energy storage technologies, and electric vehicles.

10.4.1 Blockchain for Advanced Metering Infrastructure

The advancement of Advanced Metering Infrastructure (AMI) and the widespread deployment of smart meters have revolutionized the way utility companies, customers, and manufacturers interact within the smart grid network. The two-way communication enabled by smart meters allows for a more connected and dynamic energy ecosystem. Since smart meters can collect comprehensive data on energy production, use, status, and diagnostics, they are more advanced than conventional meters. Billing, user appliance control, tracking, and debugging are all common uses for these data. These frequent data transfers yet occur across a wide area network and are kept either in a customary centralized storage system or in the cloud. The existence of the centralized system could be problematic in and of itself, with prospective alterations posing hazards, privacy issues, and single points of failure. Furthermore, more connections to a central system may result in scalability, availability, and reaction time problems. The smart grid system's electric vehicles and smart meters also generate a significant amount of payment records and data about energy use, which are often shared with other entities for billing, trading, and monitoring. However, in such a complicated system, widespread data sharing poses serious privacy risks because middlemen, brokers, and trusted partners may use the data to reveal personal information about people's identities, whereabouts, patterns of energy generation and use, energy profiles, and amounts charged or discharged. Another problem is the question of trust between producers, consumers, and centralized parties. As a result, it could be challenging for producers and consumers to accept fairness and openness from centralized parties. Understanding how to build a trustworthy, secure, and decentralized AMI system is essential.

Here, we review some pertinent Blockchain studies on AMI. The authors in [37] explore and apply Blockchain technology and smart contracts for the security and robustness of the smart grid. To reduce costs, enhance transaction rates, and improve transaction security, the contracts will act as a middleman between energy providers and consumers. After a transaction has taken place, a smart meter linked to a Blockchain network will send the record to create a new block in the distributed ledger and attach a timestamp for potential later verification.

The authors in [38] propose a permissioned Blockchain and edge-computing-based model for providing privacy protection and energy security. They utilize the permissioned Blockchain to provide the privacy of all participants and decentralized data storage and protect against malicious activities within the communication channels and central clouds. Figure 10.11 illustrates the layered activities in their proposed Blockchain system, where three entities such as edge devices, super nodes, and smart contract servers are introduced for ensuring the correctness and trustworthiness within the Blockchain system. Here, the super node is a special type of node which is authorized to choose nodes from the list of edge nodes to participate in consensus and voting process. An edge node is considered as either a regular user or a voter, which is like a node in a classic Blockchain system. Prior to joining the voting process, the super nodes need to validate the identities of edge nodes through identity authorization and covert channel authorization methods to ensure that the voting nodes are non-malicious and less likely to be compromised by the 51% attack. Smart

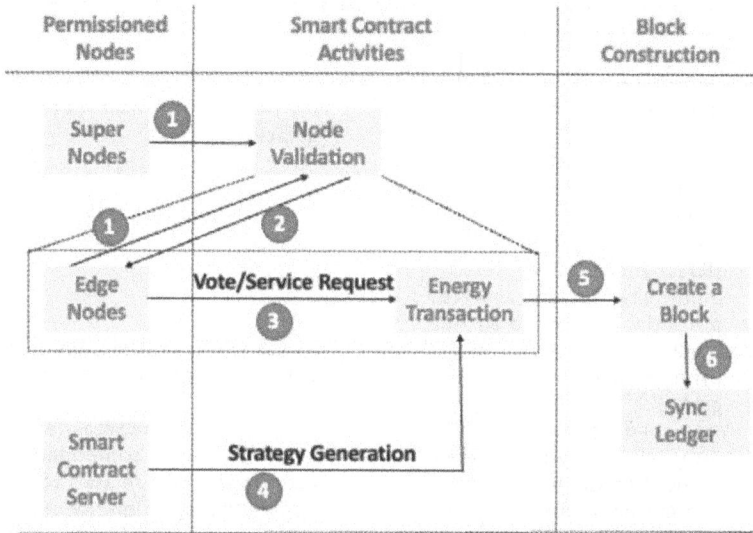

Figure 10.11 Layered activities in the proposed [38] permissioned Blockchain system.

contract is implemented in a Blockchain server, called smart contract server. This smart contract contains the optimal solution to energy resource allocation for the electricity users by considering three elements, including energy consumption, latency, and communication security. However, there will be a point of integrity concern due to compromisation of the super node [7].

In [26], the authors introduce a demand-side management paradigm for the autonomous and decentralized operation of smart energy networks. In this method, a decentralized, safe, and automated energy network is built using the Blockchain so that each of its nodes can function independently of DSO control or centralized supervision. It is also used to tamper-proof the blocks of data on energy usage that can be obtained from smart meters. The smart contract, on the other hand, is made to provide decentralized control, set up incentives or sanctions, validate demand response contracts, and implement rules for achieving a balance across demand of energy and production on the power network side. Using information from the energy generation and consumption records of UK buildings, a prototype developed on the Ethereum Blockchain platform is utilized to validate and test this concept. The findings show that this model is able to create energy flexibility levels that allow for rapid demand adjustments in close to real time and to validate all demand response contracts. But it is not made apparent how the anonymity of the energy profile was preserved on this public Blockchain. The user can be identified by looking at the publicly accessible transactions. Table 10.3 presents the survey on AMI security.

10.4.2   *Blockchain in Decentralized Energy Trading and Market*

Because energy and information may be sent in both directions, consumers can also generate. It is anticipated that the smart grid would support an increasing number of producers, consumers, and prosumers (producer + consumer) in distributed energy-trading situations.

Table 10.3 AMI Security Surveys Compared for Relevance

Ref.	Year	Objective	Remark
[39]	2020	A summary of attacks and mitigation strategies focusing just on smart meters, which are found in an AMI's hardware layer.	Only focuses on smart meter-specific attacks.
[40]	2016	Examines current academic intrusion detection systems (IDSs) and other AMI methodologies and analyzes potential dangers to an AMI.	Exclusively employs IDS as a defense strategy.
[41]	2016	Introduces several privacy and cybersecurity risks in smart grids.	Focuses solely on the confidentiality of data exchanged via AMI channels.
[42]	2015	When sending data via an SG network, reviews of current authentication mechanisms and mutual authentication are required.	Does not cover the weaknesses of the key AMI system components or the attacks that are related to them, instead concentrating primarily on authentication and access control approaches.
[43]	2015	Assessment of network vulnerabilities, attack defenses, and cybersecurity for smart grid and AMI.	Only focuses on difficulties with smart meters and the UC in an AMI about authentication and privacy.
[44]	2015	Examines software, hardware, and network configurations to identify potential cyberattack surfaces.	Solely addresses the attack surface of an AMI, and not assaults, vulnerabilities, or how to protect the AMI networks.
[45]	2014	Presents concerns around the security of smart home and smart grid.	Focuses only on the HAN, which is one tier of an AMI.

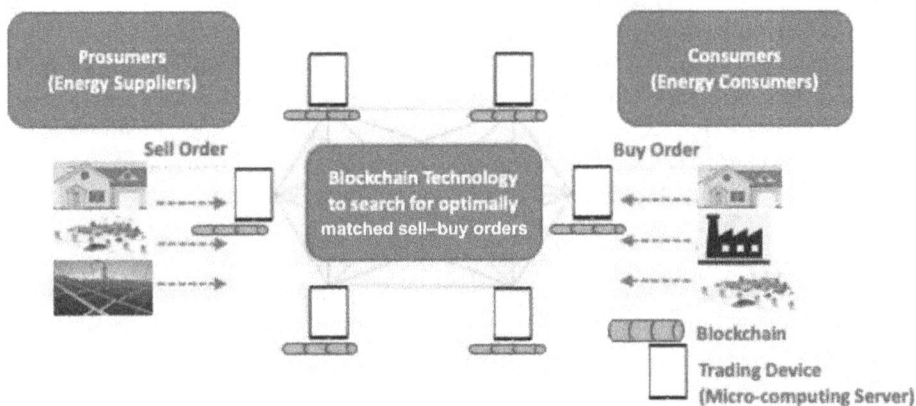

Figure 10.12 Blockchain technology to optimize P2P energy.

To achieve benefits including lowering load peaks, reducing power loss in transmission, easing the strain on the power grid to support green systems, and balancing energy supply and demand, they should be capable of trading their localized generation or more power from distributed sources with one another. The processes required for organizing bids, negotiating contracts, and carrying them out between parties must therefore be included in energy trading. Energy can also be exchanged directly and easily between consumers and producers. Without the involvement of middlemen, this direct energy trading can boost the benefits for all parties and is advantageous for the adoption of renewable energy sources. With current methods, however, customers and producers can only communicate with one another informally through a huge number of middlemen and retailers, which may provide several issues and challenges. Due to the additional operational and regulatory costs, consumers, manufacturers, and prosumers eventually pay for them. Ineffective, dishonest, or malicious intermediaries also produce a market that is not competitive, has little transparency and justice, and has monopoly incentives. A Blockchain is a valuable tool for building a more transparent and decentralized energy market and trading system due to its salient features. Figure 10.12 shows how Blockchain technology can be used to optimize P2P energy [46].

Li et al. [47] offer a peer-to-peer (P2P) energy trading system and an energy token to assure the security and decentralization of energy trade in a variety of situations, including energy harvesting, microgrids, and vehicle-to-grid networks [48]. They achieve this by combining Stackelberg game theory, credit-based payment systems, and consortium Blockchain technology. The credit-based payment mechanism is used to address the problem of transaction confirmation latency, which is very likely with PoW-based Bitcoin. Peer nodes in this system can send payments more rapidly and efficiently than Bitcoin by requesting energy coin loans from credit banks depending on their credit scores. A loan pricing approach that maximized profits for the credit banks was developed for this scheme using the Stackelberg game theory. The planned energy Blockchain prototype is not employed in this work, nor is there formal proof of the double-spending attack.

The developers of [49] introduce PriWatt, a decentralized token-based system that is based on Bitcoin and built on top of it. The purpose of PriWatt is to address the issues that arise in the smart grid energy trading system while safeguarding user identity privacy and transaction security. Multi-signature functionality, anonymous encrypted message streams, and smart contracts with Blockchain support are all part of this system. PriWatt helps buyers and sellers to conduct complex energy price bids and discussions while preventing malicious activity with the use of the contracts defined in the smart contract. For bidding and negotiating, the anonymous communications stream technique is used. The use of multiple signatures helps prevent theft, and at least two other participants must sign a transaction for it to be valid. Additionally, they use PoW for consensus like bitcoin to protect against

Table 10.4 A Comparison of State-of-the-Art Research Articles on Blockchain-Based P2P Energy Trading

Ref.	Cost and Energy Optimized	Optimization Applied	Secure against Attacks	Security Analysis	Scalable	Performance Analysis
[51]	✓	✓	X	X	X	X
[48]	✓	✓	X	X	✓	X
[52]	X	X	✓	X	✓	✓
[53]	✓	✓	X	X	✓	✓

Byzantine failures and the risks of double spending. There isn't a thorough explanation of which nodes implement Proof of Work (PoW), how miners will do so, or what the rewards are for mining correctly [50].

10.4.3 Use of Blockchain to Monitor, Measure, and Control

The current smart grid cyber-physical system (CPS) is built on a central SCADA system that is hierarchically coupled with various components like MTUs, RTUs, PMUs, and a variety of sensors. Power grids are regularly managed and monitored using the SCADA system. Once it is connected to the Internet, the SCADA system will improve large-scale distributed monitoring, measuring, and control. IoT smart devices, sensors, and PMUs regularly gather and share status information on power equipment with MTUs through RTUs, which are thought of as central repositories and control centers. The smart grid system enables intelligent control, wide-area monitoring, and governance to better manage grid safety, stability, and reliability as well as monitoring power theft and loss, making it feasible for CPS components, various grid operators, suppliers, and consumers to exchange fine-grained measurements. However, malevolent attackers or insiders can conduct cyberattacks in a variety of ways, including by modifying data in central controllers, launching an availability assault, and injecting phoney data through sensors and PMUs. As a result, the attacker can hijack control channels and issue nefarious orders. Blockchain has opened new possibilities for tracking, measuring, and managing the decentralized smart grid system [7].

Figure 10.13 shows how different Blockchains can be applied in different smart grid layers for the data protection [54]. Here, each layer of the smart grid is managed by separate Blockchain aggregator. The Blockchain offers a secure and reliable data-storing platform for protecting the smart grids. Further, it reduces the possibilities of certain types of malicious attacks by providing a higher level of anonymity and privacy to users (including producers,

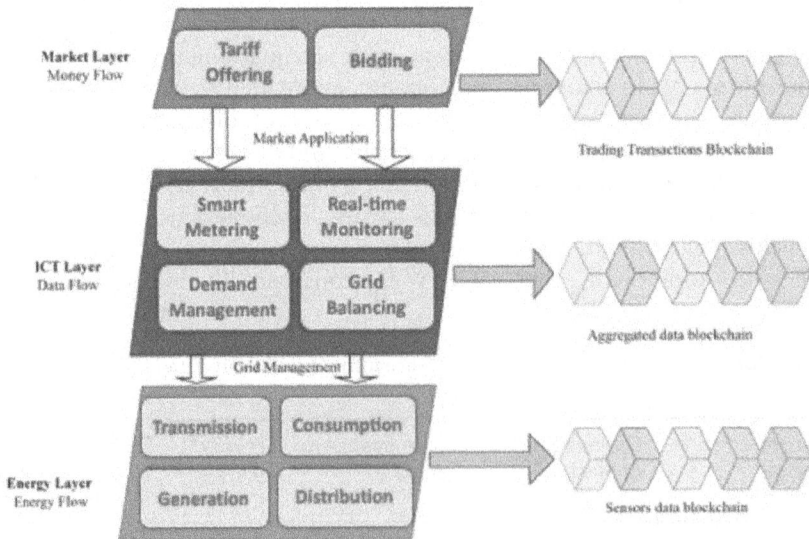

Figure 10.13 Blockchain to monitor, measure, control, and protect.

consumers, and prosumers) within the Blockchain environment. Subsequently, it increases the resilience of the electrical power system.

The focus of the authors of [55] is industrial control system data security (ICS). Plant operational data will be more secure if ICS-BlockOpS, a Blockchain-based architecture, is implemented. By employing Blockchain technology, this architecture is primarily designed to address two important ICS problems, namely immutability and redundancy. The Blockchain's tamper-proof design makes it possible to guarantee data immutability. On the other hand, a Hadoop Distributed File System-inspired efficient replication technique with assistance from Blockchain is also offered to ensure data redundancy (HDFS). However, there aren't any debates about how to handle malicious or compromised nodes that inject false data or how the Blockchain network would function with sensors and actuators that have restricted resources.

In Ref. [56], a monitoring system for smart grids based on Blockchain and smart contracts is presented to guarantee the security and transparency of energy usage. Utility companies, consensus nodes, and smart meters are three different types of nodes that make up this Blockchain network, each with a specific role. The Blockchain network frequently receives electronically measured energy information from smart meters belonging to consumers. Contrarily, the consensus nodes are in charge of maintaining the records of energy usage, verifying to produce new blocks, publishing them to the main chain, and preserving the particular subscription data supplied by utility companies. These nodes produce brief forms for each user, including meter IDs and other consumer data, prior to constructing new blocks, but before doing so. Following inspection, approval, and audit by consensus nodes, these forms are subsequently transformed into blocks. In this case, a system of immutable data records is created for utilizing the Blockchain to safeguard smart meter data records from tampering by utility companies and consumers.

To address the challenges of traditional cloud-based architecture, the authors in [57] propose a Blockchain-based decentralized CPS architecture as depicted in Figure 10.14. The sensing layer is responsible for collecting and processing the data using sensor and computing devices. The next layer is responsible for the cryptographic operations on the data, generating blocks and recording them in the distributed ledger. The third layer stores the entire Blockchain in a distributed and synchronized manner. In order to connect each layer, the fourth layer implements the underlying technologies including distributed algorithms and data storage technology. The fifth layer is responsible for the services to the users including real-time monitoring and failure prediction. The authors show that the proposed architecture offers better security and privacy as compared to the traditional cloud-based architecture. However, they do not identify the responsibilities of lightweight nodes and the nodes which are responsible for performing PoW.

10.4.4 Use of Blockchain in Microgrid

The control of microgrids is turning into a critical challenge with the integration of many distributed energy supplies (DERs). Experts are focusing on the need for demand-based microgrid control and optimal operation these days [58]. Similarly to this, Blockchain technology has been used in the industry because of its prospective advantages. A DER scheduling technique relying on Blockchain technology is described in [59]. Here, DERs can all be trusted because of the secure environment created by the usage of Blockchain technology. Although Blockchains are getting a lot of attention as a platform for distributed computation and data supervision, Münsing et al. [60] propose a Blockchain and smart-contract-assisted

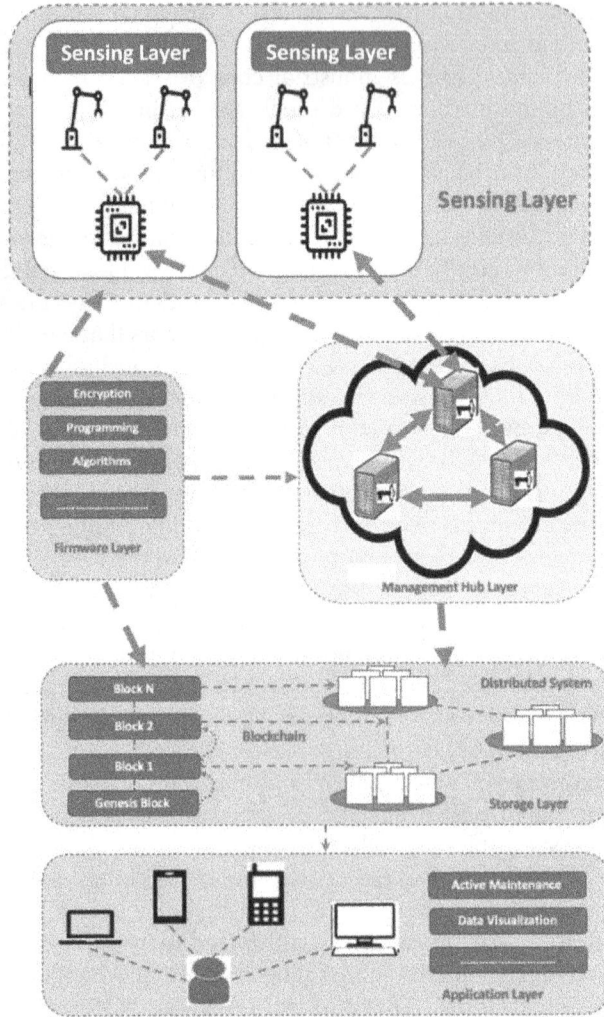

Figure 10.14 Blockchain-based CPS architecture proposed in [57].

architecture to support distributed optimization and control of DERs in microgrids where the decentralized optimal power flow (OPF) model for DERs is used. The proposed architecture utilizes the Blockchain for making the system decentralized by distributing the role of microgrid operators across all entities and providing fair energy trading without the need for a traditional utility company or microgrid operator. To schedule the battery, shapeable, and deferrable electric loads in the network, OPF model employs the alternating direction method of multipliers (ADMMs) technique. Here, the smart contract is used as an ADMM coordinator. The proposed architecture makes the payment transaction secure and automatic while it keeps the optimal schedule on the blocks. However, the authors do not consider the uncertain data and how to build trust among non-trusted entities [7].

In [61], the authors propose an autonomous Blockchain-assisted P2P and M2M energy trading system named *SynergyGrids* to improve the social utility and overall benefits of prosumers and consumers in the microgrids. The proposed system offers secure energy trading between the peers within the microgrids and the peers from different microgrids using smart contracts. The authors also present the pricing mechanism that ensures the increase in utility of the participants by considering the total available energy, the usage, and the distance between consumer and prosumer for cost calculation. As depicted in Figure 10.15, the main components of the proposed model are: (i) Prosumers/consumers:—responsible for selling/buying the energy; (ii) microgrids:—responsible for fulfilling the local energy demands; (iii) utility grids:—the main grids that facilitate energy transfer; (iv) blockchain:—responsible for decentralization.

The authors in [62] propose a distributed proportional-fairness control scheme for DERs in microgrids using Blockchain and smart contracts. The proposed scheme considers a group of DERs acting as voltage regulators and reduces the penetration and sacrifices their revenue over control periods for balancing the voltage regulations. The authors also introduce a principle based on the exchange of credits for providing incentive to the DERs for fair participation in voltage regulation. However, they do not include any punishment mechanism for the fraudulent transactions [7]. The authors in [63] also address the voltage regulation problem where they present a Blockchain-based transactive energy system (TES) and pay the attention to punish for the fraudulent activities or provide incentives for voltage regulation services. However, they do not pay any attention to the consensus mechanism [7].

In [64], the authors address the traditional microgrid transaction management problems due to centralized trading system such as (i) trust issues between transaction center and traders (ii) fairness, transparency, and effectiveness of information for transaction centers; and (iii) threats to transaction security and the interests of traders. Therefore, they propose a decentralized electricity transaction mode based on Blockchain and continuous double auction (CDA) mechanism to support independent and direct P2P transactions between distributed generations and consumers in the microgrid energy market as illustrated in Figure 10.16. In this technique, initially, buyers and sellers can present quotes to the CDA

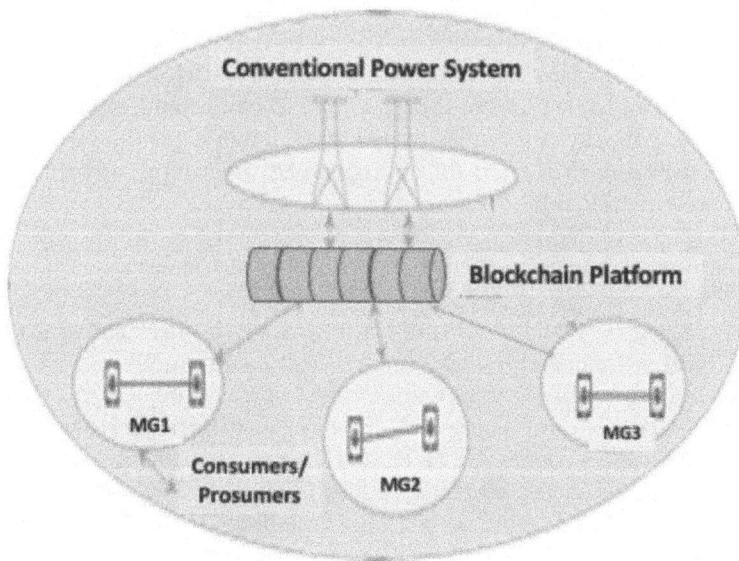

Figure 10.15 Blockchain-supported distributed microgrid energy trading proposed in [61].

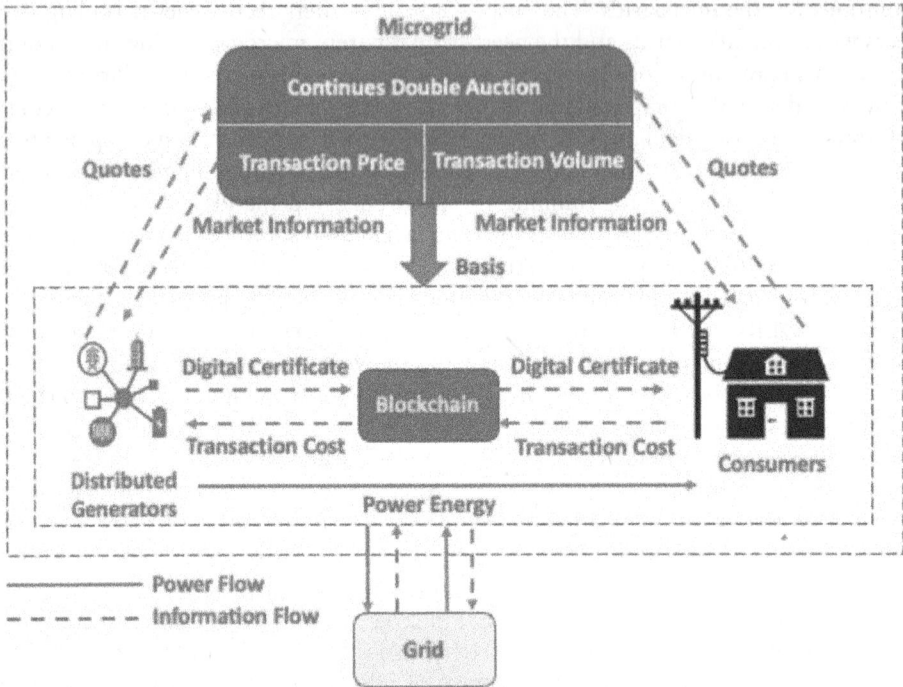

Figure 10.16 The overall structure of microgrid electricity transactions proposed in [64].

market according to an adaptive aggressiveness (AA) strategy and adjust quotes dynamically according to the market information. Afterward, the buyers and sellers accomplish the digital proof of energy trading by using multi-signature and Blockchain. Here, Blockchain provides security of the transactions whereas the multi-signature ensures protection against any manipulation of a contract between the buyer and seller. However, the authors do not consider how rich-rule problem of Proof of Stake (PoS) can be addressed [7]. We reach the stage where different Blockchain applications are available depending on the type of microgrid that is being used, such as an AC microgrid, DC microgrid, or hybrid AC-DC MG [25, 65–76] as presented in Table 10.6.

10.4.5 Blockchain Applications in Electric Vehicles

The integration of electric vehicles (EVs) into the smart grid offers numerous benefits that contribute to a more sustainable, resilient, and efficient energy future. An EV can store and exchange energy with the charging stations, home, and other neighboring EVs in P2P manner. This leads to four scenarios, i.e., Vehicle-to-Vehicle (V2V), Vehicle-to-Home (V2H), Vehicle-to-Grid (V2G), and Grid-to-Vehicle (G2V). Due to short range communications and mobility of EVs, it introduces new security and privacy issues. Figure 10.17 illustrates an overview how smart contract can be used for secure purchasing and selling the energy between them.

In [77], the authors propose a secure EV-charging framework based on smart contract that is, integrated renewable energy sources and smart grids. They utilize permissioned Blockchain system and the contract theory to design and implement optimal smart contracts and

Table 10.6 Blockchain for Sustainable Microgrids: A Summary

Ref.	Subdomain	Objectives	Solutions/Results	Technologies	Advantages/ Opportunities	Challenges
[65]	Local market for energy	Converting all banking-based transactions to cryptocurrency-based transactions	Flow of money and energy based on Blockchain	Public Blockchain	Authenticating funds and automating transaction control	Governmental policies, commercial concerns, and technological limitations
[25]	Microgrid, smart grid, and local energy market	Maximizing energy efficiency and cutting down on electricity expenditures.	Decreased electricity costs and optimized energy use, particularly during peak hours	A private Blockchain using the PoW algorithm	Optimal electricity prices for each time period, as well as a balance between local energy demand and production	Implementing scheduling algorithm for a large-community private Blockchain
[66]	Microgrid/smart grid	Reducing the price of electricity	Mechanism for a decentralized market	Private Blockchain	-	Selling oversupply
[67]	Microgrid and local energy market	P2P energy exchanges	Lower electricity prices	Public Blockchain	Control over energy flows and	Political restrictions, commercial
[68]	Microgrid and local energy market	P2P energy exchanges	Suggested frameworks are semi-centralized and decentralized.	Solc, Mocha, React. js, Next.js, Ganache Meta Mask, Ganache- cli, and Web3	Framework 1 is less flexible, more secure, and using more transactions Framework 2 is less secure and using fewer transactions.	A smart contract's restrictions
[69]	Microgrid/smart grid	Ensure security and reach agreement when cyber-attacks happen.	Proposed architecture	Either public or private Blockchain	Resistance to attacks	Security for transaction

(Continued)

Table 10.6 (Continued)

Ref.	Subdomain	Objectives	Solutions/Results	Technologies	Advantages/Opportunities	Challenges
[70]	Microgrid	Improve the microgrid's electricity flow	Model suggested/optimized energy flow	Private Blockchain	Lower import prices	Efficiency and security in communications
[71]	Renewable energy	Energy conservation	Proposed framework and methods	Blockchain can be private or public	-	Infrastructure for technology and investment costs
[72]	Local energy market/microgrid	P2P energy exchanges	Proposed trading framework for multi-agent cooperation and sharing of the energy Internet	Blockchain technology can be public or private.	Promote peer-to-peer energy transfers	Concerns about hazards and security
[73]	Microgrid and local energy market	P2P energy exchanges	Envisioned trading environment	Private Blockchain	Transparent transactions	Political restrictions, commercial concerns, and technological constraints
[74]	Hybrid AC–DC microgrid	Increased safety	Proposed structure	Public Blockchain	Increased safety	Restrictions on power injection
[75]	DC microgrid	Energy conservation	Proposed structure	Either public or private Blockchain	Maximum use of renewable energy	Political restrictions, commercial concerns, and technological constraints
[76]	Hybrid AC–DC Microgrid	Energy conservation	Proposed structure	Private Blockchain	Energy efficiency and safe transactions	Political restrictions, commercial concerns, and technological constraints

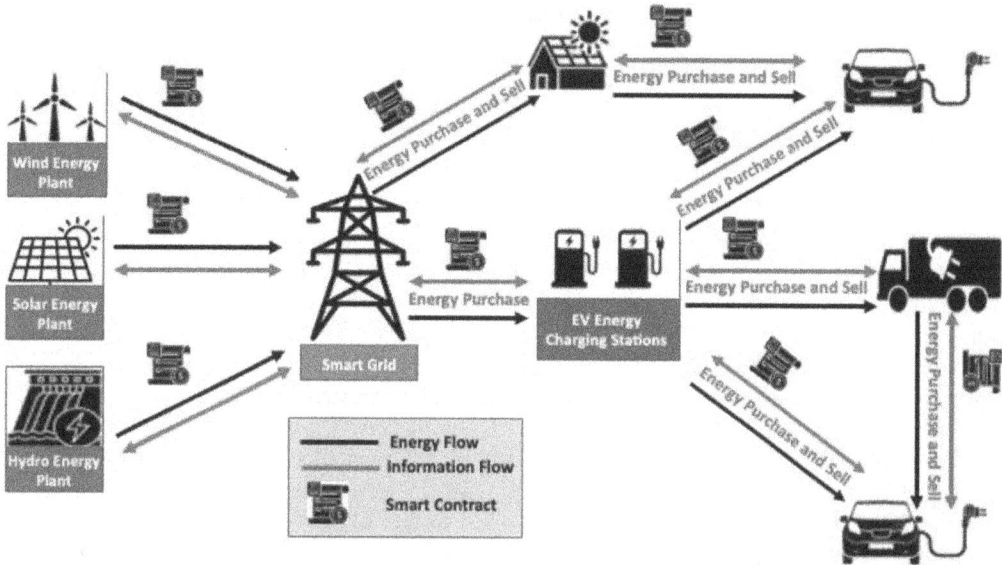

Figure 10.17 Smart contract-based secure energy trading in EVs.

a novel energy allocation algorithm. They also utilize delegated Byzantine fault tolerance (DBFT) for achieving an efficient and fast consensus. Here, the framework allows only the preselected EVs to participate in auditing and creating a new block. However, the authors do not present any discussion on who is responsible for validating the transactions issues [7]. The authors in [78] propose a Blockchain-assisted charging coordination mechanism for charging the energy storage units (ESUs) including EVs in a realizable, transparent, and decentralized manner from the utility providers. The utility providers and energy storage units are connected with the Blockchain network. The charging request containing its demand, state of charge, and time to complete the charge, needs to be sent by each ESU to the Blockchain network. However, the authors do not present any consensus mechanism [7].

The authors in [79] propose a transparent, autonomous, and privacy-preserving technique to search the cheapest and viable charging stations for EVs based on energy prices and the distance. The bid request needs to be sent to the Blockchain network. The Blockchain protects the EVs' identity privacy and hides their geographical location, makes verifiable, and increases the transparency of bidding requests. In [80], the authors present a security model for decentralized EV-charging management to provide the security of EVs. This model introduces the lightning network and Blockchain network. Initially, the lightning network is set up for the registration of the EVs, charging piles, and operator. This network is responsible for helping the Blockchain network by creating trust among the participants and ensuring the security of payments.

10.4.6 Blockchain Applications in Cyber-Physical Security

The development of the smart grid led to several flaws that make it possible to manipulate or attack several of the cyber-physical smart grid's components. A thorough analysis of the results of various attacks in [81] suggests that cyber and physical attacks are closely related and should be treated as a single entity. A further explanation of the many features and historical context of the numerous cyber-physical security problems in the smart grid is provided by Gupta et al. [82]. There are many different types, effects, and impacts of cyber-physical attacks, including time synchronization [83], GPS spoofing [84], and denial-of-service (DOS) [85] attacks.

To strengthen the grid's defense against cyber-physical threats, many Blockchain solutions have been thoroughly studied. A general summary of the various Blockchain security mechanisms is provided in [86]. The authors in [87] discuss on the use of Blockchain technology to enhance the security, dependability, and safety of the electricity grid using the meters as nodes in a distributed network that records the readings of the meters as individual blocks in the chain. In [24], the authors utilize a network built on the Blockchain so that the customers can monitor electricity use without being concerned about outside influence. The work in [37] makes use of smart contracts to improve the cyber resilience of secure transactive energy applications and smart grids. This is also highly helpful for applications that involve trading in energy. The importance of utilizing Blockchain technology to create a trustworthy network for the usage of intelligent electric car operations was examined by Kim [88]. Many cyberattacks can be avoided by his plan before they have an adverse impact. Every home has a smart meter installed as part of the smart grid, which collects real-time data on electricity consumption for use by the utilities.

In [89], the authors examine how Blockchain technology can improve smart grid security and privacy. It suggests using Blockchain technology in smart grids for peer-to-peer energy trading, data aggregation, energy distribution systems, and equipment diagnostics and maintenance. The authors review commercial Blockchain smart grid projects and explore the challenges of integrating these technologies. They also highlight the benefits of incorporating these technologies. The report analyzes Blockchain technology's smart grid and other uses.

The authors in [7] note smart grid's decentralization and ability to integrate green and renewable energy technology. A smart grid integrates green and renewable energy technology efficiently. This chapter examines how Blockchain technology may address smart grid security issues and summarizes existing Blockchain-based research in this area. The authors also summarize the key practical projects, experiments, and products; discuss major research challenges; and suggest future options for using Blockchain technology to address smart grid security issues. This work uses cutting-edge technology to help create an ecologically conscious society.

10.5   Limitations of Blockchain and Future Research Directions

Blockchain technology has brought many benefits. However, it is important to recognize that it also comes with limitations and challenges that need to be considered. Here are some limitations and challenges associated with Blockchain technology [54]:

- **Costlier:** The nodes in Blockchain receive rewards for their efforts in processing transactions and maintaining the network in the form of transaction fees or newly minted cryptocurrency tokens. However, the nodes can claim bigger rewards for completing transactions in a business that follows the law of supply and demand.
- **Scalability:** Blockchain can face scalability issues, with an increased number of nodes and transactions. As each transaction needs to be validated and included in the chain by consensus among the nodes, the process becomes slower as the network grows.

- **Energy Consumption:** Blockchain uses Proof-of-Work (PoW) consensus algorithm that requires extensive computational power and leads to high-energy consumption.
- **Data Storage:** Blockchain stores all transaction data across all nodes in the network, which can lead to challenges in terms of storage requirements, especially for Blockchains with extensive transaction histories.
- **Error Risk:** Even though the Blockchain is an extremely safe technology, there is always a risk when humans are involved.
- **Transaction Costs:** The cost of Blockchain transactions can be very high, especially during the peak times of network congestion.
- **Lack of Interoperability:** There are several Blockchain platforms with different protocols and standards, resulting in a lack of interoperability. This limits seamless data and asset transfers between different Blockchain networks, hindering broader adoption and integration.
- **Security Risks:** While Blockchain technology is known for its robust security features, it is not completely immune to security risks and vulnerabilities. Centralization risks, 51% attacks, and vulnerabilities in smart contracts can all pose threats to Blockchain networks.

Blockchain technology has the potential to reshape the energy industry and contribute significantly to global energy sustainability if we can address the following challenges in future:

- **Scalability Solutions:** We need to focus on developing scalable Blockchain architecture to manage a large number of transactions efficiently. We need to explore sharding, off-chain solutions, and layer-two protocols for enhancing transaction throughput while maintaining the network security and decentralization.
- **Interoperability:** There are numerous Blockchain platforms, each with its unique features, protocols, and standards. Ensuring interoperability between the smart grid systems and Blockchain networks is critical. We need to work on protocols and standardization before deploying Blockchain networks to the smart grids for enabling seamless data and asset transfers between heterogeneous Blockchain networks.
- **Privacy and Security Enhancements:** Advancements in cryptographic techniques and privacy-preserving technologies need to be incorporated for protecting user privacy and sensitive data while maintaining the transparency and auditability benefits of Blockchain.
- **Smart Contract Security:** We need to minimize the vulnerabilities and ensure secure execution of automated agreements by improving the security of smart contracts. Techniques like formal verification, advance cryptographic techniques, and auditing tools need to be developed for enhancing the reliability of smart contract.
- **Distributed Energy Management:** We need to design and develop advance distributed energy management systems for leveraging Blockchain technology to optimize energy distribution, consumption, and storage in a decentralized manner.
- **Energy-Efficient Consensus Mechanisms:** PoW-based Blockchains require extensive computational power and consume high energy. Future research will explore alternative consensus mechanisms that are more energy-efficient, such as Proof of Stake (PoS) variants and other novel consensus algorithms, to reduce the environmental impact of Blockchain technology smart on grids.

- **Incentive Mechanisms and Market Design:** We need to design effective incentive mechanisms for energy producers and consumers to participate actively and contribute to the stability and efficiency of the grid.
- **Real-World Implementations and Case Studies:** We need to focus on pilot projects to evaluate the effectiveness, scalability, and impact of Blockchains on energy systems while integrating Blockchains to real-world applications of smart grids.
- **Regulatory Compliance and Policy Implications:** We need to focus on how Blockchains in smart grids can adopt existing and evolving energy policies, regulations, and environmental goals. In future, the policymakers need to address the issues related to energy market regulations, data privacy, and environmental impacts.

10.6   Conclusion

The application of Blockchain technology to smart grids is an emerging and rapidly evolving area of research that is gaining significant attention from researchers, industry stakeholders, and policymakers. The integration of Blockchains with smart grid systems holds the potential to revolutionize the energy sector by addressing various challenges and unlocking new opportunities. In this chapter, a comprehensive study of Blockchain applications to smart grids has been presented. First, we have presented a background of Blockchain and its classification, working principle, characteristics, and applications. Then, a number of recent research works proposed in different literatures on Blockchain integration into smart grid systems for energy management, energy trading, security and privacy, microgrid management, and electric vehicle management have been presented. Finally, we have presented the limitations of Blockchain and future research direction of applying Blockchain in smart grids.

Acknowledgment

This work is supported by the Visiting Faculty Research Program (VFRP) with the Information Assurance Branch of the AFRL, Rome, NY, United States, and the Information Institute (II). Any opinions, findings, and conclusions or recommendations expressed in this material are those of the authors and do not necessarily reflect the views of the Air Force Research Laboratory.

Bibliography

[1] U. Ghosh, P. Chatterjee, S. Shetty, C. Kamhoua and L. Y. Njilla , "Towards Secure Software-Defined Networking Integrated Cyber-Physical Systems: Attacks and Countermeasures," in *Cybersecurity and Privacy in Cyber-Physical Systems*, CRC Press, May, 2019.

[2] U. Ghosh, P. Chatterjee and S. Shetty, "Securing SDN-enabled Smart Power Grids: SDN-enabled Smart Grid Security," in *Research Anthology on Smart Grid and Microgrid Development*, IGI Global, 2022, pp. 1028–1046.

[3] E. S. G. Challenge, "Energy Storage Grand Challenge Roadmap," [Online]. Available: www.energy.gov/energy-storage-grand-challenge/articles/energy-storage-grand-challenge-roadmap [Accessed 11 July 2023].

[4] S. Aggarwal, N. Kumar, S. Tanwar and M. Alazab, " Survey on Energy Trading in the Smart Grid: Taxonomy, Research Challenges and Solutions," *IEEE Access*, vol. 9, pp. 116231–116253, 2021.

[5] P. Research, "Smart Grid Market Size to Worth Around us$ 162.8 bn by 2030," [Online]. Available: www.globenewswire.com/Ne/news-release/2021/12/15/2352520/0/en/Smart-Grid-Market-Size-to-Worth-Around-US-162-8-Bn-by-2030.html [Accessed 11 July 2023].

[6] X. Liang, S. Shetty, D. Tosh, C. Kamhoua, K. Kwiat and L. Njilla, "ProvChain: A Blockchain-Based Data Provenance Architecture in Cloud Environment with Enhanced Privacy and Availability," in *17th IEEE/ACM International Symposium on Cluster, Cloud and Grid Computing (CCGRID)*, Madrid, Spain, 2017.

[7] M. B. Mollah, J. Zhao, D. Niyato, K.-Y. Lam, X. Zhang, A. M. Y. M. Ghias, L. H. Koh and L. Yang, "Blockchain for Future Smart Grid: A Comprehensive Survey," *IEEE Internet of Things Journal*, vol. 8, no. 1, pp. 18–43, 2021.

[8] D. Das, S. Banerjee, P. Chatterjee, U. Ghosh and U. Biswas, "Blockchain for Intelligent Transportation Systems: Applications, Challenges, and Opportunities," *IEEE Internet of Things Journal*, vol. 10, no. 21, pp. 18961-18970, 1 Nov.1, 2023, doi: 10.1109/JIOT.2023.3277923.

[9] K. K. R. Choo, U. Ghosh, D. Tosh, R. M. Parizi and A. Dehghantanha, "Introduction to the Special Issue on Decentralized Blockchain Applications and Infrastructures for Next Generation Cyber-Physical Systems," *ACM Transactions on Internet Technology*, vol. 21, no. 2, pp. 1–3, 2021.

[10] S. Nakamoto, "Bitcoin: A Peer-to-Peer Electronic Cash System," 2008. Available: https://bitcoin.org/bitcoin.pdf.

[11] B. A. Tama, B. J. Kweka, Y. Park and K.-H. Rhee, "A Critical Review of Blockchain and Its Current Applications," in *2017 International Conference on Electrical Engineering and Computer Science (ICECOS)*, Palembang, Indonesia, 2017, pp. 109-113, doi: 10.1109/ICECOS.2017.8167115.

[12] Y. Guo, Z. Wan and X. Cheng, "When Blockchain Meets Smart Grids: A Comprehensive Survey," *High-Confidence Computing*, vol. 2, no. 2, 2022.

[13] D. Das, S. Banerjee, U. Ghosh and U. Biswas, "A Decentralized Vehicle Anti-Theft System using Blockchain and Smart Contracts," *Peer-to-Peer Networking and Applications*, vol. 14, pp. 2775–2788, 2021.

[14] D. Das, S. Banerjee, P. Chatterjee, U. Ghosh and U. Biswas, "A Secure Blockchain Enabled V2V Communication System Using Smart Contracts," *IEEE Transactions on Intelligent Transportation Systems*, vol. 24, no. 4, pp. 4651–4660, 2023.

[15] A. P. Singh, R. N. Pradhan, A. K. Luhach, S. Agnihotri, N. Z. Jhanjhi, S. Verma, Kavita, U. Ghosh and D. S. Roy, "A Novel Patient-Centric Architectural Framework for Blockchain-Enabled Healthcare Applications," *IEEE Transactions on Industrial Informatics*, vol. 17, no. 8, pp. 5779–5789, 2021.

[16] Y. Ren, F. Zhu, J. Wang, P. Sharma and U. Ghosh, "Novel Vote Scheme for Decision-Making Feedback Based on Blockchain in Internet of Vehicles," *IEEE Transactions on Intelligent Transportation Systems*, vol. 23, no. 2, pp. 1639–1648, 2022.

[17] D. Das, S. Banerjee, P. Chatterjee, U. Ghosh, U. Biswas and W. Mansoor, "Security, Trust, and Privacy Management Framework in Cyber-Physical Systems using Blockchain," in *IEEE 20th Consumer Communications & Networking Conference (CCNC)*, Las Vegas, NV, 2023.

[18] A. A. Malik, D. Tosh and U. Ghosh, "Non-Intrusive Deployment of Blockchain in Establishing Cyber-Infrastructure for Smart City," in *16th Annual IEEE International Conference on Sensing, Communication, and Networking (SECON)*, Boston, MA, 2019.

[19] A. Gomez Rivera, D. K. Tosh and U. Ghosh, "Resilient Sensor Authentication in SCADA by Integrating Physical Unclonable Function and Blockchain," *Cluster Computing*, vol. 25, pp. 1869–1883, 2022.

[20] D. Das, S. Banerjee, K. Dasgupta, P. Chatterjee, U. Ghosh and U. Biswas, "Blockchain Enabled SDN Framework for Security Management in 5G Applications," in *ACM 24th International Conference on Distributed Computing and Networking (ICDCN)*, Kharagpur, India, 2023.

[21] Y. Yoldaş, A. Önen, S. Muyeen, A. V. Vasilakos and I. Alan, "Enhancing Smart Grid with Microgrids: Challenges and Opportunities," *Renewable and Sustainable Energy Reviews*, vol. 72, pp. 205–214, 2017.

[22] J. Lukić, M. Radenković, M. Despotović-Zrakić, A. Labus and Z. Bogdanović, "Supply Chain Intelligence for Electricity Markets: A Smart Grid Perspective," *Information Systems Frontiers*, vol. 19, no. 1, pp. 91–107, 2017.

[23] J. Xie, H. Tang, T. Huang, F. R. Yu, R. Xie, J. Liu and Y. Liu, "A Survey of Blockchain Technology Applied to Smart Cities: Research Issues and Challenges," *IEEE Communications Surveys & Tutorials*, vol. 21, no. 3, pp. 2794–2830, 2019.

[24] J. Gao, K. O. Asamoah, E. B. Sifah, A. Smahi, Q. Xia, H. Xia, X. Zhang and G. Dong, "Grid-Monitoring: Secured Sovereign Blockchain Based Monitoring on Smart Grid," *IEEE Access*, vol. 6, pp. 9917–9925, 2018.

[25] E. Mengelkamp, B. Notheisen, C. Beer, D. Dauer and C. Weinhardt, "A Blockchain-Based Smart Grid: Towards Sustainable Local Energy Markets," *Computer Science-Research and Development*, vol. 33, no. 1–2, pp. 207–214, 2018.

[26] C. Pop, T. Cioara, M. Antal, I. Anghel, I. Salomie and M. Bertoncini, "Blockchain-Based Decentralized Management of Demand Response Programs in Smart Energy Grids," *MDPI Sensor*, vol. 18, no. 1, 2018.

[27] F. Imbault, M. Swiatek, R. D. Beaufort and R. Plana, "The Green Blockchain: Managing Decentralized Energy Production and Consumption," in *2017 IEEE International Conference on Environment and Electrical Engineering and 2017 IEEE Industrial and Commercial Power Systems Europe (EEEIC/I&CPS Europe)*, Milan, Italy, 2017.

[28] G. Dileep, "A Survey on Smart Grid Technologies and Applications," *Renewable Energy*, vol. 146, pp. 2589–2625, 2020.

[29] S. Rusitschka, K. Eger and C. Gerdes, "Smart Grid Data Cloud: A Model for Utilizing Cloud Computing in the Smart Grid Domain," in *First IEEE International Conference on Smart Grid Communications*, Gaithersburg, MD, 2010.

[30] A. R. Metke and R. L. Ekl, "Security Technology for Smart Grid Networks," *IEEE Transactions on Smart Grid*, vol. 1, no. 1, pp. 99–107, 2010.

[31] M. Mylrea and S. N. G. Gourisetti, "Blockchain for Smart Grid Resilience: Exchanging Distributed Energy at Speed, Scale and Security," in *Resilience Week (RWS)*, Wilmington, DE, 2017.

[32] F. Ye, Y. Qian and R. Q. Hu, "An Identity-Based Security Scheme for a Big Data Driven Cloud Computing Framework in Smart Grid," in *IEEE Global Communications Conference (GLOBECOM)*, San Diego, CA, 2015.

[33] Y. Guo, Z. Wan and X. Cheng, "When Blockchain Meets Smart Grids: A Comprehensive Survey," *High-Confidence Computing*, vol. 2, no. 2, 2022.

[34] A. Ipakchi and F. Albuyeh, "Grid of the Future," *IEEE Power and Energy Magazine*, vol. 7, no. 2, pp. 52–62, 2009.

[35] H. Farhangi, "The Path of the Smart Grid," *IEEE Power and Energy Magazine*, vol. 8, no. 1, pp. 18–28, 2010.

[36] W. Su and A. Q. Huang, "The Energy Internet," *Woodhead Publishing*, Sawston, UK, 2018.

[37] M. Mylrea and S. N. G. Gourisetti, "Blockchain for Smart Grid Resilience: Exchanging Distributed Energy at Speed, Scale and Security," in *Resilience Week (RWS)*, Wilmington, DE, 2017.

[38] K. Gai, Y. Wu, L. Zhu, L. Xu and Y. Zhang, "Permissioned Blockchain and Edge Computing Empowered Privacy-Preserving Smart Grid Networks," *IEEE Internet of Things Journal*, vol. 6, no. 5, pp. 7992–8004, 2019.

[39] S. Pealy and M. A. Matin, "A Survey on Threats and Countermeasures in Smart Meter," in *IEEE International Conference on Communication, Networks and Satellite (Comnetsat)*, Batam, Indonesia, 2020.

[40] W. Tong, L. Lu, Z. Li, J. Lin and X. Jin, "A Survey on Intrusion Detection System for Advanced Metering Infrastructure," in *Sixth International Conference on Instrumentation & Measurement, Computer, Communication and Control (IMCCC)*, Harbin, China, 2016.

[41] P. Jokar, N. Arianpoo and V. Leung, "A Survey on Security Issues in Smart Grids," *Security Communication Networks*, vol. 9, pp. 262–273., 2016.

[42] N. Saxena and B. J. Choi, "State of the Art Authentication, Access Control, and Secure Integration in Smart Grid," *Energies*, vol. 8, pp. 11883–11915, 2015.

[43] A. Anzalchi and A. Sarwat, "A Survey on Security Assessment of Metering Infrastructure in Smart Grid Systems," in *SoutheastCon*, Fort Lauderdale, FL, 2015.

[44] J. C. Foreman and D. Gurugubelli, "Identifying the Cyber Attack Surface of the Advanced Metering Infrastructure," *The Electricity Journal*, vol. 28, no. 1, pp. 94–103, 2015.

[45] N. Komninos, E. Philippou and A. Pitsillides, "Survey in Smart Grid and Smart Home Security: Issues, Challenges and Countermeasures," *IEEE Communications Surveys & Tutorials*, vol. 16, no. 4, pp. 1933–1954, 2014.

[46] M. E. a. T. T. Develop, *Blockchain Technology to Optimize P2P Energy Trading*, Tokyo, Japan, 2021. Available: www.mitsubishielectric.com/news/2021/pdf/0118.pdf.

[47] Z. Li, J. Kang, R. Yu, D. Ye, Q. Deng and Y. Zhang, "Consortium Blockchain for Secure Energy Trading in Industrial Internet of Things," *IEEE Transactions on Industrial Informatics*, vol. 14, no. 8, pp. 3690–3700, 2018.

[48] J. Guerrero, A. C. Chapman and G. Verbič, "Decentralized P2P Energy Trading Under Network Constraints in a Low-Voltage Network," *IEEE Transactions on Smart Grid*, vol. 10, no. 5, pp. 5163–5173, 2019.

[49] N. Z. Aitzhan and D. Svetinovic, "Security and Privacy in Decentralized Energy Trading Through Multi-Signatures, Blockchain and Anonymous Messaging Streams," *IEEE Transactions on Dependable and Secure Computing*, vol. 15, no. 5, pp. 840–852, 2018.

[50] J. Wu and N. K. Tran, "Application of Blockchain Technology in Sustainable Energy Systems: An Overview," *Sustainability*, vol. 10, no. 9, p. 3067, 2018.

[51] C. Dang, J. Zhang, C.-P. Kwong and L. Li, "Demand Side Load Management for Big Industrial Energy Users Under Blockchain-Based Peer-to-Peer Electricity Market," *IEEE Transactions on Smart Grid*, vol. 10, no. 6, pp. 6426–6435, 2019.

[52] M. A. Ferrag and L. Maglaras, "DeepCoin: A Novel Deep Learning and Blockchain-Based Energy Exchange Framework for Smart Grids," *IEEE Transactions on Engineering Management*, vol. 67, no. 4, pp. 1285–1297, 2020.

[53] S. Wang, A. F. Taha, J. Wang, K. Kvaternik and A. Hahn, "Energy Crowdsourcing and Peer-to-Peer Energy Trading in Blockchain-Enabled Smart Grids," *IEEE Transactions on Systems, Man, and Cybernetics: Systems*, vol. 49, no. 8, pp. 1612–1623, 2019.

[54] A. S. Musleh, G. Yao and S. M. Muyeen, "Blockchain Applications in Smart Grid–Review and Frameworks," *IEEE Access*, vol. 7, pp. 86746–86757, 2019.

[55] A. Maw, S. Adepu and A. Mathur, "ICS-BlockOpS: Blockchain for Operational Data Security in Industrial Control System," *Pervasive and Mobile Computing*, vol. 59, p. 101048, 2019.

[56] J. Gao, K. O. Asamoah, E. B. Sifah, A. Smahi, Q. Xia, H. Xia, X. Zhang and G. Dong, "GridMonitoring: Secured Sovereign Blockchain Based Monitoring on Smart Grid," *IEEE Access*, vol. 6, pp. 9917–9925, 2018.

[57] J. Wan, J. Li, M. Imran and D. Li, "A Blockchain-Based Solution for Enhancing Security and Privacy in Smart Factory," *IEEE Transactions on Industrial Informatics*, vol. 15, no. 6, pp. 3652–3660, 2019.

[58] X. Zhang and M. Fan, "Blockchain-Based Secure Equipment Diagnosis Mechanism of Smart Grid," *IEEE Access*, vol. 6, pp. 66165–66177, 2018.

[59] V. Gunes, S. Peter, T. Givargis and F. Vahid, "A Survey on Concepts Applications Challenges in Cyber-Physical Systems," *KSII Transactions on Internet and Information Systems*, vol. 8, no. 12, pp. 4242–4268, 2014.

[60] E. Münsing, J. Mather and S. Moura, "Blockchains for Decentralized Optimization of Energy Resources in Microgrid Networks," in *IEEE Conference on Control Technology and Applications (CCTA)*, Maui, HI, 2017.

[61] M. Aloqaily, O. Bouachir, Ö. Özkasap and F. S. Ali, "SynergyGrids: Blockchain-Supported Distributed Microgrid Energy Trading," *Peer-to-Peer Networking Applications*, vol. 15, pp. 884–900, 2022.

[62] P. Danzi, M. Angjelichinoski, C. Stefanović and P. Popovski, "Distributed Proportional-Fairness Control in Microgrids via BLOCKCHAIN Smart Contracts," in *IEEE International Conference on Smart Grid Communications (SmartGridComm)*, Dresden, Germany, 2017.

[63] S. Saxena, H. Farag, H. Turesson and H. M. Kim, "Blockchain Based Grid Operation Services for Transactive Energy Systems," *arXiv preprint arXiv:1907.08725*, 2019.

[64] J. Wang, Q. Wang, N. Zhou and Y. Chi, "A Novel Electricity Transaction Mode of Microgrids Based on Blockchain and Continuous Double Auction," *Energies*, vol. 10, no. 12, p. 1971, 2017.

[65] A. S. Yahaya, N. Javaid, F. A. Alzahrani, A. Rehman, I. Ullah, A. Shahid and M. Shafiq, "Blockchain Based Sustainable Local Energy Trading Considering Home Energy Management and Demurrage Mechanism," *Sustainability*, vol. 12, p. 3385, 2020.

[66] C. Antal, T. Cioara, M. Antal, V. Mihailescu, D. Mitrea, I. Anghel, I. Salomie, G. Raveduto, M. Bertoncini, V. Croce, T. Bragatto, F. Carere and F. Bellesini, "Blockchain Based Decentralized Local Energy Flexibility Market," *Energy Reports*, vol. 7, pp. 5269–5288, 2021.

[67] G. Vieira and J. Zhang, "Peer-to-Peer Energy Trading in a Microgrid Leveraged by Smart Contracts," *Renewable and Sustainable Energy Reviews*, vol. 143, 2021.

[68] A. Kavousi-Fard, A. Almutairi, A. Al-Sumaiti, A. Faroughian and S. Alyami, "An Effective Secured Peer-to-Peer Energy Market Based on Blockchain Architecture for the Interconnected

Microgrid and Smart Grid," *International Journal of Electrical Power & Energy Systems*, vol. 132, 2021.

[69] G. V. Leeuwen, T. AlSkaif, M. Gibescu and W. V. Sark, "An Integrated Blockchain-Based Energy Management Platform with Bilateral Trading for Microgrid Communities," *Applied Energy*, vol. 2020, 263.

[70] A. Yildizbasi, "Blockchain and Renewable Energy: Integration Challenges in Circular Economy Era," *Renewable Energy*, vol. 176, 2021.

[71] Y. Tsao and V. Thanh, "Toward Sustainable Microgrids with Blockchain Technology-Based Peer-to-Peer Energy Trading Mechanism: A Fuzzy Meta-Heuristic Approach," *Renewable and Sustainable Energy Reviews*, vol. 2021, no. 110452, 136.

[72] X. Wang, P. Liu and Z. Ji, "Trading Platform for Cooperation and Sharing Based on Blockchain Within Multi-Agent Energy Internet," *Global Energy Interconnection*, vol. 4, no. 4, pp. 384–393, 2021.

[73] Q. Li, A. Li, T. Wang and Y. Cai, "Interconnected Hybrid AC-DC Microgrids Security Enhancement Using Blockchain Technology Considering Uncertainty," *International Journal of Electrical Power & Energy Systems*, vol. 133, no. 107324, 2021.

[74] G. S. Mahesh, G. D. Babu and V. Rakesh, "Energy Management with Blockchain Technology in DC Microgrids," *Materials Today: Proceedings*, vol. 47, no. 10, pp. 2232–2236, 2021.

[75] S. Wang, Z. Xu and J. Ha, "Secure and Decentralized Framework for Energy Management of Hybrid AC/DC Microgrids Using Blockchain for Randomized Data," *Sustainable Cities and Society*, vol. 76, no. 103419, 2022.

[76] M. Yilmaz and P. T. Krein, "Review of the Impact of Vehicle-to-Grid Technologies on Distribution Systems and Utility Interfaces," *IEEE Transactions on Power Electronics*, vol. 28, no. 12, pp. 5673–5689, 2013.

[77] Z. Su, Y. Wang, Q. Xu, M. Fei, Y. -C. Tian and N. Zhang, "A Secure Charging Scheme for Electric Vehicles with Smart Communities in Energy Blockchain," *IEEE Internet of Things Journal*, vol. 6, no. 3, pp. 4601–4613, 2019.

[78] M. Baza, M. Nabil, M. Ismail, M. Mahmoud, E. Serpe and M. A. Rahman, "Blockchain-Based Charging Coordination Mech- Anism for Smart Grid Energy Storage Units," in *IEEE International Conference on Blockchain (Blockchain)*, Atlanta, GA, 2019.

[79] F. Knirsch, A. Unterweger and D. Engel, "Privacy-Preserving Blockchain-Based Electric Vehicle Charging with Dynamic Tariff Decisions," *Computer Science – Research and Development*, vol. 33, pp. 71–79, 2018.

[80] X. Huang, C. Xu, P. Wang and H. Liu, "LNSC: A Security Model for Electric Vehicle and Charging Pile Management Based on Blockchain Ecosystem," *IEEE Access*, vol. 6, pp. 13565–13574, 2018.

[81] Y. Mo et al., "Cyber–Physical Security of a Smart Grid Infrastructure," *Proceedings of the IEEE*, vol. 100, no. 1, pp. 195–209, 2012.

[82] A. Gupta, A. Anpalagan, G. H. S. Carvalho, A. S. Khwaja and L. Guan, "Retracted: Prevailing and Emerging Cyber Threats and Security Practices in IoT-Enabled Smart Grids: A Survey," *Journal of Network and Computer Applications*, vol. 132, pp. 118–148, 2019.

[83] Z. Zhang, S. Gong, A. D. Dimitrovski and H. Li, "Time Synchronization Attack in Smart Grid: Impact and Analysis," *IEEE Transactions on Smart Grid*, vol. 4, no. 1, pp. 87–98, 2013.

[84] C. Konstantinou, M. Sazos, A. S. Musleh, A. Keliris, A. Al-Durra and M. Maniatakos, "GPS Spoofing Effect on Phase Angle Monitoring and Control in a Real-Time Digital Simulator-Based Hardware-in-the-Loop Environment," *ET Cyber-Physical Systems: Theory & Applications*, vol. 2, pp. 180–187, 2017.

[85] S. Liu, X. P. Liu and A. E. Saddik, "Denial-of-Service (DoS) Attacks on Load Frequency Control in Smart Grids," in *IEEE PES Innovative Smart Grid Technologies Conference (ISGT)*, Washington, DC, 2013.

[86] D. Minoli and B. Occhiogrosso, "Blockchain Mechanisms for IoT Security," *Internet of Things*, vols. 1–2, pp. 1–13, 2018.

[87] G. Liang, S. R. Weller, F. Luo, J. Zhao and Z. Y. Dong, "Distributed Blockchain-Based Data Protection Framework for Modern Power Systems Against Cyber Attacks," *IEEE Transactions on Smart Grid*, vol. 10, no. 3, pp. 3162–3173, 2019.

[88] S. Kim, "Chapter Two—Blockchain for a Trust Network Among Intelligent Vehicles," *Advances in Computers*, vol. 111, pp. 43–68, 2018.

[89] T. Alladi, V. Chamola, J. J. P. C. Rodrigues and S. A. Kozlov, "Blockchain in Smart Grids: A Review on Different Use Cases," *Sensors*, vol. 19, no. 22, p. 4862, 2019.

[90] V. Gunes, S. Peter, T. Givargis and F. Vahid, "A Survey on Concepts, Applications, and Challenges in Cyber-Physical Systems," *KSII Transactions on Internet and Information Systems*, vol. 8, no. 12, pp. 4242–4268, 2014.

[91] R. Rajkumar, I. Lee, L. Sha and J. Stankovic, "Cyber-physical Systems: The Next Computing Revolution," in *Design Automation Conference, Anaheim, Anaheim*, CA, 2010.

[92] R. Baheti and H. Gill, "Cyber-Physical Systems," *The Impact of Control Technology*, pp. 161–166, 2011.

[93] "Cyber-Physical Systems—A Concept Map," [Online]. Available: https://ptolemy.berkeley.edu/projects/cps/ [Accessed 24 July 2023].

[94] Q. Yang, J. Yang, W. Yu, D. An, N. Zhang and W. Zhao, "On False Data-Injection Attacks against Power System State Estimation: Modeling and Countermeasures," *IEEE Transactions on Parallel and Distributed Systems*, vol. 25, no. 3, pp. 717–729, 2014.

For Product Safety Concerns and Information please contact our EU
representative GPSR@taylorandfrancis.com
Taylor & Francis Verlag GmbH, Kaufingerstraße 24, 80331 München, Germany

www.ingramcontent.com/pod-product-compliance
Lightning Source LLC
Chambersburg PA
CBHW061401210326
41598CB00035B/6054